OBSERVATIONAL EVIDENCE FOR BLACK HOLES IN THE UNIVERSE

ASTROPHYSICS AND SPACE SCIENCE LIBRARY

VOLUME 234

Executive Committee

W. B. BURTON, *Sterrewacht, Leiden, The Netherlands*
J. M. E. KUIJPERS, *Faculty of Science, Nijmegen, The Netherlands*
E. P. J. VAN DEN HEUVEL, *Astronomical Institute, University of Amsterdam, The Netherlands*
H. VAN DER LAAN, *Astronomical Institute, University of Utrecht, The Netherlands*

Editorial Board

I. APPENZELLER, *Landessternwarte Heidelberg-Königstuhl, Germany*
J. N. BAHCALL, *The Institute for Advanced Study, Princeton, U.S.A.*
F. BERTOLA, *Università di Padova, Italy*
W. B. BURTON, *Sterrewacht, Leiden, The Netherlands*
J. P. CASSINELLI, *University of Wisconsin, Madison, U.S.A.*
C. J. CESARSKY, *Centre d'Etudes de Saclay, Gif-sur-Yvette Cedex, France*
O. ENGVOLD, *Institute of Theoretical Astrophysics, University of Oslo, Norway*
J. M. E. KUIJPERS, *Faculty of Science, Nijmegen, The Netherlands*
R. McCRAY, *University of Colorado, JILA, Boulder, U.S.A.*
P. G. MURDIN, *Royal Greenwich Observatory, Cambridge, U.K.*
F. PACINI, *Istituto Astronomia Arcetri, Firenze, Italy*
V. RADHAKRISHNAN, *Raman Research Institute, Bangalore, India*
K. SATO, *School of Science, The University of Tokyo, Japan*
F. H. SHU, *University of California, Berkeley, U.S.A.*
B. V. SOMOV, *Astronomical Institute, Moscow State University, Russia*
R. A. SUNYAEV, *Space Research Institute, Moscow, Russia*
Y. TANAKA, *Institute of Space & Astronautical Science, Kanagawa, Japan*
S. TREMAINE, *Princeton University, U.S.A.*
E. P. J. VAN DEN HEUVEL, *Astronomical Institute, University of Amsterdam, The Netherlands*
H. VAN DER LAAN, *Astronomical Institute, University of Utrecht, The Netherlands*
N. O. WEISS, *University of Cambridge, U.K.*

OBSERVATIONAL EVIDENCE FOR BLACK HOLES IN THE UNIVERSE

Proceedings of a Conference
held in Calcutta, India,
January 10–17, 1998

Edited by

SANDIP K. CHAKRABARTI

*S.N. Bose National Center for Basic Sciences,
Calcutta, India*

KLUWER ACADEMIC PUBLISHERS

DORDRECHT / BOSTON / LONDON

A C.I.P. Catalogue record for this book is available from the Library of Congress.

ISBN 0-7923-5298-X

Published by Kluwer Academic Publishers,
P.O. Box 17, 3300 AA Dordrecht, The Netherlands.

Sold and distributed in North, Central and South America
by Kluwer Academic Publishers,
101 Philip Drive, Norwell, MA 02061, U.S.A.

In all other countries, sold and distributed
by Kluwer Academic Publishers,
P.O. Box 322, 3300 AH Dordrecht, The Netherlands.

Present estimate of the enclosed mass at the center of the Milky Way
as a function of the projected distance (Ghez et al., this volume).
In terms of the mass density of the Compact object, this is the strongest
evidence of a black hole so far (Ho, this volume).

Printed on acid-free paper

All Rights Reserved
©1999 Kluwer Academic Publishers
No part of the material protected by this copyright notice may be reproduced or
utilized in any form or by any means, electronic or mechanical,
including photocopying, recording or by any information storage and
retrieval system, without written permission from the copyright owner

Printed in the Netherlands.

Table of Contents

Organizing Committees viii

Preface ix

ACCRETION DISK THEORY:
FROM THE STANDARD MODEL UNTIL ADVECTION
G.S. Bisnovatyi-Kogan 1

ACCRETION DISKS AROUND BLACK HOLES:
TWENTY FIVE YEARS LATER
Sandip K. Chakrabarti 19

VISCOSITY IN ACCRETION DISKS
P.J. Wiita 49

SHOCK FORMATION IN ADIABATIC ACCRETION
FLOWS AROUND A KERR BLACK HOLE
Ju-Fu Lu 61

ROTATING ACCRETION FLOWS NEAR A
BLACK HOLE: A NUMERICAL STUDY
Dongsu Ryu 73

SIMULATIONS OF SHOCKS WITH SMOOTHED
PARTICLES HYDRODYNAMICS METHOD
D. Molteni, G. Gerardi, M.A. Valenza & G. Lanzafame ... 83

NUCLEOSYNTHESIS IN ADVECTIVE ACCRETION
DISKS AROUND GALACTIC AND EXTRA-GALACTIC
BLACK HOLES
B. Mukhopadhyay 105

COMPUTATION OF MASS-OUTFLOW RATES FROM
ADVECTIVE ACCRETION DISKS AROUND
BLACK HOLES
Tapas K. Das 113

BLACK HOLE SOLUTIONS OF EINSTEIN'S
EQUATIONS AN OVERVIEW
Thomas Zannias 123

WATER MEGA MASERS IN NGC4258
M. Miyoshi *141*

SUPERMASSIVE BLACK HOLES IN GALACTIC
NUCLEI – OBSERVATIONAL EVIDENCE AND
SOME ASTROPHYSICAL CONSEQUENCES
Luis C. Ho *157*

TIDAL DISRUPTION OF A STAR BY A MASSIVE
BLACK HOLE
Hyung Mok Lee *187*

MULTIFREQUENCY MONITORING OF BLAZARS
Leo O. Takalo *203*

THE OJ287 SUPERMASSIVE BINARY BLACK HOLE
MODEL AND THE NEW UNIFIED SCHEME FOR
THE AGNs
Aimo K. Sillanpää *209*

HIGH ENERGY GAMMA-RAY EMISSION FROM
BLAZARS: EGRET OBSERVATIONS
R. Mukherjee *215*

STRONG GRAVITY AND X-RAY SPECTROSCOPY
A. Maciołek-Niedźwiecki & P. Magdziarz *231*

MY INVOLVEMENT IN THE EARLY YEARS OF
RADIO ASTRONOMY
M.K. Das Gupta *241*

RADIO OBSERVATIONS OF ACTIVE GALACTIC
NUCLEI: EVIDENCE FOR DISKS AND
BLACK HOLES
D.J. Saikia *247*

HIGH PROPER MOTION STARS IN THE VICINITY
OF Sgr A*: EVIDENCE FOR A SUPERMASSIVE
BLACK HOLE AT THE CENTER OF OUR GALAXY
A.M. Ghez, B.L. Klein, M. Morris & E.E. Becklin *265*

BLACK HOLES IN OUR GALAXY
DYNAMICAL EVIDENCE
Phil Charles *279*

OUTBURSTS IN BLACK HOLE X-RAY TRANSIENTS:
CLUES FROM MULTIWAVELENGTH OBSERVATIONS
Carole A. Haswell 293

THE HIGH-ENERGY SPECTRA OF ACCRETING
BLACK HOLES: VIEWING THE MATTER AS IT
DISAPPEARS DOWN THE SCHWARZSCHILD DRAIN
*L.G. Titarchuk, C.R. Shrader, S. Trudolyubov,
M. Revnivtsev & K. Borozdin* 309

X-RAY SPECTRAL VARIABILITY OF
BLACK HOLE BINARIES
M. Gilfanov, E. Churazov & R. Sunyaev 319

QUASI-PERIODIC OSCILLATIONS IN THE X-RAY
FLUX FROM THE BLACK HOLE CANDIDATES
Tadayasu Dotani 341

X-RAY PROPERTIES OF GRS 1915+105
B. Paul 351

X-RAY OBSERVATION OF BLACK HOLE NOVAE
S. Kitamoto 369

A TRANSITION DISK MODEL FIT TO CYGNUS X-1
R. Misra 385

ARE THE X-RAYS COMING FROM THE INSIDE OF
THE ACCRETION DISK OF THE BLACK HOLE?
H. Negoro 389

List of Participants 395

Author Index 399

Scientific Advisory Committee

Prof. Roger D. Blandford (Caltech, USA)
Prof. Matio Livio (STScI, USA)
Prof. Paul J. Wiita (GSU, USA)
Prof. Lev G. Titarchuk (NASA/GSFC, USA)
Prof. Hyung Mok Lee (Pusan, Korea)
Prof. Prahlad C. Agrawal (TIFR, India)
Prof. Chanchal K. Majumdar (SNBNCBS, India)
Prof. Arun K. Sen (ECRA, India)

Local Organizing Committee:

Prof. Sandip .K. Chakrabarti (SNBNCBS, Convener)
Dr. Tushar K. Das (ECRA)
Dr. Narayan Banerjee (Jadavpur University)
Dr. Amitabha Lahiri (SNBNCBS)
Mr. Tapas K. Das (SNBNCBS)
Mr. Indranil Chattopadhyay (SNBNCBS)
Mr. Arnab Ray (SNBNCBS)
Mr. Banibrata Mukhopadhyay (SNBNCBS)

PREFACE

Black Holes are probably the most enigmatic objects in the whole of spacetime. They are simple to describe, yet most difficult to detect. We do not know for sure if we heard the 'birth cry' of a black hole (we are probably not even sure how the 'cry' would 'sound' like). We have not witnessed the 'agony' of a dying black hole. Neither have we detected a 'living' one with a hundred percent certainty. But we are sure that they should exist simply because the fundamental laws of Nature tell us so. Our common faith and conviction in these laws bring us together in various conferences and symposia to exchange notes and ideas to see how far we have progressed to detect these enigmatic objects. With the advent of new theoretical understanding of the nature of hydrodynamic and radiative solutions close to the black holes, together with recent observations with very high resolution instruments aboard spacecrafts, it was felt that it is about time that these new results be looked at more closely. With this idea the conference on 'Observational Evidence for Black Holes in the Universe' was organized to seriously debate and discuss on the sole issue of whether black holes have been detected or not. This was the first occasion where expert Astrophysicists from around the world gathered with this one point agenda. This *Edited Volume* is the fruit of this gathering.

The year of 1998, the present year, is very special. This year celebrates one hundred years of discovery of electrons, seventy-five years of discovery of Compton effect, fifty years of pion discovery, twenty-five years of asymptotic freedom, and more importantly (to us, astrophysicists at least) twenty five years of the publications of two very important papers which defined standard accretion disks. These sentiments have been reflected in some of the articles in this book. Especially, it is noted that perhaps a special type of Compton scattering from rapidly moving matter brings the signatures of the disappearing matter into a black hole to us.

The venue of the conference, namely Calcutta, is also very special. On 20th of June, 1756, in a hot summer day, an incident took place in a dark, 18 feet long cubic chamber in Calcutta. Around 40 British soldiers were suffocated to their end. Since then this incident is known as the 'Black Hole of Calcutta incident' in English literature. When the 'collapsars' were termed as 'Black Holes' in 1967, J.A. Wheeler of

Princeton probably had this *one way chamber* in mind. It is no coincidence that people in Calcutta would be curious about the existence of celestial black holes.

The conference was the third and final part of a three-part workshop entitled 'Multiwavelength Studies of Stars and Compact Objects' which took place during January 1-17, 1998. The lectures were arranged in a make-shift lecture hall in a partially complete building of S.N. Bose National Centre for Basic Sciences, located at a newly developed area of Calcutta metropolis (Salt Lake) where the Astrophysics department was barely one year old. The participants, thirty-two from abroad and rest from various parts of India stayed in the guest house of the Centre. The living and dining arrangements were modest. But all these did not dampen the spirit of the conference. The conference was covered by at least half-a-dozen national newspapers. All India Radio and national and provincial television stations broadcast interviews and news-worthy clips from the conference. There was a very lively debate on whether black holes exist or not in a panel consisting of G.B. Kogan (Chair), L.G. Titarchuk, M. Miyoshi, M. Gilfanov, S.K. Chakrabarti and A. Sillanpää. There was also an official press conference at the Press Club of India. Towards the end, the participants enjoyed a launch-trip to Sundarbans, the land of Royal Bengal Tigers. In the concluding session on the launch, after being satisfied with only the fresh footsteps of the tigers, one participant suggested that the search for tigers is no different from the search for black holes – one knows that they are there, but one can't really find them! Close encounter to either of these two objects is clearly one way and one does not really want to come close to either of them. In both the cases, participants had to be satisfied with indirect evidences only. I mention this witty remark (made in the midst of wilderness) of the participant to impress on the fact that the participants never actually lost sight of the original goal of the conference!

That a large number of new results were presented at the conference could be judged from the quality of the papers in this book. Most of the speakers of the third part of the Workshop (namely, the conference) and a few of the selected ones from the first two parts (whose talks were relevant to the subject matter of this book) contributed to this book. Dr. Ghez, Dr. Charles, Dr. Dotani and Dr. Saikia could not attend the conference despite their strong wishes. They were requested to contribute and they kindly obliged. Dr. M.K. Das Gupta's pioneering observation of the double radio structure of Cygnus-A in the early fifties may be considered to be the first circumstantial evidence for extragalactic black holes. At our request, Dr. Das Gupta contributed a short paper recalling the early years of Radio Astronomy. Every state-of-the-art aspect of black hole astrophysics has been covered, from theory to space- and ground-based observations of galactic and extra-galactic black holes.

The authors were requested to keep in mind that this is an 'Edited Volume', and not just another proceedings. Most of the authors did keep this in mind and the readers are expected to learn, not just the results, but the means by which the results were obtained. I thank all the participants for attending the meeting and the contributors for the trouble they took in preparing the manuscripts.

The conference was directly sponsored by three organizations: S.N. Bose National Center for Basic Sciences (funded by Department of Science and Technology, Government of India), Eastern Center for Research in Astrophysics (funded by University Grants Commission, Government of India) and International Centre for Theoretical Physics, Italy. However, indirectly, a numerous organizations throughout the world supported the conference by deputing the participants to the meeting. I thank all these organizations most sincerely.

Monetary support is only a part of a major event such as this meeting. The infra-structural facilities of the entire S.N. Bose Centre, including the guest house, were generously offered to us by the Director, Prof. C.K. Majumdar. All the non-academic staff members of the Centre were involved in smoothly running the proceedings. Four graduate students of astrophysics, namely, A. Ray, I. Chattopadhyay, B. Mukhopadhyay and T.K. Das and a visitor S. Dasgupta had to shuttle between the Centre and the Airport to welcome the delegates at the Airport and to see them off. I thank them all. Special thanks are due to two helpers Mr. B. Naskar and Mr. S. Ghose who stayed at the Centre for the entire period of the workshop to keep xerox and library facilities available to the participants round the clock.

My effort would be of some success if this book inspires a few students in future.

Calcutta: 30.5.1998   Sandip K. Chakrabarti, Editor

Some participants of the Part Three of the Workshop

ACCRETION DISC THEORY: FROM THE STANDARD MODEL UNTIL ADVECTION

G. S. BISNOVATYI-KOGAN
IKI, Profsoyuznaya 84/32, Moscow 117810 Russia

Abstract. Accretion disc theory was first developed as a theory with the local heat balance, where the whole energy produced by a viscous heating was emitted to the sides of the disc. One of the most important new invention of this theory was a phenomenological treatment of the turbulent viscosity, known as "alpha" prescription, when the $(r\phi)$ component of the stress tensor was approximated by (αP) with a unknown constant α. This prescription played the role in the accretion disc theory as well important as the mixing-length theory of convection for stellar evolution. Sources of turbulence in the accretion disc are discussed, including nonlinear hydrodynamical turbulence, convection and magnetic field role. In parallel to the optically thick geometrically thin accretion disc models, a new branch of the optically thin accretion disc models was discovered, with a larger thickness for the same total luminosity. The choice between these solutions should be done of the base of a stability analysis. The ideas underlying the necessity to include advection into the accretion disc theory are presented and first models with advection are reviewed. The present status of the solution for a low-luminous optically thin accretion disc model with advection is discussed and the limits for an advection dominated accretion flows (ADAF) imposed by the presence of magnetic field are analysed.

1. Introduction

Accretion is served as a source of energy in many astrophysical objects, including different types of binary stars, binary X-ray sources, most probably quasars and active galactic nuclei (AGN). While first development of accretion theory started long time ago (Bondi and Hoyle, 1944; Bondi, 1952), the intensive development of this theory began after discovery of first X ray sources (Giacconi et al, 1962) and quasars (Schmidt, 1963). Accretion into stars, including neutron stars, is ended by a collision with an inner boundary, which may be a stellar surface, or outer boundary of a magnetosphere for strongly magnetized stars. We may be sure in this case, that all gravitational energy of the falling matter will be transformed into heat and radiated outward.

Situation is quite different for sources containing black holes, which are discovered in some binary X-ray sources in the galaxy, as well as in many AGN. Here matter is falling to the horizon, from where no radiation arrives, so all luminosity is formed on the way to it. The efficiency of accretion is not known from the beginning, contrary to the accretion into a star, and depends strongly on such factors, like angular momentum of the falling matter, and magnetic field embedded into it. It was first shown by Schwarzman (1971), that during spherical accretion of nonmagnetized gas the efficiency may be as small as 10^{-8} for sufficiently low mass fluxes. He had shown that presence of magnetic field in the accretion flux matter increase the efficiency up to about 10%, and account of heating of matter due to magnetic field annihilation in the flux rises the efficiency up to about 30% (Bisnovatyi-Kogan & Ruzmaikin, 1974). In the case of a thin disc accretion when matter has large angular momentum, the efficiency is about 1/2 of the efficiency of accretion into a star with a radius equal to the radius of the last stable orbit. Matter cannot emit all the gravitational energy, part of which is absorbed by the black hole. In the case of geometrically thick and optically thin accretion discs the situation is approaching the case of spherical symmetry, and a presence of a magnetic field plays also a critical role.

Here we consider a development of the theory of a disk accretion, starting from creation of a so called "standard model", and discuss recent trends, connected with a presence of advection.

2. Development of the standard model of the disc accretion into a black hole

Matter falling into a black hole is gathered into a disc when its angular momentum is sufficiently high. It happens when the matter falling into a black hole comes from the neighbouring ordinary star companion in the binary, or when the matter appears as a result of a tidal disruption of the star which trajectory of motion approaches sufficiently close to the black hole, so that forces of self gravity could be overcome. The first situation is observed in many galactic X-ray sources containing a stellar mass black hole (Cherepashchuk, 1996). A tidal disruption happens in quasars and active galactic nuclei (AGN), if the model of supermassive black hole surrounded by a dense stellar cluster of Lynden-Bell (1969) is true for these objects.

The models of the accretion disc structure around a black hole had been investigated by Lynden-Bell (1969), Pringle and Rees (1972). The modern "standard" theory of the disc accretion was formulated in the papers of Shakura (1972), Novikov and Thorne (1973) and Shakura and Sunyaev (1973). It is important to note, that all authors of the accretion disc theory from USSR were students (N.I.Shakura) or collaborators (I.D.Novikov and R.A.Sunyaev) of academician Ya.B.Zeldovich, who was not among the authors, but whose influence on them hardly could be overestimated.

The equations of the standard disc accretion theory were first formulated by Shakura (1972); some corrections and generalization to general relativity (GR) were done by Novikov and Thorne (1973), see also correction to their equations in GR made by Riffert & Herold (1995). The main idea of this theory is to describe a geometrically thin non-self-gravitating disc of the mass M_d, which is much smaller then the mass of the black hole M, by hydrodynamic equations averaged over the disc thickness $2h$.

2.1. EQUILIBRIUM EQUATIONS

The small thickness of the disc in comparison with its radius $h \ll r$ indicate to small importance of the pressure gradient ∇P in comparison with gravity and inertia forces. That leads to a simple radial equilibrium equation denoting the balance between the last two forces occurring when the angular velocity of the disc Ω is equal to the Keplerian one Ω_K,

$$\Omega = \Omega_K = \left(\frac{GM}{r^3}\right)^{1/2}. \qquad (1)$$

Note, just before a last stable orbit around a black hole, and of course inside it, this suggestion fails, but in the "standard" accretion disc model the relation (1) is suggested to be fulfilled all over the disc, with an inner boundary at the last stable orbit.

The equilibrium equation in the vertical z-direction is determined by a balance between the gravitational force and pressure gradient

$$\frac{dP}{dz} = -\rho \frac{GMz}{r^3} \qquad (2)$$

For a thin disc this differential equation is substituted by an algebraic one, determining the half-thickness of the disc in the form

$$h \approx \frac{1}{\Omega_K}\left(2\frac{P}{\rho}\right)^{1/2}. \qquad (3)$$

The balance of angular momentum, related to the ϕ component of the Euler equation has an integral in a stationary case written as

$$\dot{M}(j - j_{in}) = -2\pi r^2 \, 2h t_{r\phi}, \quad t_{r\phi} = \eta r \frac{d\Omega}{dr}. \quad (4)$$

Here $j = v_\phi r = \Omega r^2$ is a specific angular momentum, $t_{r\phi}$ is a component of the viscous stress tensor, $\dot{M} > 0$ is a mass flux per unit time into a black hole, j_0 is an integration constant having, after multiplication by $\dot{M}$, a physical sense of difference between viscous and advective flux of the angular momentum, when j_{in} itself is equal to the specific angular momentum of matter falling into a black hole. In the standard theory the value of j_{in} is determined separately, from physical considerations. For the accretion into a black hole it is suggested, that on the last stable orbit the gradient of the angular velocity is zero, corresponding to zero viscous momentum flux. In that case

$$j_{in} = \Omega_K r_{in}^2, \quad (5)$$

corresponding to the Keplerian angular momentum of the matter on the last stable orbit. During accretion into a slowly rotating star which angular velocity is smaller than a Keplerian velocity on the inner edge of the disc, there is a maximum of the angular velocity close to its surface, where viscous flux is zero, and there is a boundary layer between this point and stellar surface. In that case (5) remains to be valid. The situation is different for accretion discs around rapidly rotating stars with a critical Keplerian speed on the equator. Here there is no extremum of the angular velocity of the disc which smoothly joins the star. In stationary self-consistent situation when the accreting star remains to rotate critically during the process of a disc accretion, the specific angular momentum of matter joining the star is determined by a relation (Bisnovatyi-Kogan, 1993): $j_{in} = \frac{dJ}{dM}|_{crit}$, where the derivative is taken along the states of the star having a Keplerian equatorial speed. For stars with a polytropic structure, corresponding to equation of state $P = K\rho^{1+\frac{1}{n}}$, this derivative is calculated numerically giving the value $j_{in} = 0.176 \Omega_K r_{in}^2$ for $n = 1.5$; 0 for $n = 2.5$; and negative values of j_{in} for larger n.

Note, that in the pioneering paper of Shakura (1972) the integration constant j_{in} was found as in (5), but was taken zero in his subsequent formulae. Importance of using j_{in} in the form (5) was noticed by Novikov and Thorne (1973), and became a feature of the standard model.

2.2. VISCOSITY

The choice of the viscosity coefficient is the most difficult and speculative poblem of the accretion disc theory. In the laminar case of microscopic (atomic or plasma) viscosity, which is very low, the stationary accretion disc must be very massive and very thick, and before its formation the matter is collected by disc leading to a small flux inside. It contradicts to observations of X-ray binaries, where a considerable matter flux along the accretion disc may be explained only when viscosity coefficient is much larger then the microscopic one. In the paper of Shakura (1972) it was suggested, that matter in the disc is turbulent, what determines a turbulent viscous stress tensor, parameterized by a pressure

$$t_{r\phi} = -\alpha \rho v_s^2 = -\alpha P, \qquad (6)$$

where v_s is a sound speed in the matter. This simple presentation comes out from a relation for a turbulent viscosity coefficient $\eta_t \approx \rho v_t l$ with an average turbulent velocity v_t and mean free path of the turbulent element l. It follows from the definition of $t_{r\phi}$ in (4), when we take $l \approx h$ from (3)

$$t_{r\phi} = \rho v_t h r \frac{d\Omega}{dr} \approx \rho v_t v_s = -\alpha \rho v_s^2, \qquad (7)$$

where a coefficient $\alpha < 1$ is connecting the turbulent and sound speeds $v_t = \alpha v_s$. Presentations of $t_{r\phi}$ in (6) and (7) are equivalent, and only when the angular velocity differs considerably from the Keplerian one the first relation to the right in (7) is more preferable. That does not appear (by definition) in the standard theory, but may happen when advective terms are included.

Development of a turbulence in the accretion disc cannot be justified simply, because a Keplerian disc is stable in linear approximation to the development of perturbations. It was suggested by Ya.B.Zeldovich, that in presence of very large Reynolds number Re $= \frac{\rho v l}{\eta}$ the amplitude of perturbations at which nonlinear effects become important is very low, so in this situation a turbulence may develop due to nonlinear instability even when the disc is stable in linear approximation. Another source of viscous stresses may arise from a magnetic field, but it was suggested by Shakura (1972), that magnetic stresses cannot exceed the turbulent ones.

Magnetic plasma instability as a source of the turbulence in the accretion discs has been studied extensively in last years (see review of Balbus and Hawley, 1998). They used an instability of the uniform magnetic field parallel to the axis in differentially rotating disc, discovered by Velikhov (1959). It could be really important in absence of any other source of the turbulence, but it is hard to believe that there is no radial or azimuthal component of the magnetic field in matter flowing into the accretion disc from the companion star. In that case the field amplification due to twisting by a differential rotation take place without necessity of any kind of instability.

It was shown by Bisnovatyi-Kogan and Blinnikov (1976, 1977), that inner regions of a highly luminous accretion discs where pressure is dominated by radiation, are unstable to vertical convection. Development of this convection produce a turbulence, needed for a high viscosity. Other regions of a standard accretion disc should be stable to development of a vertical convection, so other ways of a turbulence excitation are needed there. With alpha- prescription of viscosity the equation of angular momentum conservation is written in the plane of the disc as

$$\dot{M}(j - j_{in}) = 4\pi r^2 \alpha P_0 h. \tag{8}$$

When angular velocity is far from Keplerian the relation (4) is valid with a coefficient of a turbulent viscosity

$$\eta = \alpha \rho_0 v_{s0} h, \tag{9}$$

where values with the index "0" denote the plane of the disc.

2.3. HEAT BALANCE

In the standard theory a heat balance is local, what means that all heat produced by viscosity in the ring between r and $r + dr$ is radiated through the sides of disc at the same r. The heat production rate Q_+ related to the surface unit of the disc is written as

$$Q_+ = h\, t_{r\phi} r \frac{d\Omega}{dr} = \frac{3}{8\pi}\dot{M}\frac{GM}{r^3}\left(1 - \frac{j_{in}}{j}\right). \tag{10}$$

Heat losses by a disc depend on its optical depth. The first standard disc model of Shakura (1972) considered a geometrically thin disc as an optically thick in a vertical direction. That implies energy losses Q_- from the disc due to a radiative conductivity, after a substitution of the differential equation of a heat transfer by an algebraic relation

$$Q_- \approx \frac{4}{3}\frac{acT^4}{\kappa \Sigma}. \tag{11}$$

Here a is a constant of a radiation energy density, c is a speed of light, T is a temperature in the disc plane, κ is a matter opacity, and a surface density $\Sigma = 2\rho h$. Here and below ρ, T, P without the index "0" are related to the disc plane. The heat balance equation is represented by a relation

$$Q_+ = Q_-, \qquad (12)$$

A continuity equation in the standard model of the stationary accretion flow is used for finding of a radial velocity v_r

$$v_r = \frac{\dot{M}}{4\pi r h \rho} = \frac{\dot{M}}{2\pi r \Sigma}. \qquad (13)$$

Equations (1),(3),(8), (12), completed by an equation of state $P(\rho, T)$ and relation for the opacity $\kappa = \kappa(\rho, T)$ represent a full set of equations for a standard disc model. For power low equations of state of an ideal gas $P = P_g = \rho \mathcal{R} T$ ($\mathcal{R}$ is a gas constant), or radiation pressure $P = P_r = \frac{aT^4}{3}$, and opacity in the form of electron scattering κ_e, or Karammers formulae κ_k, the solution of a standard disc accretion theory is obtained analytically (Shakura, 1972; Novikov & Thorne, 1973; Shakura & Sunyaev, 1973). Checking the suggestion of a large optical thickness confirms a self-consistency of the model. One of the shortcoming of the analytical solutions of the standard model lay in the fact, that solutions for different regions of the disc with different equation of states and opacities are not matched to each other.

2.4. OPTICALLY THIN SOLUTION

Few years after appearance of the standard model it was found that in addition to the optically thick disc solution there is another branch of the solution for the disc structure with the same input parameters M, $\dot{M}$, α which is also self-consistent and has a small optical thickness (Shapiro, Lightman & Eardley, 1976). Suggestion of the small optical thickness implies another equation of energy losses, determined by a volume emission $Q_- \approx q\rho h$, where due to the Kirchoff law the emissivity of the unit of a volume q is connected with a Planckian averaged opacity κ_p by an approximate relation $q \approx acT_0^4 \kappa_p$. Note, that Krammers formulae for opacity are obtained after Rosseland averaging of the frequency dependent absorption coefficient. In the optically thin limit the pressure is determined by a gas $P = P_g$. Analytical solutions are obtained here as well, from the same equations with volume losses and gas pressure. In the optically thin solution the thickness of the disc is larger then in the optically thick one, and density is lower.

While heating by viscosity is determined mainly by heavy ions, and cooling is determined by electrons, the rate of the energy exchange between them is important for a thermal structure of the disc. The energy balance equations are written separately for ions and electrons. For small accretion rates and lower matter density the rate of energy exchange due to binary collisions is so slow, that in the thermal balance the ions are much hotter then the electrons. That also implies a high disc thickness and brings the standard accretion theory to the border of its applicability. Nevertheless, in the highly turbulent plasma the energy exchange between ions and electrons may be strongly enhanced due to presence of fluctuating electrical fields, where electrons and ions gain the same energy. In such conditions difference of temperatures between ions and electrons may be negligible. Regretfully, the theory of relaxation in the turbulent plasma is not completed, but there are indications to a large enhancement of the relaxation in presence of plasma turbulence, in comparison with the binary collisions (Quataert,. 1997).

2.5. ACCRETION DISC STRUCTURE FROM EQUATIONS DESCRIBING CONTINUOUSLY OPTICALLY THIN AND OPTICALLY THICK DISC REGIONS

In order to find equations of the disc structure valid in both limiting cases of optically thick and optically thin disc, and smoothly describing transition between them, Eddington approximation had been used for obtaining formulae for a heat flux and for a radiation pressure (Artemoma et al., 1996). The following expressions had been obtained for the vertical energy flux from the disc F_0, and the radiation pressure in the symmetry plane

$$F_0 = \frac{2acT_0^4}{3\tau_0}\left(1 + \frac{4}{3\tau_0} + \frac{2}{3\tau_*^2}\right)^{-1}, \quad P_{rad,0} = \frac{aT_0^4}{3}\frac{1 + \frac{4}{3\tau_0}}{1 + \frac{4}{3\tau_0} + \frac{2}{3\tau_*^2}}, \quad (14)$$

where $\tau_0 = \kappa_e \rho h$, $\tau_* = (\tau_0 \tau_{\alpha 0})^{1/2}$, $\tau_{\alpha 0} \approx \kappa_p \rho h$. At $\tau_0 \gg \tau_* \gg 1$ we have (11) from (14). In the optically thin limit $\tau_* \ll \tau_0 \ll 1$ we get

$$F_0 = acT_0^4 \tau_{\alpha 0}, \quad P_{rad,0} = \frac{2}{3} acT_0^4 \tau_{\alpha 0}. \quad (15)$$

Using F_0 instead of Q_- and equation of state $P = \rho \mathcal{R} T + P_{rad,0}$, the equations of accretion disc structure together with equation $Q_+ = F_0$, with Q_+ from (10), have been solved numerically by Artemova et al. (1996). It occurs that two solutions, optically thick and optically thin, exist separately when luminosity is not very large. Two solutions intersect at $\dot{m} = \dot{m}_b$ and there is no global solution for accretion disc at $\dot{m} > \dot{m}_b$ (see Fig.1). It was concluded by Artemova et al (1996), that in order to obtain a global physically meaningful solution at $\dot{m} > \dot{m}_b$, account of advection is needed.

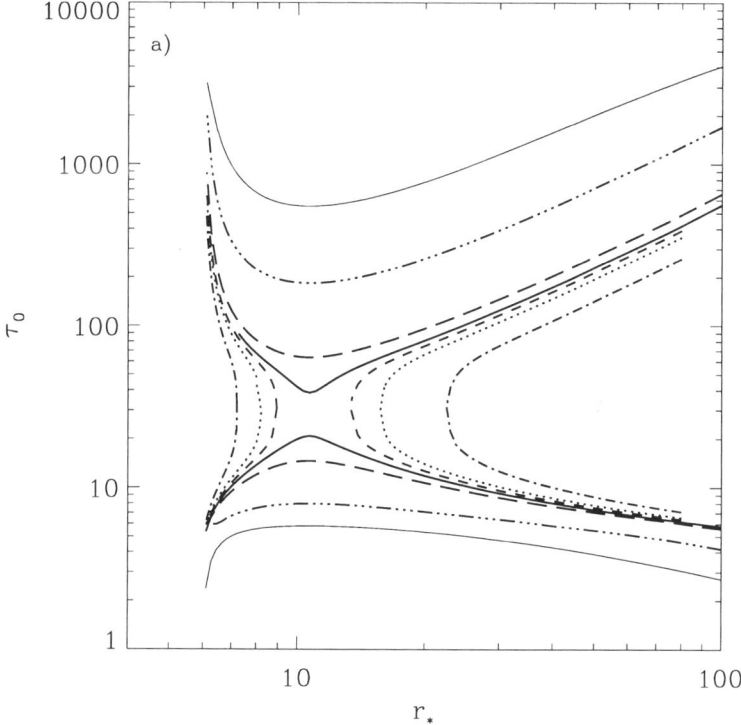

Figure 1. The dependences of the optical depth τ_0 on radius, $r_* = r/r_g$, for the case $M_{BH} = 10^8 \ M_\odot$, $\alpha = 1.0$ and different values of $\dot m$. The thin solid, dot-triple dash, long dashed, heavy solid, short dashed, dotted and dot-dashed curves correspond to $\dot m = 1.0, 3.0, 8.0, 9.35, 10.0, 11.0, 15.0$, respectively. The upper curves correspond to the optically thick family, lower curves correspond to the optically thin family.

3. Accretion discs with advection

Standard model gives somewhat nonphysical behaviour near the inner edge of the accretion disc around a black hole. For high mass fluxes when central regions are radiation-dominated ($P \approx P_r$, $\kappa \approx \kappa_e$), the radial dependence follows relations (Shakura & Sunyaev, 1973)

$$\rho \sim r^{3/2} \mathcal{J}^{-2} \to \infty, \quad T \sim r^{-3/8}, \tag{16}$$

$$h \sim \mathcal{J} \to 0, \quad \Sigma \sim r^{3/2} \mathcal{J}^{-1} \to \infty, \quad v_r \sim r^{-5/2} \mathcal{J} \to 0,$$

where limits relate to the inner edge of the disc with $r = r_{in}$, $\mathcal{J} = 1 - \frac{j_{in}}{j} = 1 - \sqrt{\frac{r_{in}}{r}}$. At smaller $\dot M$, when near the inner edge $P \approx P_g$, $\kappa \approx \kappa_e$, there are different type of singularities

$$\rho \sim r^{-33/20} \mathcal{J}^{2/5} \to 0, \quad T \sim r^{-9/10} \mathcal{J}^{2/5} \to 0, \tag{17}$$

$$h \sim r^{21/20} \mathcal{J}^{1/5} \to 0, \quad \Sigma \sim r^{-3/5} \mathcal{J}^{3/5} \to 0, \quad v_r \sim r^{-2/5} \mathcal{J}^{-3/5} \to \infty.$$

This results from the local form of the equation of the thermal balance (12). It is clear from physical ground, that when a local heat production due to viscosity goes to zero, the heat brought by radial motion of matter along the accretion disc becomes more important. In presence of this advective heating (or cooling term, depending on the radial entropy S gradient) written as

$$Q_{adv} = \frac{\dot{M}}{2\pi r} T \frac{dS}{dr}, \tag{18}$$

the equation of a heat balance is modified to $Q_+ + Q_{adv} = Q_-$. In order to describe self-consistently the structure of the accretion disc we should also modify the radial disc equilibrium, including pressure and inertia terms

$$r(\Omega^2 - \Omega_K^2) = \frac{1}{\rho} \frac{dP}{dr} - v_r \frac{dv_r}{dr}. \tag{19}$$

Appearance of inertia term leads to transonic radial flow with a singular point. Conditions of a continuous passing of the solution through a critical point choose a unique value of the integration constant j_{in}. First approximate solution for the advective disc structure have been obtained by Paczynski and Bisnovatyi-Kogan (1981), but a corresponding set of equations had been discussed earlier (Hoshi and Shibazaki, 1977; Liang and Thompson, 1980). Attempts to find a solution for advective disc structure (see e.g. Matsumoto et al., 1984; Abramovicz et al., 1988) gave the following results. For moderate values of $\dot{M}$ a unique continuous transonic solution was found, passing through singular points, and corresponding to a unique value of j_{in}. The number of critical points in the radial flow happens always to be more then unity. This is connected with two reasons. First, the gravitational potential ϕ_g in papers dealing with advective disc solutions was different from Newtonian one (Paczyński and Wiita, 1980): $\phi_g = \frac{GM}{r-r_g}$, $r_g = \frac{2GM}{c^2}$. The advantage of this potential is a realistic approximation of the general relativistic (GR) effects, namely, infinitive gravitational attraction at a gravitational radius r_g, and existence of the stable circular orbits only up to $r = 3r_g$, like in exact GR. Appearance of two critical points for a radial flow in this potential was analysed by Chakrabarti and Molteni (1993). The second reason of multiplicity of singular points is connected with using of equations averaged over a thickness of the disc. That changes a structure of hydrodynamic equations, leading to a position of singular points not coinciding with a unit Mach number point, and increasing a number of critical points. For the potential of Paczynski-Wiita the multiplicity of the singular points, coinciding with the sonic ones, happens also in full hydrodynamical description without averaging of equations over the disc thickness (Chakrabarti & Molteni, 1993).

When $\dot{M}$ is becoming so high, that radiation pressure starts to be important, still unresolved problems appear in a construction of the advective disc model. These problems are connected with increasing of a number of a critical points from one side, and loss of uniqueness of the transonic solution from another. So, with increasing of $\dot{M}$ the solution becomes non-unique at some parameters, or was not found at all (see Matsumoto et al., 1984; Abramovicz et al., 1988; Artemova et al., 1996a). At high $\dot{M}$ the integral curves are very sensitive to input conditions: form of viscosity stresses (4) or (6), choice of boundary conditions etc. The system of equations has a very small resource of stability, so it cannot be excluded, that the failures are connected with an improper choice of a numerical method and development of numerical instabilities prevents of finding a unique physical solution. In addition to continuous solutions, solutions with standing shock waves have been investigated (Chakrabarti, 1996).

3.1. TWO-TEMPERATURE ADVECTIVE DISCS

In the optically thin accretion discs at low mass fluxes the density of the matter is low and energy exchange between electrons and ions due to binary collisions is slow. In this situation, due to different mechanisms of heating and cooling for electrons and ions, they may have different temperatures. First it was realized by Shapiro, Lightman & Eardley (1976) where advection was not included. It was noticed by Narayan and Yu (1995), that advection in this case is becoming extremely important. It may carry the main energy flux into a black hole, leaving rather low efficiency of the accretion up to $10^{-3} - 10^{-4}$ (advective dominated accretion flows - ADAF). This conclusion is valid only when the effects, connected with magnetic field annihilation and heating of matter due to it are neglected.

In the ADAF solution the ion temperature is about a virial one $kT_i \sim GMm_i/r$, what means that even at high initial angular momentum the disc becomes very thick, forming practically a quasi-spherical accretion flow. It is connected also with an "alpha" prescription of viscosity. At high ion temperatures, connected with a strong viscous heating, the ionic pressure becomes high, making the viscosity very effective. So, due to suggestion of "alpha" viscosity in the situation, when energy losses by ions are very low, some kind of a "thermo-viscous" instability is developed, because heating increases a viscosity, and viscosity increases a heating. Development of this instability leads to formation of ADAF.

A full account of the processes, connected with a presence of magnetic field in the flow, is changing considerably the picture of ADAF. It was requires, that expressions for a magnetic heating of the matter $\mathcal{H}_B$, obtained from the condition of stationarity of the flow (20), and from the Ohm's law (25), should be identical. That gives some restrictions for the choice of a characteristic velocity v_E. Comparison between (20) and (25) shows the identity of these two expressions at $v_E = v_r$, , $\frac{\alpha}{\mathcal{J}\alpha_m} = \frac{3\sqrt{2}}{4}$. So, the model is becoming self-consistent at the reasonable choice of the parameters. Note, that in the advective models $\mathcal{J}$ is substituted by another function which is not zero at the inner edge of the disc. The heating due to magnetic field reconnection $\mathcal{H}_B$ in the equations (26), (27), may be written as

$$\mathcal{H}_B = \frac{3}{16\pi}\frac{B^2}{r\rho}v_r = \frac{1}{2\mathcal{J}}\mathcal{H}_{\eta i}\left(\frac{v_B}{v_K}\right). \qquad (30)$$

So, at $v_B = v_K$ the expressions for viscous and magnetic heating are almost identical. The distribution of the magnetic heating between electrons and ions has a critical influence on the model, if we neglect the influence of a plasma turbulence on the energy relaxation, and take into account only the energy exchange by binary collisions from (29). Observations of the magnetic field reconnection in the solar flares show (Tsuneta, 1996), that electronic heating prevails.

It follows from the physical picture of the field reconnection, that transformation of the magnetic energy into a heat is connected with the change of the magnetic flux, generation of the vortex electrical field, accelerating the particles. This vortex field has a scale of the turbulent element and suffers rapid and chaotic changes. The accelerating forces on electrons and protons in this fields are identical, but accelerations themselves differ ~ 2000 times, so during a sufficiently short time of the turbulent pulsation the electron may gain much larger energy, then the protons. Additional particle acceleration and heating happens on the shock fronts, appearing around turbulent cells, where reconnection happens. In this process acceleration of the electrons is also more effective than of the protons. In the paper of Bisnovatyi-Kogan and Lovelace (1997) the equations (26), (27) have been solved in the approximation of nonrelativistic electrons, $v_B = v_K$, what permitted to unite a viscous and magnetic heating into a unique formula. The combined heating of the electrons and ions were taken as $\mathcal{H}_e = (2-g)\mathcal{H}_{\eta i}$, $\mathcal{H}_e = g\mathcal{H}_{\eta i}$. In the expression for a cyclotron emission self-absorption was taken into account according to Trubnikov (1973). The results of calculations for $g = 0.5 \div 1$ show that almost all energy of the electrons is radiated, so the relative efficiency of the two-temperature, optically thin disc accretion cannot become lower then 0.25. Note again that accurate account of a plasma turbulence for a thermal relaxation and corresponding increase of the term Q_{ie} may restore the relative efficiency to its unity value, corresponding to the optically thick discs.

4. Discussion

Observational evidences for existence of black holes inside our Galaxy and in the active galactic nuclei (Cherepashchuk, 1996; Ho, 1998) make necessary to revise theoretical models of the disc accretion. Large part of high energy radiation indicates to its origin close to the black hole, where standard accretion disc model is not a appropriate. The improvements of a model are connected with account of advective terms and more accurate treatment of the magnetic field effects. Conclusions about existence of ADAF solution for an optically thin accretion disc at low mass flux are connected with an incomplete account of the effects connected with magnetic field annihilation. Their account does not permit to make a relative efficiency of the accretion lower then ~ 0.25 from the standard value. It is expected that more accurate treatment of the relaxation connected with the plasma turbulence will even more increase the efficiency, making it close to unity (see also Fabian and Rees, 1995).

Some observational data which were interpreted as an evidence for the existence of the ADAF regime have disappeared after additional accumulation of data. The most interesting example of this sort is connected with the claim of the proof of the existence of event horizon of the black holes due to manifestation of the ADAF regime of accretion (Narayan et al., 1997). Analysis of the more complete set of the observational data (Chen et al., 1997) had shown disappearance of the statistical effect claimed as an evidence for ADAF. This example shows how dangerous is to base a proof of the theoretical model on the preliminary observational data. It is even more dangerous, when the model is physically not fully consistent. Then even a reliable set of the observational data cannot serve as a proof of the model. The classical example from astrophysics of this kind gives the theory of the origin of the elements presented in the famous book of G.Gamov (1952), where the model of the hot universe was developed. In addition to rich advantages of this model, the author also wanted to explained the origin of heavy elements in the primordial explosion, neglecting the problems connected with an absence of the stable elements with the number of baryons equal to 5 and 8. G. Gamov considered a good coincidence of his calculations, where the mentioned problem was neglected, and the observational curve, as a best proof of his theory of the origin of the elements. The farther developments have shown that his outstanding theory explains lot of things, except the origin of the heavy elements, which are produced due to stellar evolution.

It looks like it is difficult to use ADAF for solution of the problem of existence of underluminous AGN, where the observed flux of the energy is smaller, then the expected from the standard accretion disc models. Two possible ways may be suggested. One is based on a more accurate estimations of the accretion mass flow into the black hole, which could be overestimated. Another, more attractive possibility, is based on existence of another mechanisms of the energy losses in the form of accelerated particles, like in the radio-pulsars, where their losses exceed strongly a radiation losses. This is very probable to happen in a presence of a large scale magnetic field which may be also responsible for a formation of the observed jets. To extend this line, we may suggest, that underlumilnous AGN loose main part of their energy to the formation of jets. The search of the correlation between existence of jets and lack of the luminosity could be very informative.

References

Abramovicz, M.A., Czerny, B., Lasota, J.P., Szuszkiewicz, E. 1988, ApJ, 332, 646
Artemova, I.V., Bisnovatyi-Kogan, G.S., Björnsson, G., Novikov, I.D. 1996, ApJ, 456, 119
Artemova, I.V., Bisnovatyi-Kogan, G.S., Björnsson, G., Novikov, I.D. 1996a, Preprint TAC 1996-029
Balbus, S.A. & Hawley, J.F. 1998, Rev. Mod. Phys. 70, 1
Bisnovatyi-Kogan, G.S. 1993, A& A, 274, 796
Bisnovatyi-Kogan, G.S. and Blinnikov, S.I. 1976, Pisma Astron. Zh., 2, 489
Bisnovatyi-Kogan, G.S. and Blinnikov, S.I. 1977, A&A, 59, 111
Bisnovatyi-Kogan, G.S. and Lovelace, R.V.L. 1997, ApJ, 486, L43
Bisnovatyi-Kogan, G.S., & Ruzmaikin, A.A., 1974, Astrophys. and Space Sci., 28, 45
Bisnovatyi-Kogan, G.S., & Ruzmaikin, A.A., 1976, Astrophys. and Space Sci., 42, 401
Bondi, H. 1952, MN R.A.S., 112, 195
Bondi, H., Hoyle, F. 1944, MN R.A.S., 104, 273
Chakrabarti, S.K. 1996, ApJ, 464,1996
Chakrabarti, S.K., Molteni, D. 1993, ApJ, 417, 671
Chen, W., Cui, W., Frank, J., King, A., Livio, M., Zhang, S.N. 1997, Talk on High Energy Ap. Division Meeting. November
Cherepashchuk, A.M. 1996, Uspekhi Fiz. Nauk, 166, 809
Fabian, A.C. and Rees, M.J. 1995, Month. Not. R.A.S., 277, L55.
Gamov, G. 1952. The Creation of the Universe. Viking Press. NY.
Giacconi, R., Gursky, H, Paolini, F.R., and Rossi, B.B. 1962, Phys. Rev. Lett. 9, 439
Ho, L. 1998, this volume
Hōshi, R., Shibazaki, N. 1977, Prog. Theor. Phys., 58, 1759
Landau, L.D. 1937, Zh. Exp. Theor. Phys. 7, 203.
Liang, E.P.T., Thompson, K.A. 1980, ApJ, 240, 271
Lynden-Bell, D. 1969, Nature, 223, 690
Matsumoto, R., Sato, Sh., Fukue, J., Okazaki, A.T. 1984, Publ. Astron. Soc. Japan, 36, 71
Narayan, R., Garcia, M.R., McClintock, J.E. 1997, astro-ph/9701139
Narayan, R., Yu, I. 1995, ApJ, 452, 710
Novikov, I.D. & Thorne, K.S. 1973, in Black Holes eds. C.DeWitt & B.DeWitt (New York: Gordon & Breach), p.345
Paczyńsci, B., Bisnovatyi-Kogan, G.S. 1981, Acta Astron., 31, 283
Paczyńsci, B., Wiita, P.J. 1980, A&A, 88, 23
Pringle, J.E. & Rees, M.J. 1972, A&A, 21, 1
Quataert, E. 1997, astro-ph/9710127

Riffert, H., Herold, H. 1995, ApJ, 450, 508
Schmidt, M. 1963, ApJ, 136, 164
Schwartsman, V.F. 1971, Soviet Astron., 15, 377
Shakura, N.I. 1972, Astron. Zh., 49, 921 (1973, Sov. Astron., 16, 756)
Shakura, N.I., & Sunyaev, R.A. 1973, A&A, 24, 337
Shapiro, S.L., Lightman, A.P., Eardley, D.M. 1976, ApJ, 204, 187
Spitzer, L. 1940, MNRAS, 100, 396
Trubnikov B.A. 1973, Voprosy Teorii Plasmy, 7, 274
Tsuneta, S. 1996, ApJ, 456, 840.
Velikhov, E.P. 1959, Sov. Phys. - JETP, 36, 995

G. S. Bisnovatyi-Kogan

S. K. Chakrabarti

ACCRETION DISKS AROUND BLACK HOLES: TWENTY FIVE YEARS LATER

SANDIP K. CHAKRABARTI
S. N. Bose National Centre For Basic Sciences
JD Block, Salt Lake, Sector-III, Calcutta-700091, India
email: chakraba@boson.bose.res.in

1. Introduction

Study of accretion processes onto stars began by the works of Hoyle & Lyttleton (1939), almost sixty years ago. They computed the rate at which pressure-less matter would be accreted on a moving star. Subsequently, pressure was included and the spherical flow solution was perfected by Bondi (1952). However, emission from rapidly infalling matter was not found to be strong enough to explain high luminosities of quasars and AGNs. Suggestions to improve the luminosity by magnetic dissipation were then put forth (Shvartsman, 1971; Shapiro, 1973ab). Indeed, efforts to improve luminosity of a spherical flow are on even in recent days (Chang & Ostriker, 1985; Babul, Ostriker & Mészáros, 1989; Nobili, Turolla & Zampieri, 1991). Meanwhile, possible evidence of disklike structures around one of the binary components were found (Kraft, 1963) and some tentative suggestions that matter should accrete in the form of accretion disks were put forward (Burbidge & Prendergast, 1968; Lynden-Bell, 1969). More quantitative studies were made by Shakura (1972). However, the beginning of modern accretion disk physics is traditionally attributed to the two classical articles, one by Shakura & Sunyaev (1973, hereafter referred to as SS73) and the other by Novikov & Thorne (1973, hereafter referred to as NT73), both of which were published exactly twenty five years ago.

Early history of the development of the accretion disk model has been provided by Bisnovatyi-Kogan (1998) in this volume and we shall not elaborate on that here. Instead, we shall dwell on the major developments of this subject, including high points and excitements since 1973. SS73 and NT73 compute the structure of the accretion disk assuming the angular momentum distribution is Keplerian, independent of how matter is supplied and

independent of the nature of the viscous processes. Strictly speaking, no *accreting* disk of *finite temperature* can be perfectly Keplerian at all points. To see this, we have to go back to the basics. Let's start with a general set of equations (see Chakrabarti, 1996a, hereafter referred to as C96a) which the flow satisfies on the equatorial plane around a Schwarzschild black hole (modeled here as an object around which matter moves in a pseudo-Newtonian potential of Paczynski-Wiita (1980)):
(a) The radial momentum equation:

$$v\frac{dv}{dx} + \frac{1}{\rho}\frac{dp}{dx} + \frac{l^2_{Kep} - l^2}{x^3} = 0 \quad (1a)$$

(b) The continuity equation:

$$\frac{d}{dx}(\Sigma x v) = 0 \quad (1b)$$

(c) The azimuthal momentum equation:

$$v\frac{dl(x)}{dx} - \frac{1}{\Sigma x}\frac{d}{dx}(x^2 W_{x\phi}) = 0 \quad (1c)$$

(d) The entropy equation:

$$\Sigma v T \frac{ds}{dx} = \frac{h(x)v}{\Gamma_3 - 1}(\frac{dp}{dx} - \Gamma_1 \frac{p}{\rho}\frac{d\rho}{dx}) = Q^+_{mag} + Q^+_{nuc} + Q^+_{vis} - Q^-$$

$$= Q^+ - g(x, \dot{m})q^+ = f(\alpha, x, \dot{m})q^+. \quad (1d)$$

Here, x, v and l are the dimensionless distance, velocity and specific angular momentum (measured in units of $2GM/c^2$, c and $2GM/c$ respectively) from the black hole. p and ρ are the dimensionless isotropic pressure and mass density respectively, Σ, W and $W_{x\phi}$ are the mass density, pressure and viscous stress, integrated over the vertical height $h(x)$ (assuming thin disk approximation $h(x) << x$). T is the proton and electron temperature (assuming strong coupling between electrons and protons, but see, Chakrabarti & Titarchuk, 1995, hereafter CT95, where two temperature flow was studied without the magnetic heating effects). In the entropy equation, we have included the possibility of magnetic heating (due to stochastic field) and nuclear energy release as well. On the right hand side, we wrote Q^+ collectively proportional to the cooling term for simplicity (purely on dimensional grounds). The quantity f is a measure of cooling efficiency of the flow. Also,

$$\Gamma_3 = 1 + \frac{\Gamma_1 - \beta}{4 - 3\beta}; \quad \Gamma_1 = \beta + \frac{(4 - 3\beta)^2(\gamma - 1)}{\beta + 12(\gamma - 1)(1 - \beta)} \quad (2)$$

and $\beta(x)$ is the ratio of gas pressure to total (gas plus magnetic plus radiation) pressure:

$$\beta(x) = \frac{\rho kT/\mu m_p}{\rho kT/\mu m_p + \bar{a}T^4/3 + B^2(x)/4\pi} \tag{3}$$

$B(x)$ is the strength of magnetic field in the flow, $\bar{a}$ is the Stefan's constant, k is the Boltzmann constant, μ is the electron number per particle (and is generally a function of x in case of strong nucleosynthesis effects), m_p is the mass of the proton.

Now if a disk is strictly Keplerian, $l = l_{Kep}$, and eq. 1a is satisfied only if,

$$v\frac{dv}{dx} + \frac{1}{\rho}\frac{dp}{dx} = 0$$

at all the points. For a polytropic flow, $p = p(\rho)$, and the integral of the above equation gives,

$$\frac{1}{2}v^2 - W(p) = W_0 \tag{4}$$

where, W_0 is the value of the potential $W = -\int \frac{dp}{\rho}$ at $v = 0$ surface. Now, since the potential must be negative for a bound flow, we see that above equation cannot be satisfied unless $W = 0 = v$ everywhere, i.e., when the flow is strictly non-accreting.

1.1. TWO CORRECTIONS OF STANDARD KEPLERIAN DISK MODEL

Before we go into the more general state-of-the-art advective disk model in next Section, we wish to discuss about two corrections to the 'standard' notion of a Keplerian disk. First one concerns the angular momentum distribution equation. Novikov & Thorne (1973) and all the papers which followed it too closely (recent one being, Lasota, 1994 which has many other errors) used $\tilde{l} = u_\phi$ to be the conserved specific angular momentum. This is not true. For a fluid with specific enthalpy h, the conserved specific angular momentum is $l = hu_\phi$. This correction in the angular momentum equation was first introduced in Chakrabarti (1996b, 1996c; hereafter referred to as C96b, and C96c respectively). See, eq. (8) of C96b.

The second major correction is done in the expression of viscous stress $W_{x\phi} = -\alpha p$. Normally, in a strictly Keplerian disk where radial velocity is negligible, this form is alright. But when the inertial pressure or ram pressure ρv^2 is significant, then one has to add this on the right hand side,

$$W_{x\phi} = -\alpha_\pi(p + \rho v^2) = -\alpha_\pi \Pi.$$

The effect of ρv^2 is not just cosmetic or rescaling of α, but it has a deeper significance. Π is conserved across a discontinuity, or shocks, for instance.

Thus, with this definition, viscous stress would also remain continuous across the shock. This ensures that no undue transport of angular momentum takes place in the disk. In a smooth flow this definition should be used, particularly, when infall velocity is significant as in an advective disk.

2. Need and Attempts for a Disk Model Alternative to Keplerian

In a black hole accretion, the specific binding energy on an equatorial plane ($\theta = \pi/2$) is give by (C96b, C96c),

$$u_t = -\left[\frac{\Delta}{(1-V^2)(1-\Omega l)(g_{\phi\phi} + lg_{t\phi})}\right]^{1/2}, \qquad (5)$$

where, $g_{\mu\nu}$ is the metric coefficient, and,

$$\Delta = r^2 - 2r + a^2; \qquad \omega = \frac{2ar}{A}; \qquad A = r^4 + r^2 a^2 + 2ra^2$$

with a is the Kerr parameter and r is the radial coordinate. V is the radial velocity measured in the rotating frame, l is the conserved specific angular momentum and Ω is the angular velocity of the orbiting matter. On the horizon, $\Delta = 0$, hence for a flow with finite binding energy on the horizon, $V = 1$. Since for causality, the speed of sound, $a_s < 1$, flow must be supersonic on the horizon (Chakrabarti, 1990; C96b). When a flow, strictly Keplerian at a large distance, enters through the horizon with $V = 1$, it must cross a sonic point at an intermediate distance. The flow must start having large radial velocities as it approaches the horizon, i.e., the flow must be advective. It can carry energy, entropy along with matter when advecting. Abramowicz & Zurek (1981) computes the sound speed at the sonic point to be,

$$a_s^2 = \frac{1}{x_c^2}[l_k^2(x_c) - l^2] \qquad (6)$$

and concludes that the flow must be sub-Keplerian at that point, since, by definition, $a_s^2 > 0$. However, this is not generally true when viscous flow with cooling, heating, nuclear energy release etc. are considered (C96a). The flow could be Keplerian or even super-Keplerian at the sonic point. What is definitely true, however, is that irrespective of any heating or cooling effects, the flow must be *sub-Keplerian* as well as supersonic at the horizon.

This necessitates the study of non-Keplerian flows more seriously. Indeed, even the observations from *Uhuru* back in the seventies of Cyg X-1 required a non-Keplerian component. This component was thought to be Compton clouds in models of Zdziarski (1986) or a coronal layers with possible holes in two-phase models of Haardt & Maraschi (1991). Recently,

Chakrabarti (1995), CT95, Chakrabarti (1997a; hereafter C97a) looked into this problem including advection and bulk motion and the theoretical predictions of the advective flows were fully put to test.

2.1. ATTEMPTS TO NON-KEPLERIAN FLOW STUDY: THICK ACCRETION DISKS

The pressure effect ($\frac{1}{\rho}\frac{dp}{dx}$ of eq. 1a) was first included by Paczyński and collaborators and others in a series of papers (Abramowicz et al. 1978; Koźlowski et al. 1978; Jaroszyński et al, 1980; Paczyński et al., 1980; Chakrabarti, 1985). Pressure from radiation causes the flow to be non-Keplerian, and fattens the disk geometrically. Matter with angular momentum refuses to come closer to the axis of the disk, and a funnel wall is also produced. All these were very exciting, since the origin and formation of jets was a real problem, and the funnel wall, with its super-Eddington luminosity seems to be helpful to push matter out along the axis. Meanwhile, Rees et al. (1982) also suggested (albeit using qualitative considerations) that the thick accretion disks are possible for very low accretion rate, since the gas would remain very hot with virial temperature $T \sim GMm_p/rk$ (where, k is the Boltzman constant). However, these disks did not have any radial velocity and early attempts to include this (see next Section) were not very successful. Normal proceedings was perturbed by the so-called Papaloizou-Pringle (1985) instability which was not found to be so fatal after all (e.g., Kojima, 1986). After two dimensional advective disks came about these thick disks are no longer studied in isolation, since in presence of centrifugal barrier, these 'classical' thick disks are special cases of advective disks.

2.2. ADVECTION MODELS: UPS AND DOWNS

Meanwhile, development was going on to include advection term ($v\frac{dv}{dx}$ of eq. 1a) as well. A large number of workers, since 1980s realized that accretion disks, both thin and thick, need a face-lifting by addition of radial velocity. However, earlier works had a partial success. Works of Liang & Thomson (1980), Abramowicz & Zurek (1981) Paczyński & Bisnovatyi-Kogan (1982), Paczyński & Muchotrzeb (1982), Muchotrzeb (1983) etc. deserve some mention in this respect. Several conclusions drawn in these works were incorrect, specially the existence of a special α parameter in Muchotrzeb (1983) was contested by Abramowicz & Kato (1989) as an artifact of finite distance of the boundary. Secondly, six B parameters describing vertical averaging was narrowed down in slim disk model of Abramowicz et al. (1988). However, while correcting a set of errors new errors and wrong concepts were introduced. (This is quite normal in the developmental phase of a subject.) One conclusion which excited the community temporarily was that the black holes should allow multiple solutions, just because there are multiple sonic points (Abramowicz & Zurek, 1981). Today we know that the entropies of the flow at these two points are completely different, and this usually means two completely different flows pass through two different sonic points. In fact, by topological reason one of the solutions cannot enter into the black hole at all. On the other hand, these two flows with two different entropies could be connected by standing shocks to obtain a unique steady solution (C96a, Chakrabarti, 1996d; hereafter C96d) since irreversible increase of entropy takes place at the shock. Matsumoto et al. (1984) tried to repair the inner edge of a strictly Keplerian flow by introducing transonic flows with nodal type sonic points. Similarly, both the methods and the global solutions of slim disks (including later solutions named advection dominated flows which follow this approach) are not correct. For instance, by choosing arbitrary initial velocity, sound speed, angular momentum etc., the solutions obtained were found to be globally incorrect. Fig. 3 of Abramowicz et al. 1988 suggests that (a) angular momentum should deviate away from a Keplerian disk as it approaches a Keplerian disk (for $r_{out} = 10^2$ case), and (b) the flow with accretion rate 50 times the *critical rate*, i.e., 800 times the Eddington rate was found to be deviating from a Keplerian disk at $r_{out} = 10^5 R_g$, where $R_g = 2GM/c^2$, the Schwarzschild radius. Today, we know that both these solutions are wrong. Both of these errors were coming from the initial condition. All the errors in slim disk approach have propagated into its recent re-incarnation of advection dominated flows (ADAF) which are supposed to deviate from a Keplerian disk by efficient evaporation typically at a million Schwarzschild radius (Narayan & Yi, 1994, 1995 and Narayan 1997). It is easily shown that this is impossible especially for a high viscosity flow which ADAF uses.

3. Fundamentals of Advective disks

An advective disk is the one which advects, or carry 'something', namely, mass, entropy, energy etc. Since this fundamentally means that radial velocity must be present, I define advective disks as those which have finite radial velocity which may even reach the velocity of light (e.g., on the horizon). Whether they actually advect energy or not will depend on the accretion rate and viscosity which in turn decide the cooling and heating efficiencies. These disks are the most general which are studied so far. If slim disks (high accretion rate, optically thick) or ADAF (low accretion rate and high viscosity, with very low radiative efficiency) solution really exists they would automatically come out of the advective disk solutions. For a black hole accretion, advective disks are the same as the transonic disks.

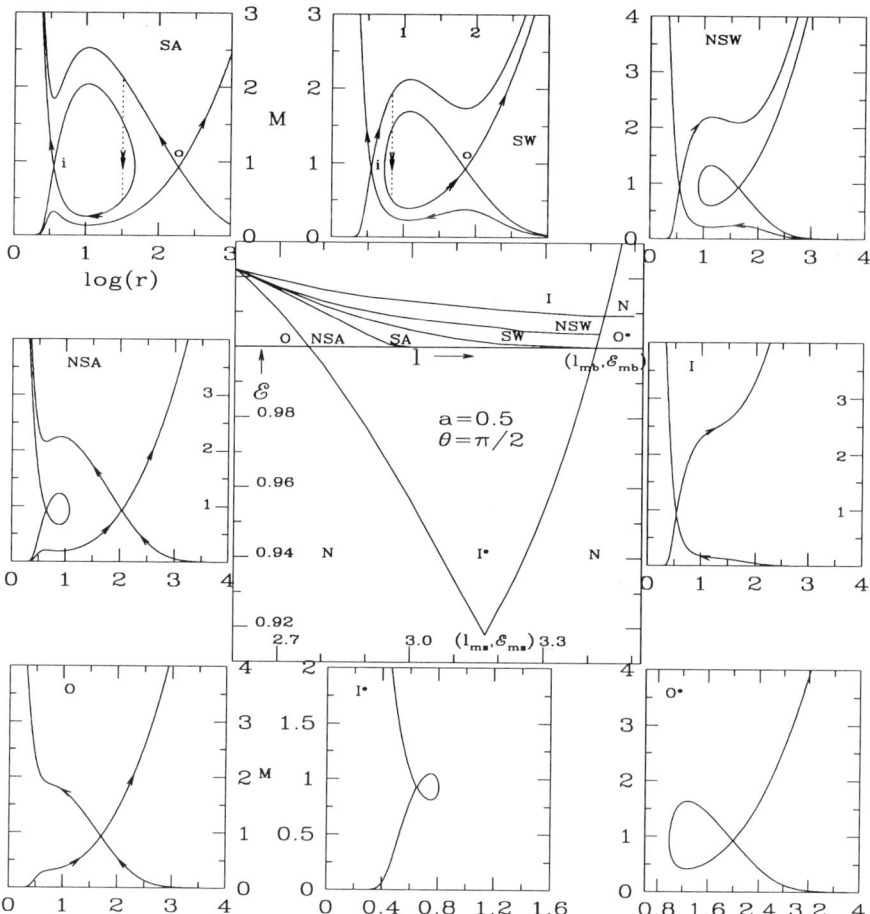

Fig. 1: Classification of the parameter space (central box) in the energy-angular momentum plane in terms of various topology of the black hole accretion. Eight surrounding boxes show the solutions from each of the independent regions of the parameter space. Each small box shows Mach number M against the logarithmic radial distance r (measured in units of GM_{BH}/c^2) Contours are of constant entropy accretion rate $\dot{\mathcal{M}}$. Similar classification is possible for all adiabatic index $\gamma < 1.5$. For $\gamma > 1.5$, only the inner sonic point is possible other than an unphysical 'O' type point [C96b].

For a neutron star accretion this is not necessarily so, as the neutron star accretion could be completely subsonic as well. Some of the recent reviews of the advective disks are in C96d and Chakrabarti (1998).

Before the full fledge advective flow is presented, consider only the inviscid, constant angular momentum thin flow in a Kerr geometry. Figure.

1, taken from C96b (also see, Chakrabarti, 1989; hereafter C89) shows all possible solutions and non-solutions. Mach number is plotted against the logarithmic radial distance (measured in units of GM/c^2). If the matter is bound, i.e., when the specific energy $\mathcal{E}$ is less than the rest mass of the flow c^2 (where c is the velocity of light, which is taken to be unity here), then there is *no* complete solution which becomes transonic (see, region $\mathcal{E} \leq 1$ in the Figure). See, Ryu (this volume) for further discussion on this classification. Thus a cool, Keplerian flow which is bound everywhere, does not have a way to enter into a black hole. Only if enough viscosity is present, then closed topology of I^* opens up (Fig. 3 below) and the originally Keplerian flow enters through the inner sonic point. However, these transonic flows would join with the Keplerian disk *very close* to the black hole (roughly around the inner sonic point). These are not ADAF solutions. In ADAF (Narayan & Yi, 1994, 1995) the flow starts to deviate from Keplerian disk (due to evaporation) from a million Schwarzschild radii.

When the flow is away from the equatorial plane, or energized by magnetic flares or other coronal effects, or in the extreme case, when the Keplerian disk itself is *very hot*, so that the specific energy is greater than 1 (rest mass), flow would deviate from a Keplerian disk and pass through outer (O, NSA), or inner (I, NSW), or both (NS, SW) sonic points. Flow may (SA, SW) or may not (NSA, NSW) have a standing shock in the flow. Only very high energy flows with weak viscosity (I) or very low energy flows with high viscosity (I^*) pass through the inner sonic point. This figure is drawn for a flow in vertical equilibrium and for adiabatic index $\gamma = 4/3$. It is to be noted that only one sonic point would be present if the polytropic index is greater than $\gamma = 1.5$ so the subdivision of the parameter space would look different and the question of shocks do not arise (Note a typographical error in C97 where it was mistakenly stated that $\gamma < 1.5$ would not have shocks.).

This classification of solution, though done for inviscid flows, is the backbone of the advective disk physics. Viscosity and cooling modify these solutions by changing the topologies in a very predictable way (see Fig. 3 below). But close to a black hole, where the infall timescale $r/v(r)$ is short compared to the viscous timescale (unless $\alpha \geq 1$) viscosity does not do much. Angular momentum remains roughly constant (and therefore behaves like an inviscid flow in some sense) in the last few to a couple of tens of Schwarzschild radii. Constant angular momentum flow introduces large centrifugal force which forms a dense region around a black hole (CEN-BOL). This is the centrifugal pressure supported boundary layer of the black hole. If the viscosity is small, this barrier is prominent, and even standing shocks may form, but when the viscosity is very large, the barrier may disappear and matter virtually falls freely near the horizon. This

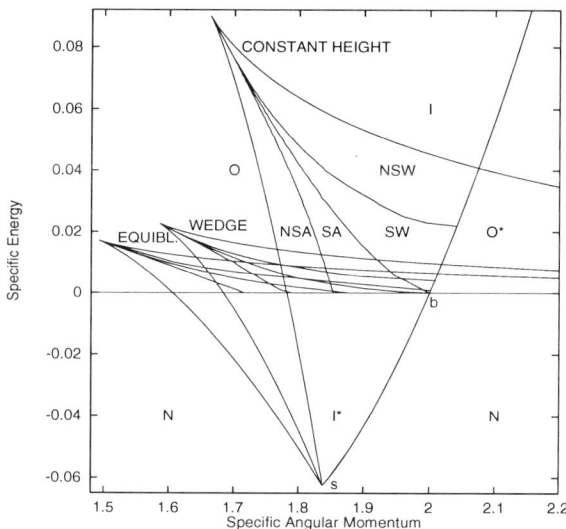

Fig. 2 : Same as in the central box of Fig. 1, but for flows of various models in pseudo-Newtonian geometry. When the models are changed as marked, the number of regions, and therefore topologies, do not change. Only the boundaries vary. The points s and b which represent the marginally stable and marginally bound quantities of the geometry remain invariant under change of models.

unique region produces the power-law hard tail in the soft states of black holes through *bulk motion Comptonization* as described in CT95.

The classification presented in Fig. 1 is generic. The total number of topological variation of the solutions does not depend on any cosmetic changes of the model, such as vertical averaging. Instead, the boundaries of the region will vary. Figure 2 shows the classification in a Schwarzschild geometry for a (a) thin disk with constant height, (b) conical wedge flow and (c) a flow in vertical equilibrium. Note that in each of the models the point s and b are fixed. They represent the marginally stable and marginally bound quantities and are functions of the geometry only, and therefore model independent. However, since it is a pseudo-Newtonian model, the point s occurs at $\mathcal{E} = -0.0625$ rather than at -0.057, valid for Schwarzschild geometry. Note also that since it is a Newtonian computation, rest mass has been subtracted from the specific energy. Thus, cool Keplerian flows have negative specific energy in this notation.

Because the flow is inviscid, the angular momentum is constant and the flow cannot join with a Keplerian disk within a finite distance at all (i.e., joins at infinity!). The energy is conserved, so the entire energy is advected towards the black hole. In presence of shocks (which form in regions SA and

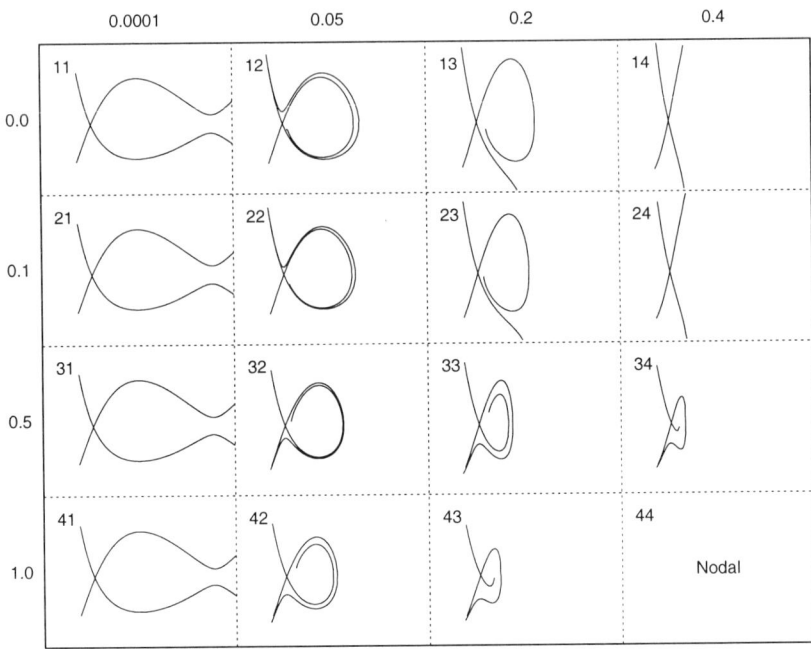

Fig. 3: Changes in advective disk topologies when viscosity parameter α_Π (marked on upper axis) and the cooling factors f (along left axis) are varied. x_{in} and l_{in} are kept fixed. ADAF type solutions ($f \sim 1$) are impossible for high viscosity. Logarithmic distance along x-axis and Mach number is along y-axis. See text for details.

SW in accretions and winds respectively) entropy can be generated due to turbulent viscosity (for instance) at the shock and the entropy generated is also advected away completely (C89, Chakrabarti, 1998). Thus, a truly advection dominated flow (completely radiatively inefficient) is the most weakly viscous flow which joins with a Keplerian disk very far away. But, then, the accretion rate should be excruciatingly low so that the energy remains roughly conserved.

This understanding is fully confirmed when one deals with a general flow as described in C96a. Figure 3 (Fig. 2a of C96a) shows a class of solutions placed in grids (grid numbers are marked in each box) of size 0 to $50R_g$ along x axis (logarithmic distance) and 0 to 2.5 (Mach number along y) along y axis. f and α_π are written on the left and upper axis respectively. All the other parameters are the same: the location of the sonic point $x_{in} = 2.795$, and the angular momentum at the inner edge $l_{in} = 1.65$. $\gamma = 4/3$. $f = 1$ solutions are the correct ADAF solutions. Unlike Narayan and collaborators model, our solutions show that ADAF (indeed all the solutions with $f > 0.5$ in this particular set of parameters) solutions exist only for small viscosity. There are no shocks. This is also confirmed by works of Bisnovatyi-Kogan (this volume) who shows that ADAF solutions for $f > 0.75$ are not possible. Non-ADAF solutions ($f \neq 1$, especially $f \leq 0.5$) are more promising, however. We note that there are two critical viscosities at which the topology changes dramatically (this behaviour of topologies revolutionized our understanding of the accretion flows in black holes). For very low viscosities the flow passes through the inner sonic point and can join a Keplerian flow (subsonic branch with $l = l_K$) very far away. For very high viscosities, the subsonic branch touches a Keplerian disk near by. For the intermediate viscosities, topologies are semi-closed, but they can be reached using the outer sonic point (discussed in C96a) when shock conditions are satisfied. If shock conditions are not satisfied then the flow has to enter through outer sonic point only, much like the original Bondi flow.

Several authors have lamented that they could not find shocks (Chen et al, 1997; Narayan et al. 1997; Narayan, 1997). The solutions of these groups are not correct, as they use the slim disk approach by specifying too many parameters at the launching point in a Keplerian disk. We have already mentioned in §1 that an accreting flow cannot be strictly Keplerian. Thus, ADAF solutions published by these groups, all of them, in my view, are incorrect. It is therefore immaterial whether these solutions contained shocks or not. Only mathematically correct solution is that of Narayan & Yi (1994) where self-similar solution was studied. But a black hole accretion is not at all self-similar. Exact solutions of the same set of equations (Fig. 3 above, for example) show very rich behaviour, with Mach number going up and down several times, while a self-similar flow has a constant Mach number. Some more reasons are written by Lu (this volume).

The correct approach is to start from a sonic point and integrate backward till the Keplerian value is reached, and forward till the horizon is reached. That way all the sonic points are used properly (unlike in slim disk approach where at most one sonic point may be used, if integrated properly). There are a large number of independent groups (Lu et al., 1996; Yang & Kafatos, 1994; Nobuta & Hanawa, 1994 etc.) who have found shocks in accretion flows. Shock study was also common in winds from stars (e.g., Habbal & Rosner, 1984, and references therein). While shocks and centrifugal barrier supported boundary layer (CENBOL) would remain the most important ingredients in accretion flows close to a black hole, the problem would still remain at the junction point when the viscosities and cooling parameters are kept constant in the advective region. As Chakrabarti et al. (1996) showed using a detailed numerical simulations, at the junction the disk becomes super-Keplerian. We conjecture that either the viscosity and cooling parameters would have to smoothly vary at the junction to connect

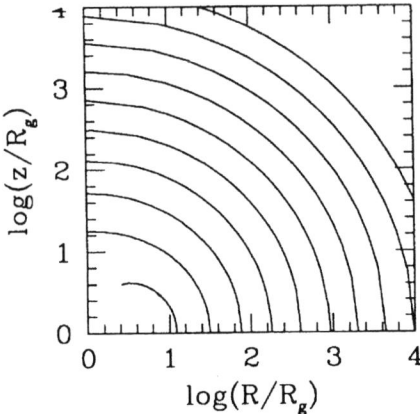

Fig. 4a : Isodensity contours in an ADAF solution. Rotating flow also moves in on the jet axis without forming any centrifugal barrier (taken from Narayan, 1997; courtesy of the Astronomical Society of Pacific Conference Series).

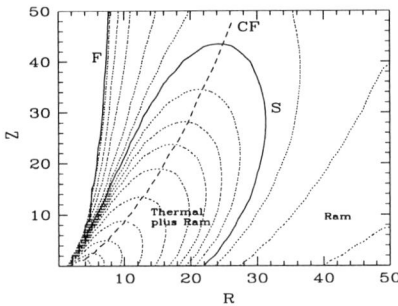

Fig. 4b : Isobaric (isodensity for polytropic flows) contours for advective flows which have centrifugal barrier CF, funnel wall F and possible shocks S (C96c).

the advective disk with a Keplerian disk, or, the advective disk would be a bit unsteady to try to match with a Keplerian disk. This may even cause quasi-periodic oscillations.

We end this Section with a comparison of a 'solution' of ADAF and the solution of an advective disk. Figure 4a is supposed to be a two dimensional flow density contours of ADAF (Narayan, 1997) and Fig. 4b shows contours of thermal pressure inside the shock (marked S) and thermal plus ram pressure outside the shock (taken from C96c). In ADAF 'solution', even when the matter has angular momentum, it has no problem on the axis. This is fundamentally incorrect. In Fig. 4b, the funnel wall (F) and the centrifugal barrier (CF) are formed as expected. If, for the sake of argument one assumed that in ADAF all angular momentum is removed

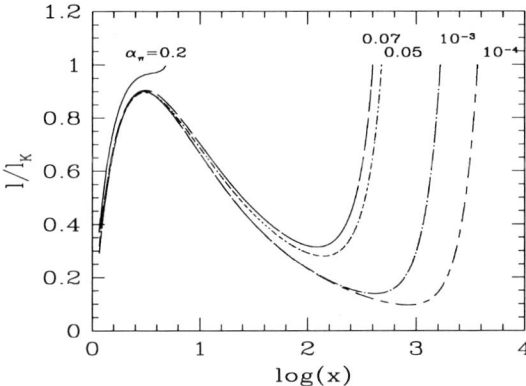

Fig. 5: Ratio of disk angular momentum to the Keplerian angular momentum of several advective disk solutions whose viscosity parameters (α_Π) are marked. At or near $l/l_K = 1$, the disk may join with a Keplerian flow if energetics are right. Note that size of the advective region inversely varies with α_Π.

completely, from Fig. 2, one notes that there is no solution close to a black hole which simultaneously has zero angular momentum and at the same time has negative energy (coming from Keplerian disk). These does not imply that the original philosophy of Rees et al. (1982) is incorrect. Disks with very low accretion rate might be possible, only if the magnetic heating were negligible (see, Bisnovatyi-Kogan, this volume).

4. What Should a Realistic Accretion Disk Look Like?

Figure 5 shows the angular momentum distribution of a collection of solutions with $\gamma = 4/3$ for different choices of the viscosity parameter α_Π (marked on each curve). Each distribution touches a location x_K where $l/l_K = 1$, where, roughly speaking, one would expect the advective region to join a Keplerian disk. First note that when other parameters (basically, specific angular momentum and the location of the inner sonic point) remain roughly the same, x_K changes inversely with α_Π. If one assumes, as CT95 and C97a did, that alpha viscosity parameter *decreases* with vertical height, then it is clear from the general behaviour of Fig. 5 above that x_K would go up with height. The disk will then look like a sandwitch with higher viscosity matter flowing along the equatorial plane with Keplerian disk closest to the black hole. This fact that the inner edge of the disk should move in and out when the black hole goes in soft or hard state (e.g.,

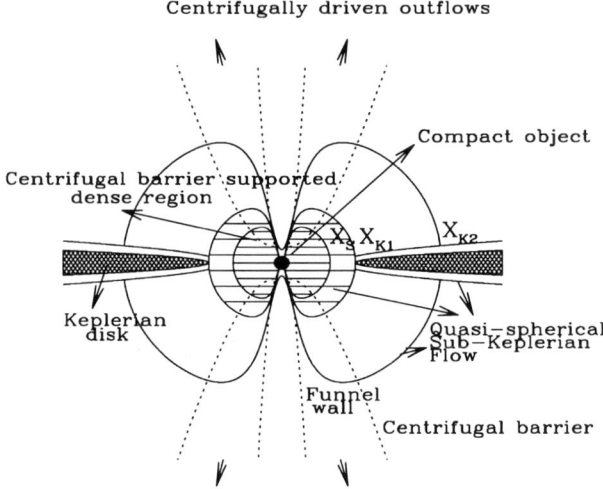

Fig. 6: Schematic diagram of an unmagnetized advective disk with all its components. Compact object is surrounded by a centrifugal barrier supported boundary layer (CENBOL), which in turn is surrounded by an advective region. The entire flow may emerge from a Keplerian disk or from a mixture of Keplerian and a sub-Keplerian flow. Jets and outflows are originated from the CENBOL.

Gilfanov, Churazov & Sunyaev, 1997; Zhang et al., 1997) are thus naturally established from this advective disk solution. Note the presence of two groups of solutions, one with a small viscosity, and the other with a large viscosity. These two groups correspond to two sets of solutions presented in Fig. 3. In the intermediate viscosity, if the shock condition is not satisfied, and there is no solution which passes through the outer sonic point, there is no continuous solution connecting the Keplerian disk with horizon. In that case, the only possible alternative is to produce an unsteady flow.

Figure 6 shows the schematic diagram of a disk as we perceive it close to a black hole. There is, as yet, *no single global solution* for the whole system. What we now have are the bits and pieces of the behaviours of thin flows under various parametric variations. After combining them, the whole picture emerges. This was first presented in Texas Symposium in 1994 (Chakrabarti, 1995), and later was used to compute spectra of black hole candidates (CT95, C97a). If the inflow parameters are such that shocks form, then x_s is the shock boundary, otherwise it is simply the boundary of the centrifugal barrier when pressure effects are included. Without the pressure effects one could have the funnel wall and the centrifugal barrier (dotted curves) in between which outflows are likely to emerge.

In Section 6, we shall establish that the general observational results agree with such a picture, even when the spectrum is non-stationary.

5. Progresses in Numerical simulation works

Last two decades saw tremendous progress in numerical simulation work, mostly due to the advent of faster computers with larger memory and due to better numerical algorithms. In 1978, Wilson showed that an accretion with significant angular momentum was accompanied by shock waves which traveled outwards. This code was later improved upon, with number of grid points as well as the evolution time orders of magnitude higher. A series of very important simulations were made with this code to show that thick accretion disks can indeed form in inviscid flows (Hawley, Smarr & Wilson, 1984, 1985). These simulations also confirm the results of Wilson (1978) that non-steady shock waves are formed which travel outward. From the post-shock flow, a very strong wind is generated which is hollow in nature which 'hugs' the funnel wall. Due to inviscid nature of the flow, centrifugal force kept it away from the axis of symmetry.

However, there was one problem: since in the contemporary period, only available theoretical work on non-Keplerian disk model was that of a thick disk, the numerical results of the thermodynamic quantities could not be compared properly. Figures 7(a-b) show examples of two numerical solutions of Hawley (1984) which are compared with thick disk solutions. There is only *qualitative* similarity between the theoretical work and the numerical work. This discrepancy was mostly due to the fact that the thick disk model was not *advective* while the numerical simulation allowed the matter to rush to the black hole as fast it wanted! Of course, to a very smaller extent, the discrepancy was due to numerical error, because of numerical diffusion of energy and angular momentum, but it could be ignored in the present discussion. Today, we have a complete theoretical solution on advective disks, and a large number of numerical simulations (Chakrabarti & Molteni, 1993; Molteni, Lanzafame & Chakrabarti, 1994; Molteni, Ryu & Chakrabarti, 1995; see also, Ryu, this volume; Molteni, this volume) show how accurately the numerical simulation results match with the theoretical solution. Indeed, the analytical works could now be used to test a code in a spherical coordinates. The progress in this field is clearly obvious.

A new understanding has emerged recently regarding the time-dependent behaviour of the accretion flows. First, Chakrabarti & Molteni (1995) showed how a highly viscous flow can re-distribute angular momentum inside a flow to form a Keplerian disk even when the inflow at the outer boundary

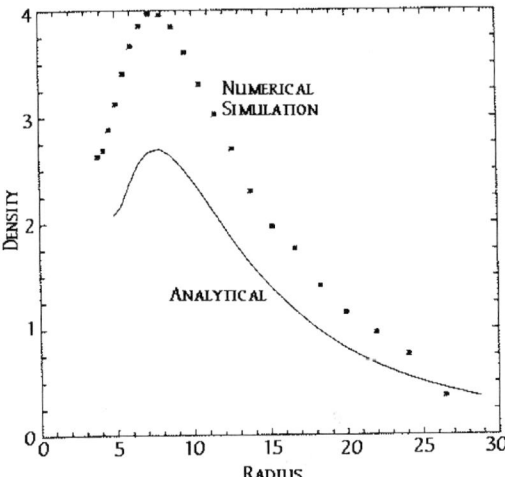

Fig. 7a : Numerical results of density variation along the equatorial plane of an inviscid flow of specific angular momentum $l = 3.77GM/c$ is compared with the analytical solution from thick accretion disk. Agreement is only qualitative. (Adapted from Hawley, 1984).

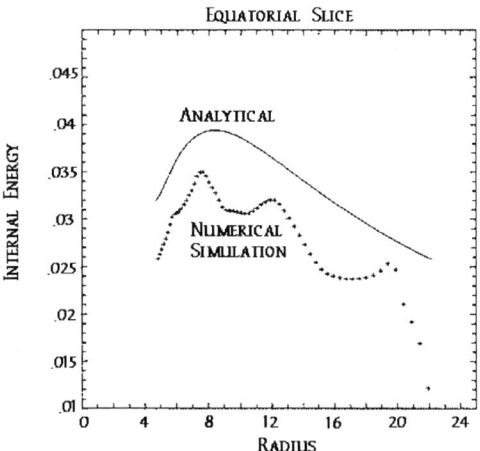

Fig. 7b : Numerical results of internal energy (a measure of temperature) variation along the equatorial plane of an inviscid flow of specific angular momentum $l = 3.8GM/c$ is compared with the analytical solution from thick accretion disk. Details do not agree. (Adapted from Hawley, 1984).

is highly sub-Keplerian. Second, in some regions of the parameter space, particularly in NSA and NSW of Fig. 2, though steady solutions which pass only though the outer sonic points were predicted, the flow chose to produce shocks which started oscillating with frequencies similar to that observed in quasi-periodic oscillations. This is because the solution topology (Fig. 1) has an inner sonic point through which higher entropy flow enters into the black hole. Since no stable shock was predicted, unsteady shocks are produced (see, Ryu, this volume). On the other hand, in presence of cooling effects, shocks oscillate (even when a steady solution is predicted) when the cooling time scale and the infall time scale roughly match (see, Molteni, this volume). Whereas the first type of oscillation weakly depends on the accretion rates, especially when the rate is very low (i.e., even in the absence of cooling), the second type of oscillation strongly depends on cooling effects, and therefore accretion rates. Because such oscillations not only explain the frequencies, but also the amplitude and other timing properties (e.g., Paul, this volume), it is a good candidate for the explanation of the quasi-periodic oscillations observed in the black hole candidates.

6. Progresses in Modeling Observational Results Which Required Accretion Disks

One can list a large number of interesting spectral features that are observed from a black hole candidate (see, Chakrabarti, 1998 for a review). The spectrum could be hard or soft or it could have a power law hard tail even in soft states (Tanaka & Lewin, 1995) or it could show X-ray novae behaviour where different components have completely different timing properties (Orosz et al., 1997) or it could show quasi-periodic oscillation (Dotani, this volume) or it could be in a quiescent state for years (McClintock, Horne & Remillard, 1995) or etc. We discuss them briefly to appreciate how they demanded the deviation from a standard Keplerian disk of SS73 and NT73.

Black holes are being fundamentally black, their proper identifications must necessarily include quantification of very special spectral signatures of radiating matter entering in them. Since the inner boundary condition of the flow must be unique, advective disk solutions (which achieve this boundary value automatically) are the only disk solutions, which, predict spectral features most self-consistently. A single global solution for a multidimensional flow is still missing and some details such as magnetic field, have not been incorporated self-consistently yet. Thus, the discussion on the spectral properties can only be a bit qualitative. This is easily compensated for by the support of a well developed theory which satisfactorily explains stationary and non-stationary features around galactic as well as extra-galactic black holes using a single framework.

6.1. TRIGGERING OF HARD AND SOFT STATES

Black holes are known to show hard and soft states (see, Ebisawa, Titarchuk & Chakrabarti, 1996; hereafter ETC96). When the viscosity of the inflow changes, the Keplerian and sub-Keplerian components (Fig. 6) redistribute matter among themselves depending on viscosity of the flow which at the same time, also change the inner-edge of the Keplerian component. Sudden rise in viscosity would bring more matter to the Keplerian component (with rate $\dot{m}_d$) and bring the Keplerian edge closer to the black hole (see, Chakrabarti & Molteni, 1995 for numerical simulations) and sudden fall of viscosity would bring more matter to the sub-Keplerian halo component (with rate $\dot{m}_h$) and the Keplerian component would go farther out. Disk component $\dot{m}_d$ not only governs the soft X-ray intensity directly coming to the observer, it also provides soft photons to be inverse Comptonized by sub-Keplerian CENBOL electrons. The CENBOL (comprised of matter coming from $\dot{m}_d$ and $\dot{m}_h$) will remain hot and emit power law (energy spectral index, $F_\nu \sim \nu^{-\alpha}$, $\alpha \sim 0.5 - 0.7$) hard X-rays only when its intercepted soft photons from the Keplerian disk are insufficient, i.e., when $\dot{m}_d << 1$ to $\dot{m}_d \sim 0.1$ or so, while $\dot{m}_h$ is much higher. For $\dot{m}_d \sim 0.1 - 0.5$ (with $\dot{m}_h \sim 1$), CENBOL cools catastrophically and no power law is seen (this is sometimes called a high state). With somewhat larger $\dot{m}_d$, the power law due to the bulk motion of electrons (CT95; Titarchuk, Mastichiadis & Kylafis, 1997) is formed at around $\alpha \sim 1.5$ (this is sometimes called a very high state). Such hard/soft transitions are regularly seen in black hole candidates (Dolan et al, 1979; Ebisawa et al., 1994; Zhang et al., 1997). This α may weakly depend on the flow angular momentum (Chakrabarti, Titarchuk, Kazanas & Ebisawa, 1996).

6.2. CONSTANCY OF SLOPES IN HARD AND SOFT STATES

Observations indicate that in hard states, power law slopes remain almost constant with luminosity (e.g. Sunyaev et al., 1994; Ebisawa et al., 1996; CT95; Kuznetsov et al., 1997; Grove et al. 1998). Advective disks also show this property (CT95; C97a). Particularly important is the weak power law in the soft state as this is not observed in neutron star candidates. In advective disks, matter behave democratically ($V = 1$) close to a horizon independent of its history. This unique fact produces unique spectra through bulk motion Comptonization and readily explains the weak power law tail in the soft states (CT95; Titarchuk, Mastichiadis & Kylafis, 1997). Since this part of the spectra is universal, it should have been seen even in hard states, had the dominant spectra due to thermal Comptonization been somehow subtracted. In the intermediate states both the power laws (hard and soft) are seen (Ling et al., 1997).

6.3. VARIATION OF INNER EDGE OF THE KEPLERIAN COMPONENT

Observations indicate that the Keplerian disk component varies with accretion rates (e.g. Gilfanov, Churazov & Sunyaev, 1997). In advection dominated models of Narayan & Yi (1994, 1995) such variations are achieved by evaporation and condensation of the disks by unknown fundamental physics. However, such variation is a natural property of the advective disks (Fig. 5). As the viscosity increases, x_K becomes smaller in viscous time scale, at the same time more matter is added to the Keplerian component.

6.4. RISE AND FALL OF X-RAY NOVAE

X-rays novae (e.g., A0620-00, GS2000+25, GS1124-68, V404 Cygni etc.) produce bursts of intense X-rays which decay with time (decay time is typically 30d). This phenomenon may be repeated every tens to hundreds of years. While in persistent black hole candidates (such as, Cyg X-1, LMC X-1, LMC X-3) Keplerian and sub-Keplerian matter may partially redistribute to change states, in X-ray novae candidates the net mass accretion rate may indeed decrease with time after the outburst, even if some redistribution may actually take place. First qualitative explanation of the change of states in X-ray novae in terms of the advective disk model was put forward by ETC96. The biggest advantage of the advective solution is that it automatically moves the inner edge of the Keplerian disk as viscosity is varied. Similar to the dwarf novae outbursts, where the Keplerian disk instability is triggered far away (e.g., Cannizzo, 1993) here also the instability may develop and cause the viscosity to increase, and the resulting Keplerian disk with higher accretion rate moves forward. In Fig. 8a we show the spectral evolution of a two component advective disk whose inner edge (marked on each curve) is approaching towards the black hole due to rise in viscosity. The component accretion rates have been kept fixed at $\dot{m}_d = 0.01$ and $\dot{m}_h = 1.0$ respectively. No shock is assumed but the centrifugal barrier has a similar effect (C97a). Here photon numbers are plotted against their energy. Assuming that this evolution is the cause of the spectral variation in GRO J1655-40 as reported by Orosz et al. (1997) and Haswell (this volume), we can have an idea of viscosity working in that disk. In Fig. 8b, upper panel, we show the variation of the photon rate with time (in arbitrary units). Along X-axis, we plotted $R^{3/2}$ (where R is the inner edge of the Keplerian disk in units of R_g) which is a measure of infall time $R/v(R)$. If the viscosity is such that it causes four days of delay

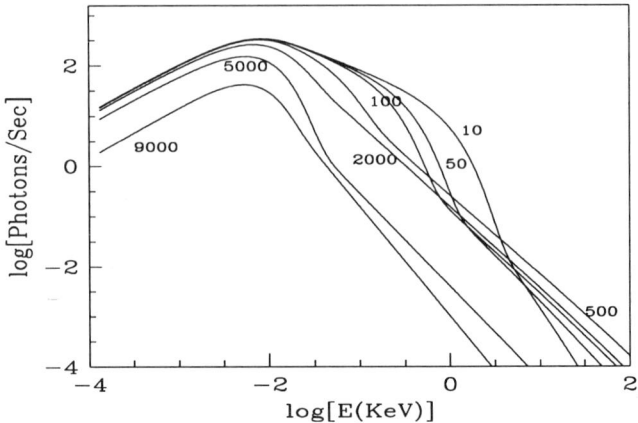

Fig. 8a : Photon numbers are plotted against photon energy. Location of the inner edge of the Keplerian component of the disk is marked on each curve.

between the soft X-rays and optical (as reported in Orosz et al. 1997) then it seems that whole of the rising phase takes around nine days as reported. Note also that the slope of B waveband is higher as compared to that of I as expected. Here I, R & B are chosen to be in energy bands of 1.24 − 2.2eV, 2.2 − 3.16eV and 3.16 − 4.13eV respectively, and soft and hard X-rays are in 2 − 12keV and 12 − 1000KeV ranges respectively. The lower panel shows the hardness ratio variation with time. First the optical band rises keeping hard to soft almost fixed. This is marked as the *horizontal* branch. Then along the *diagonal* branch the hard component rises faster than the soft. Finally, towards the end of the novae rising phase, the soft component rises rapidly, keeping the optical ratio nearly constant. This is marked as the *vertical* component. Such detailed predictions should be verifiable with observations. Small modifications or parameters, such as the angular momentum of the advective region, and the variation of the accretion rate in the two components may be necessary when actually fitting the data.

6.5. QUIESCENT STATES OF X-RAY NOVAE CANDIDATES

After years of X-ray bursts, the novae becomes very faint and hardly detectable in X-rays. This is called the quiescent state. This property is built into advective disk models. As already demonstrated (Fig. 5) x_K recedes from the black holes as viscosity is decreased. With the decrease of viscosity, less matter goes to the Keplerian component (Chakrabarti & Molteni, 1995) i.e., $\dot{m}_d$ goes down. Since the inner edge of the Keplerian disk does

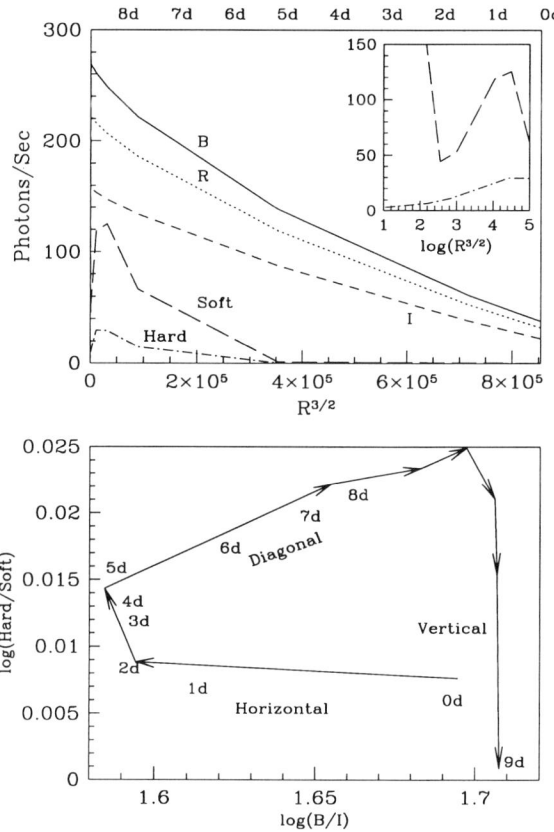

Fig. 8b : Upper panel shows the rising phase of an X-ray novae while the lower panel shows the hardness ratio. Photon numbers are plotted against $R^{3/2}$, a measure of infall time $R/v(R)$. Upper axis gives the number of days that are passed. Inset: details in the last day of the rising phase. Note that rise of B band is sharper than the rise of I band; and soft X-ray rise is delayed by four days. In lower panel, one sees three distinct phases: horizontal, diagonal and vertical branches.

not go all the way to the last stable orbit, optical radiation is weaker in comparison with what it would have been predicted by a SS73 model. This behaviour is seen in V404 Cyg (Wagner et al. 1994) and A0620-00 (McClintock et al., 1995). The deviated component from the Keplerian disk almost resembles a constant energy rotating flow described in detail in C89. It is also possible that our own galactic center may have this low viscosity, low accretion rate with almost zero emission efficiency global advective disks as mentioned in C96d.

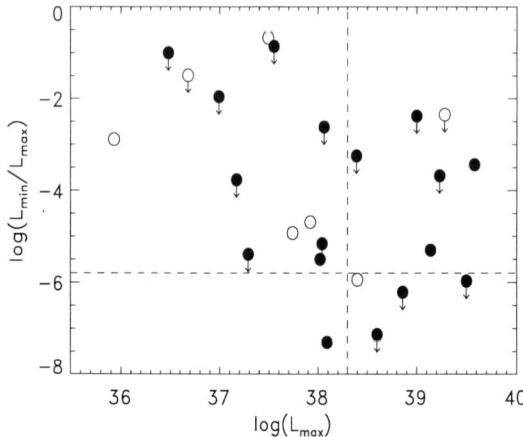

Fig. 9 : Ratio of minimum and maximum luminosities from a collection of compact objects believed to be black holes (filled circles) and neutron stars (open circles) [taken from Chen et al 1998]. ADAF predicts that black holes should lie below the dashed line while neutron stars should be above it.

Workers of ADAF 'solution' claims to fit these states (Narayan et al., 1997) well. In this model highly viscous ($\alpha \sim 0.1 - 0.5$) quasi-spherical flow resulted from Keplerian disk evaporation (which is also in equipartition with magnetic field at all radii!) was used. Typically one data point in the hard X-ray region was used and the fits are poor. On the contrary, the advective disk solution we advocate does not require such evaporation, and the advective ion torus of low mass accretion rate comes most naturally out of the governing equations only for very low viscosity case (Fig. 5).

While we are dealing with low luminosity phase of black hole candidates, we may as well mention a recent observation by a team of experts (Chen et al., 1998, also see, Chen, Shrader & Livio, 1997). They find that it is not possible to distinguish between a black hole and a neutron star purely based on observations of total luminosity. Figure 9 (taken from Chen et al. 1998) shows the ratio of minimum and maximum luminosities of a collection of neutron stars (open circles) and black holes (filled circles). The result goes against the prediction by the ADAF solution that all the black holes should lie below dashed line and all the neutron stars should lie above (Garcia, McClintock & Narayan, 1997).

6.6. QUASI-PERIODIC OSCILLATIONS OF X-RAYS

In some large region of the parameter space the solutions of the governing equations 1(a-d) are inherently time-dependent. Just as a pendulum inherently oscillates, the physical quantities of the advective disks also show oscillations of the CENBOL region for some range in parameter space. This oscillation is triggered by competitions among various time scales (such as infall time scale, cooling time scales by different processes). Thus, even if black holes do not have hard surfaces, quasi-periodic oscillations could be produced. Although any number of physical processes such as acoustic oscillations (Taam, Chen & Swank, 1997), disko-seismology (Nowak & Wagoner, 1993), trapped oscillations (Kato, Honma & Matsumoto, 1988) could produce such oscillation frequencies, modulation of 10 − 100 per cent or above cannot be achieved without bringing in the dynamical participation of the hard X-ray emitting region, namely, the CENBOL. By expanding back and forth (and puffing up and collapsing, alternatively) CENBOL intercepts variable amount of soft photons and reprocesses them. Some of the typical observational results are presented in Dotani (this volume), Halpern & Marshall (1996); Cui et al. (1997). Recently more complex behaviour has been seen in GRS 1915+105 (Morgan, Remillard & Greiner, 1997; Paul et al., 1998; Paul, this volume), which may be understood by considering several cooling mechanisms simultaneously. This will be reported elsewhere.

7. Outflows from the Advective Disks

Problem with properly explaining observed outflows and jets from black hole candidates is that they have to originate not despite of the accretion flows, but because of it. This is because unlike stellar surface, black holes do not have hard surfaces and atmospheres from where winds could independently come out. Thus physical processes must exist close to the black hole to join topologies of wind and accretion. In Fig. 1 above, we see that all the wind type solutions must have specific energy $\mathcal{E} \geq 1$ at the sonic point (C89). $\mathcal{E}$ is higher if the flow has to be supersonic very close to the black hole (region I in Fig. 1). In C89, it was shown that entropy measure $\dot{\mathcal{M}}$ must be higher in order that flows may emerge through inner sonic point. Thus we need to search for a physical mechanism which dumps more entropy to a little amount of energized matter. One possible, natural source of entropy is a stationary or non-stationary shock wave.

Chakrabarti (1997b) suggested one simple method to compute the outflow rate assuming that the inflow and outflows are both conical. Assume for the sake of argument that our system is made up of the infalling gas, the dense boundary layer of the compact object, and the outflowing wind. The sub-Keplerian, hot and dense, quasi-spherical region forms either due to centrifugal barrier or due to pair plasma pressure or pre-heating effects. The accretion rate of the incoming flow is given by,

$$\dot{M}_{in} = \Theta_{in}\rho\vartheta r^2. \tag{7}$$

Here, Θ_{in} is the solid angle subtended by the inflow, ρ and ϑ are the density and velocity respectively, and r is the radial distance in units of GM/c^2. In this unit, for a freely falling gas,

$$\vartheta(r) = [\frac{1-\Gamma}{r}]^{1/2} \quad \text{and} \quad \rho(r) = \frac{\dot{M}_{in}}{\Theta_{in}}(1-\Gamma)^{-1/2}r^{-3/2} \tag{8}$$

Here, Γ/r^2 (with Γ assumed to be a constant) is the outward radiative force.

We assume that the outer boundary of CENBOL is at $r = r_s$ where the inflow gas is compressed. The compression could be abrupt due to standing shock or gradual as in a shock-free flow with angular momentum (C97a). This details are irrelevant. At this barrier, then

$$\rho_+(r_s) = R\rho_-(r_s) \quad \text{and} \quad \vartheta_+(r_s) = R^{-1}\vartheta_-(r_s) \tag{9}$$

where, R is the compression ratio. Exact value of the compression ratio is a function of the flow parameters, such as the specific energy and the angular momentum. Here, the subscripts $-$ and $+$ denote the pre-shock and post-shock quantities respectively. At the shock surface, the total pressure (thermal pressure plus ram pressure) is balanced.

$$P_-(r_s) + \rho_-(r_s)\vartheta_-^2(r_s) = P_+(r_s) + \rho_+(r_s)\vartheta_+^2(r_s). \tag{10}$$

Assuming that the thermal pressure of the pre-shock incoming flow is negligible compared to the ram pressure, using eqs. (10) we find,

$$P_+(r_s) = \frac{R-1}{R}\rho_-(r_s)\vartheta_-^2(r_s). \tag{11}$$

The isothermal sound speed in the post-shock region is then,

$$C_s^2 = \frac{P_+}{\rho_+} = \frac{(R-1)(1-\Gamma)}{R^2}\frac{1}{r_s} = \frac{(1-\Gamma)}{f_0 r_s} \tag{12}$$

where, $f_0 = R^2/(R-1)$. An outflow is expected to be subsonic close to the black hole and supersonic far away. In the subsonic region, the pressure and density are expected to be almost constant and thus it is customary to assume isothermality condition up to the sonic point (Tarafdar, 1988). The sonic point conditions are computed from the radial momentum equation,

$$\vartheta\frac{d\vartheta}{dr} + \frac{1}{\rho}\frac{dP}{dr} + \frac{1-\Gamma}{r^2} = 0. \tag{13}$$

and the continuity equation

$$\frac{1}{r^2}\frac{d(\rho\vartheta r^2)}{dr} = 0 \tag{14}$$

in the usual way, i.e., by eliminating $d\rho/dr$,

$$\frac{d\vartheta}{dr} = \frac{N}{D} \tag{15}$$

where,

$$N = \frac{2C_s^2}{r} - \frac{1-\Gamma}{r^2} \quad \text{and} \quad D = \vartheta - \frac{C_s^2}{\vartheta} \tag{16}$$

and putting $N = 0$ and $D = 0$ conditions. These conditions yield, at the sonic point $r = r_c$, for an isothermal flow,

$$\vartheta(r_c) = C_s, \quad \text{and} \quad r_c = \frac{1-\Gamma}{2C_s^2} = \frac{f_0 r_s}{2} \tag{17}$$

where, we have utilized eq. (12) to substitute for C_s.

The constancy of the integral of the radial momentum equation (eq. 13) in an isothermal flow gives:

$$C_s^2 \ln \rho_+ - \frac{1-\Gamma}{r_s} = \frac{1}{2}C_s^2 + C_s^2 \ln \rho_c - \frac{1-\Gamma}{r_c} \tag{18}$$

where, we have ignored the initial value of the outflowing radial velocity $\vartheta(r_s)$ at the dense region boundary, and also used eq. (17a). We have also put $\rho(r_c) = \rho_c$ and $\rho(r_s) = \rho_+$. Upon simplification, we obtain,

$$\rho_c = \rho_+ exp(-f) \quad \text{where,} \quad f = f_0 - \frac{3}{2}. \tag{19}$$

Thus, the outflow rate is given by,

$$\dot{M}_{out} = \Theta_{out}\rho_c \vartheta_c r_c^2 \tag{20}$$

where, Θ_{out} is the solid angle subtended by the outflowing cone. Upon substitution, one obtains,

$$\frac{\dot{M}_{out}}{\dot{M}_{in}} = R_{\dot{m}} = \frac{\Theta_{out}}{\Theta_{in}} \frac{R}{4} f_0^{3/2} exp(-f) \tag{21}$$

which, explicitly depends only on the compression ratio:

$$\frac{\dot{M}_{out}}{\dot{M}_{in}} = R_{\dot{m}} = \frac{\Theta_{out}}{\Theta_{in}} \frac{R}{4} [\frac{R^2}{R-1}]^{3/2} exp(\frac{3}{2} - \frac{R^2}{R-1}) \tag{22}$$

apart from the geometric factors. This simple result is independent of the size of the dense cloud or the outward radiation force constant Γ. This is because the gravitational and radiation force have very simple forms ($\propto 1/r^2$).

Also, outward driving centrifugal force was ignored. Similarly, the ratio is independent of the mass accretion rate which should be valid only for low luminosity objects. For high luminosity flows, Comptonization would cool the dense region completely (CT95) and the mass loss will be negligible. In reality there would be a dependence (probably weak) on these quantities when full general relativistic considerations of the rotating flows are made. Exact and detailed computations using both the transonic inflow and outflow (where the compression ratio R is also computed self-consistently) are in Das & Chakrabarti (1998).

Figures 10(a-b) contain the basic results. Figure 10a shows the ratio $R_{\dot{m}}$ as a function of the compression ratio R (plotted from 1 to 7), and Figure 10b shows the same quantity as a function of the polytropic constant $n = (\gamma - 1)^{-1}$ (drawn from $n = 3/2$ to 3), γ being the adiabatic index. In Fig. 10a, the curve is drawn for any generic compression ratio. and in Fig. 10b, the curve is drawn assuming the strong shock limit only: $R = (\gamma + 1)/(\gamma - 1) = 2n + 1$. In both the cases, $\Theta_{out} \sim \Theta_{in}$ has been assumed for simplicity. Note that if the compression does not take place (namely, if the denser region does not exist), then there is no outflow in this model. Indeed for, $R = 1$, the ratio $R_{\dot{m}}$ is zero as expected. Since compression in shocks goes along with entropy generation, the outflows are associated with large entropy generation.

In a relativistic inflow or for a radiation dominated inflow, $n = 3$ and $\gamma = 4/3$. In the strong shock limit, the compression ratio is $R = 7$ and the ratio of inflow and outflow rates becomes,

$$R_{\dot{m}} = 0.052 \, \frac{\Theta_{out}}{\Theta_{in}}. \qquad (23a)$$

For the inflow of a mono-atomic ionized gas $n = 3/2$ and $\gamma = 5/3$. The compression ratio is $R = 4$, and the ratio in this case becomes,

$$R_{\dot{m}} = 0.266 \, \frac{\Theta_{out}}{\Theta_{in}}. \qquad (23b)$$

Since f_0 is smaller for $\gamma = 5/3$ case, the density at the sonic point in the outflow is much higher (due to exponential dependence of density on f_0, see, eq. 19) which causes the higher outflow rate, even when the actual jump in density in the postshock region, the location of the sonic point and the velocity of the flow at the sonic point are much lower. It is to be noted that generally for $\gamma > 1.5$ shocks are not expected (Chakrabarti, 1990), but the centrifugal barrier supported dense region would still exist. As is clear, the entire behavior of the outflow depends only on the compression ratio, R and the collimating property of the outflow Θ_{out}/Θ_{in}.

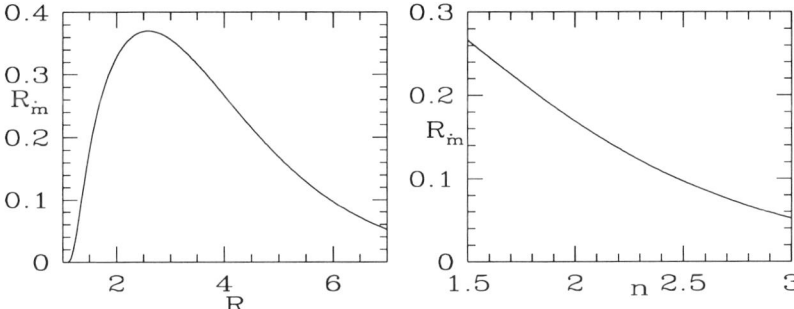

Fig. 10 (a-b) : Ratio $R_{\dot{m}}$ of the outflow rate and the inflow rate as a function of the compression ratio of the gas at the dense region boundary (a) and its variation with the polytropic constant n in the *strong shock limit* (b). Solid angles subtended by the inflow and the outflow are assumed to be comparable.

Outflows are usually concentrated near the vertical axis, while the inflow is near the equatorial plane. Assuming a half angle of $10°$ in each case, we obtain,

$$\Theta_{in} = \frac{2\pi^2}{9}; \quad \Theta_{out} = \frac{\pi^3}{162}; \quad \frac{\Theta_{out}}{\Theta_{in}} = \frac{\pi}{36}. \quad (24)$$

The ratios of the rates for $\gamma = 4/3$ and $\gamma = 5/3$ are then

$$R_{\dot{m}} = 0.0045 \quad \text{and} \quad R_{\dot{m}} = 0.023 \quad (25)$$

respectively. Thus, in quasi-spherical systems, in the case of strong shock limit, the outflow rate is at the most a couple of percent of the inflow.

It is to be noted that although the existence of outflows are well known, their rates are not. The only definite candidate whose outflow rate is known with any certainty is probably SS433 (a neutron star?) whose mass outflow rate was estimated to be $\dot{M}_{out} \gtrsim 1.6 \times 10^{-6} f^{-1} n_{13}^{-1} D_5^2 M_\odot$ yr^{-1} (Watson et al. 1986), where f is the volume filling factor, n_{13} is the electron density n_e in units of 10^{13} cm^{-3}, D_5 is the distance of SS433 in units of 5kpc. Considering a central object of mass $3M_\odot$, the Eddington rate is $\dot{M}_{Ed} \sim 0.6 \times 10^{-8} M_\odot$ yr^{-1} and assuming an efficiency of conversion of rest mass into gravitational energy $\eta \sim 0.1$, the critical rate would be roughly $\dot{M}_{crit} = \dot{M}_{Ed}/\eta \sim 6 \times 10^{-8} M_\odot$ yr^{-1}. Thus, in order to produce the outflow rate mentioned above even with our highest possible estimated $R_{\dot{m}} \sim 0.4$ (see, Fig. 10a), one must have $\dot{M}_{in} \sim 67 \dot{M}_{crit}$ which is very high indeed. One possible reason why the above rate might have been over-estimated would be that below 10^{12}cm from the central mass (Watson et al. 1986), $n_{13} >> 1$ because of the existence of the dense region at the base of the outflow.

In numerical simulations the ratio of the outflow and inflow has been computed in several occasions (Eggum, Coroniti & Katz, 1985; Molteni, Lanzafame & Chakrabarti, 1994). Eggum et al. (1985) found the ratio to be $R_{\dot{m}} \sim 0.004$ for a radiation pressure dominated flow. This is generally comparable to what we found above (eq. 25). In Molteni et al. (1994) the centrifugally driven outflowing wind generated a ratio of $R_{\dot{m}} \sim 0.1$. Here, the angular momentum was present in both inflow as well as outflow, and the shock was not very strong. Thus, the result is again comparable with what we find here. The detailed computations are presented in Das & Chakrabarti (1998).

8. Future Directions

The advective disks brought in a lot of possibilities and we are only beginning to understand them. Advective disks being hotter, a significant nucleosynthesis can take place inside them. Chakrabarti, Jin and Arnett (1987) and Jin, Arnett & Chakrabarti (1989) initiated the studies of nucleosynthesis in black hole accretion disks. These studies were made using 'then contemporary' accretion disks, namely, thick disks. In the recent studies, Mukhopadhyay & Chakrabarti (1998, see also Mukhopadhyay, this volume) finds that nuclear composition change could be significant and what is more, disks may even show instability as various elements are depleted in different radial distances. The theoretical understanding of the stability of the disk in presence of nucleosynthesis is a subject of intense study (Ray & Chakrabarti, 1998) and results would be reported elsewhere. Another important direction is to see if the two temperature character of the flow is maintained in presence of magnetic heating as recently pointed out by Bisnovatyi-Kogan (this volume). Most certainly, decoupling of protons and electrons as required by ADAF is not possible.

Twenty five years have passed by to appreciate the fact that the inertial force of matter in an accreting flow could be important to alter the basic topologies of disk accretion. Last two decades, new disks 'models' mushroomed every so often that it became customary to use one model in one occasion and another model in another occasion. Lately, what people have realized we have to go back to basics: the governing equations are unique but the solutions would depend on input parameters. One should be more concerned about the nature of the solutions rather than ad hoc models. These zoo of solutions hold the key to solving diverse problems in black hole astrophysics. Since advection can be thought to be synonymous with accretion, and it is no surprise that advective disks can resolve most of the outstanding issues very naturally. Future of this subject depends on understanding these new solutions.

References

Abramowicz, M.A. et al. (1988), *Astrophys. J.* **332**, 646
Abramowicz, M.A., Jaroszynski, M. & Sikora, M. (1978), *Astron. Ap.* **63**, 221
Abramowicz, M.A. & Kato, S. (1989), *Astrophys. J.* **336**, 304
Abramowicz, M.A. & Zurek, W.H. (1981), *Astrophys. J.* **246**, 314
Babul, A., Ostriker, J.P. & Mészáros, P. (1989), *Astrophys. J.* **347**, 59
Bishnovatyi-Kogan, G. (1998), this volume
Bondi, H. (1952), *M.N.R.A.S.* **112**, 195
Cannizzo, J. (1993) in *Accretion Disks in Compact Stellar Systems* (Ed.) J. C. Wheeler, World Scientific: Singapore
Chakrabarti, S.K. (1985), *Astrophys. J.* **288**, 1
Chakrabarti, S.K. (1989), *Astrophys. J.* **347**, 365
Chakrabarti, S.K. (1990), Theory of Transonic Astrophysical Flows (World Scientific: Singapore)
Chakrabarti, S.K. (1995), in Proc. of 17th Texas Symp. (N.Y. Acad. Sc.: New York).
Chakrabarti, S.K. (1996a), *Astrophys. J.* **464**, 623
Chakrabarti, S.K. (1996b), *MNRAS* **283**, 325
Chakrabarti, S.K. (1996c), Astrophys. J., 471, 237
Chakrabarti, S.K. (1996d), *Physics Reports* **266**, No. 5 & 6, 229
Chakrabarti, S.K. (1997a), *Astrophys. J.* **484**, 313
Chakrabarti, S.K. (1997b), *Astrophys. J.* (submitted) astro-ph/9801079
Chakrabarti, S.K. (1998), *Ind. J. Phys.* **72B**, 183, astro-ph/9803227
Chakrabarti, S.K., Jin, L. & Arnett, W.D. (1987), *Astrophys. J.* **313**, 674
Chakrabarti, S.K. & Molteni D. (1993), *Astrophys. J.* **417**, 671
Chakrabarti, S.K. & Molteni, D. (1995), *M.N.R.A.S.* **272**, 80
Chakrabarti, S.K. et al. (1996) in the *Proceedings of the IAU Asia-Pacific regional meeting* (Eds.) H.M. Lee, S.S. Kim, K.S. Kim, *J. Korean Astron. Soc.* **29**, 229
Chakrabarti, S.K., & Titarchuk, L.G. (1995) *Astrophys. J.* **455**, 623
Chakrabarti, S.K., Titarchuk, L.G., Kazanas, D. & Ebisawa, K. (1996) *A & A Suppl. Ser.* **120**, 163
Chang, K.M. & Ostriker, J. P. (1985), *Astrophys. J.* **288**, 428
Chen. W., Shrader, C.R. & Livio, M. (1997), *Astrophys. J.* **491**, 312
Chen. W. et al. (1998), in *Accretion Processes in Astrophysical Systems: Some like it hot*, Ed. S. Holt, (New York: AIP)
Chen, X.M., Abramowicz, M.A., Lasota, J.-P. (1997), *Astrophys. J.* **476**, 61
Cui, W. et al. (1997), *Astrophys. J.* **484**, 383
Das, T.K. & Chakrabarti, S.K. (1998) *Astrophys. J.* (submitted)
Dolan, J.F. et al. (1979) *Astrophys. J.* **230**, 551
Dotani, Y. (1992) in *Frontiers in X-ray Astronomy*, Eds. Y. Tanaka, & K. Koyama, K. p. 152 Universal Academy Press, Tokyo
Ebisawa, K. et al. (1994) *P.A.S.J.* **46**, 375
Ebisawa, K., Titarchuk, L. & Chakrabarti, S.K. (1996), *P.A.S.J.* **48**, No. 1
Eggum, G. E., Coroniti, F. V., Katz, J. I. (1985), *Astrophys. J.* **298**, L41
Garcia, M.R., McClintock, J.E. & Narayan, R. (1997), *Astrophys. J.* **478**, L79
Gilfanov, M., Churazov, E. & Sunyaev, R.A. (1997), in *Accretion Disks - New Aspects*, Eds. E. Meyer-Hofmeister & H. Spruit, Springer (Heidelberg).
Grove, J.E. et al. (1998), *Astrophys. J.* **500**, 899
Habbal, S.R. & Rosner, R. (1984), *J. Geophys. Res.* **89**, 10645
Halpern, J. & Marshall, H.L. (1996), *Astrophys. J.* **464**, 760
Haardt, F.& Maraschi, L. (1991), *Astrophys. J.* **380**, L51
Hawley, J.F. (1984), Ph.D. Thesis at the University of Illinois, Urbana Champaign.
Hawley, J.F., Smarr, L.L. & Wilson, J.R. (1984), *Astrophys. J.* **277**, 296
Hawley, J.F., Smarr, L.L., & Wilson, J.R. (1985), *Astrophys. J. Suppl.* **55**, 211
Hoyle, F. & Lyttleton, R.A. (1939), *Proc. Camb. Phil. Soc.* **35**, 592.

Jaroszynski, M., Abramowicz, M. & Paczyński, B. (1980), *Acta Astron.* **30**, 1
Kato, S., Honma, F. & Matsumoto, R. (1988) *P.A.S.J.* **40**, 709
Kozlowski, M., Jaroszynski, M. & Abramowicz, M.A. (1978). *Astron. Ap.* **63**, 209
Kojima, Y. (1986), *Prog. Theor. Phys.* **75** 251
Kraft, R. (1963), *Advances in Astron. & Ap*, Ed. Z. Kopal, **2**, 43.
Kuznetsov, S., et al. (1997), *M.N.R.A.S.* **292**, 651
Liang, E.P.T. & Thompson, K.A. (1980) *Astrophys. J.* **240**, 271
Ling, J. et al. (1997), *Astrophys. J.* **484**, 375
Lynden-Bell, D. (1969), *Nature* **223**, 690.
Lasota, J.P., (1994), in *Theory of Accretion Disks 2*, eds. Duschl W.J. et al. (Kluwer: Dordrecht).
Lu, J.F. et al. (1996), *Astron. Ap.* **321**, 665
Matsumoto, R., Kato, S., Fukue, J.& Okazaki, A.T. (1984), *P.A.S.J.* **36**, 71
McClintock, J.E., Horne, K. & Remillard, R.A. (1995), *Astrophys. J.* **442**, 358
Molteni, D., Lanzafame, G., & Chakrabarti, S.K. (1994). *Astrophys. J.* **425**, 161
Molteni, D., Ryu, D., & Chakrabarti, S.K. (1997), *Astrophys. J.* **470**, 460,
Morgan, E.H., Remillard, R.A., Greiner, J. (1997), *Astrophys. J.* **482**, 993
Muchotrzeb, B. & Paczyński, B. (1982), *Acta Astron.* **32**, 1
Mukhopadhyay, B. & Chakrabarti, S.K. (1998) (submitted)
Narayan, R. & Yi, I. (1994), *Astrophys. J.* **428**, L13
Narayan, R. & Yi, I. (1995), *Astrophys. J.* **444**, 231
Narayan, R. (1997) in *Accretion Phenomenon and Related Outflows*, Eds. D. Wickramasinghe, L. Ferrario & G. Bicknell (PASP Conf. Ser v. 121: San Francisco)
Narayan, R., Kato, S. & Honma, F. (1997), *Astrophys. J.* **476**, 49
Nobili, L., Turolla, R. & Zampieri, L. (1991), *Astrophys. J.* **383**, 250.
Nobuta, K. & Hanawa, T. (1994), *P.A.S.J.* **46**, 257
Novikov, I. & Thorne, K.S. (1973) in: *Black Holes* (eds.) C. DeWitt and B. DeWitt, Gordon and Breach, New York
Nowak, M. & Wagoner, R.V. (1993) *Astrophys. J.* **418**, 183
Orosz, J.A. et al. (1997), *Astrophys. J.* **478**, 830
Paczyński, B. & Bishnovatyi-Kogan, G. (1981), *Acta Astron.* **31**, 283
Paczyński, B. & Wiita, P.J. (1980), *Astron. Ap.* **88**, 23
Papaloizou, J.C.B. & Pringle, J.E. (1985) *M.N.R.A.S.* **213**, 799.
Paul, B., Agrawal, P.C., Rao, A.R., et al. (1998a), *Astrophys. J.* **492**, L63
Prendergast, K.H. & Burbidge, G.R. (1968), *Astrophys. J.* **151**, L83.
Ray, A. & Chakrabarti, S.K. (1998) (in preparation)
Rees, M.J., et al. (1982), *Nature* **295**, 17
Shakura, N.I. (1972), *Astron. Zh.* **49**, 652.
Shakura, N.I. & Sunyaev, R.A. (1973), *Astron. Ap.* **24**, 337
Shapiro, S.L. (1973a), Astrophys. J. **180**, 531.
Shapiro, S.L. (1973b), *Astrophys. J.* **185**, 69.
Shvartsman, V.F. (1971), *Sov. Astron. A.J.* **15**, 377
Sunyaev, R.A., et al. (1994) *Astron. Lett.* **20**, 777
Taam, R., Chen, X.M. & Swank, J. (1997) *Astrophys. J.* **485**, L83
Tanaka, Y. & Lewin, W.H.G. (1995) in *X Ray Binaries*, Eds. W.H.G. Lewin, J. Van Paradijs, E.P.J. Van-den Heuvel, p. 126 Cambridge University Press
Tarafdar, S.P. (1988), *Astrophys. J.* **331**, 932
Titarchuk, L. G., Mastichiadis, A., & Kylafis, N. G. (1997) *Astrophys. J.* **487**, 834
Wagner, R.M. et al. (1994) *Astrophys. J.* **401**, L97
Watson, M.G., Stewart, G. C., Brinkann, W., & King, A. R. (1986) *M.N.R.A.S.* **222**, 261
Wilson, J.R. (1978), *Astrophys. J.* **173**, 431
Yang, R., & Kafatos, M. (1995), *Astron. Ap.* **295**, 238
Zdziarski, A.A. (1986), *Astrophys. J.* **305**, 45
Zhang, S.N. et al. (1997), *Astrophys. J.* **477**, L95

VISCOSITY IN ACCRETION DISKS

PAUL J. WIITA

Department of Physics & Astronomy, Georgia State University
Atlanta, Georgia, USA

1. Introduction

The nature of the viscosity within accretion disks remains one of the greatest uncertainties afflicting models for inflows onto compact objects. Given the presence of even a modicum of angular momentum in the accreting fluid, the strength of the viscosity will have a substantial influence upon the structure of the accretion disk. Unfortunately, under nearly all astrophysically relevant circumstances, all of the well understood microscopic transverse momentum transport mechanisms such as ionic, molecular, and radiative viscosity, are extremely small. Observations with direct relevance to the nature and strength of macroscopic viscosity mechanisms are very difficult to make and their interpretations are model dependent. In the case where a black hole is the compact accretor, such observations have not yet been possible.

Therefore, advances in understanding disk viscosity must combine theoretical analyses and numerical techniques. In this article we first review the simple α parameterization, where the shear stress is proportional to the total pressure in the disk; this was introduced by Shakura & Sunyaev (1973, hereafter SS73) in their classic paper 25 years ago, which effectively launched the subject. Thermal instabilities in such α disks led to the consideration of so-called β disks, where the shear stress is assumed to couple to only the gas pressure, and this modification is discussed next. Various ways that have been proposed to produce disk viscosity are then summarized. Over the past several years an explanation of viscosity in terms of linear magnetic instabilities (the Velikhov-Chandrasekhar-Balbus-Hawley instability) has been investigated intensely, and a summary of the numerical experiments which support this hypothesis is provided before conclusions are presented.

2. The α Parameterization

The "standard thin-disk" model involves little more than the equations of mass, angular momentum, and energy conservation. Within the approximate version introduced by SS73, as well as the fully general relativistic version put forward by Novikov & Thorne (1973), the only clearly open question is how to treat the torque acting between different radial zones of the disk. If we specialize to the steady Keplerian approximation of SS73, in which inertial terms and the pressure gradient are ignored (this is adequate for a discussion of the nature of the viscosity, though a major oversimplification with respect to the structure of the inner disk region), these conservation laws become:

$$\dot{M} = -2\pi r \Sigma v_r = \text{const}, \tag{1}$$

$$\dot{M} r^2 \Omega = -2\pi r^2 W_{r\phi} + \dot{J}, \tag{2}$$

$$\frac{d}{dr}\left[\dot{M}\left(\frac{1}{2}v_\phi^2 - \frac{GM}{r}\right) + 2\pi r^2 W_{r\phi}\Omega\right] = 2\pi r Q^-. \tag{3}$$

In the above equations, $\dot{M}$ is the mass accretion rate, $\Sigma(r) \equiv \int \rho(r,z)dz$ is the column density, Ω is the angular velocity, $\dot{J}$ is the boundary flux of angular momentum, M is the mass of the central accretor, and $Q^-(r)$ is the rate at which heat is radiated from the disk. The integrated shear-stress which transfers the angular momentum outward through the disk as the mass spirals inward, is

$$W_{r\phi} \equiv \int t_{r\phi} dz = -\frac{3}{2}\Omega \int \eta dz, \tag{4}$$

in the Keplerian approximation, where η is the dynamical viscosity. The dominant local shear stress component (carrying azimuthal motions into radial ones) can be more generally written as $t_{r\phi} = \eta r \frac{\partial}{\partial r}\left(\frac{v_\phi}{r}\right)$.

To take the models further demands the inclusion of an equation of state, usually taken to be the sum of gas and radiation pressure contributions,

$$P = P_{\text{g}} + P_{\text{r}} = \frac{k}{\mu m_{\text{H}}}\rho T + \frac{a}{3}T^4; \tag{5}$$

however, possible contributions from magnetic fields, degenerate electrons and bulk motions (ram pressure) cannot be ignored under some circumstances. Then equations for the vertical structure can be derived (SS73), though we will not give them here. The heat generated in a column through the disk is

$$Q^+ = \int t_{r\phi} r \frac{\partial}{\partial r}\left(\frac{v_\phi}{r}\right) dz = W_{r\phi} r \frac{\partial}{\partial r}\left(\frac{v_\phi}{r}\right);$$

we also define, $W \equiv \int P dz$.

The viscosity prescription is "derived" by noting that in the Navier-Stokes equations the turbulent contribution to the force, $F_{\text{vis}} \approx \eta \frac{v_{\text{turb}}}{L_{\text{turb}}}$, in terms of the dynamical viscosity and the typical eddy velocity and size. In that very high Reynolds numbers are found for any plausible astrophysical fluids, the turbulence should be strong. Therefore, $F_{\text{vis}} \approx F_{\text{inertial}} \simeq \rho v_{\text{turb}}^2$, which implies $\eta_{\text{turb}} \approx \rho v_{\text{turb}} L_{\text{turb}}$. Shakura & Sunyaev argue that if the turbulence became supersonic, shocks would ensue between colliding elements, thereby directly converting bulk kinetic energy to heat; this plausible assumption yields the constraint, $v_{\text{turb}} < c_s$, the sound speed. Further, assuming isotropic turbulence, the maximal eddy size is limited by the half-thickness of the disk, z_0. Therefore,

$$-t_{r\phi,\text{turb}} \simeq \eta_{\text{turb}} \Omega < \rho c_s z_0 \Omega \approx \rho c_s^2. \tag{6}$$

A similar limit can be placed on the shear stress carried by local magnetic fields (SS73):

$$-t_{r\phi,\text{mag}} < P_{\text{mag}} = B^2/8\pi < P \approx \rho c_s^2. \tag{7}$$

Together, equations (6) and (7) lead to the α-disk prescription,

$$|t_{r\phi}| = \alpha P, \quad \text{or} \quad W_{r\phi} = \alpha W, \tag{8}$$

using equation (4). Within the Keplerian approximation this implies,

$$\eta = \frac{2}{3}\alpha P \left(\frac{GM}{r^3}\right)^{-1/2}, \tag{9}$$

with the α parameter (clearly $\alpha \leq 1$) usually assumed to be a constant throughout the disk, though this constancy is obviously unlikely to hold in reality. Various physical parameters of a disk are then functions of α, though fortunately, most of the dependences are rather weak. For example, in the so-called "middle" region of a thin disk, where $P_g > P_r$, and the dominant opacity is from electron scattering: $(z_0/r) \propto \alpha^{-1/10}$, the disk's central temperature $\propto \alpha^{-1/5}$, while the disk's central density $\propto \alpha^{-7/10}$ (SS73). Despite the obvious crudeness of this α-disk model, its simplicity has proven to be wildly popular, and certainly over 90% of the more than 6000 papers written on accretion disks over the past 25 years have employed this approximation.

3. The β Parameterization

Shortly after the α-models were proposed, a problem with them was noticed: the "inner region", where $P_r > P_g$ is unstable (Lightman & Eardley

1974, hereafter LE; Shakura & Sunyaev 1976, hereafter SS76). These instabilities result when a time-dependent analysis is performed. We shall not present much of the mathematics, but will merely summarize the results. It is also worth keeping in mind that it is in this inner region where the Keplerian approximation, in particular its assumption that pressure gradient forces can be ignored, is most likely to fail, thereby providing the impetus for studies of more general accretion flows (e.g., Paczyński & Wiita 1980; Chakrabarti 1990).

When $P_r \gg P_g$, P is independent of ρ, but the equation of vertical pressure balance against the out-of-plane gravitational forces of the black hole yields $P \approx 0.5 z_0 \Omega^2 \Sigma$. Combining this with equation (4) yields the peculiar result (LE) that $W_{r,\phi}(\Sigma, r) \propto \Sigma^{-1}$. Because the integrated stress is a decreasing function of the column density, the non-linear diffusion equation for Σ, which is (LE),

$$\frac{\partial \Sigma}{\partial t} = \frac{\partial}{\partial r}\Big[\frac{d(\Omega r^2)}{dr}\Big]^{-1} \frac{\partial}{\partial r}[r^2 W_{r,\phi}(\Sigma, r)], \qquad (10)$$

has a *negative* effective diffusion coefficient. Physically, this means that an initially stationary inner region will start to break up into rings (with $\Delta r \geq z_0$); one will have high-Σ and low-$W_{r,\phi}$ while its inner- and outer-neighbors will have low-Σ and high-$W_{r,\phi}$. Thus, the matter gets pushed into rings of low viscous stress while the alternate rings become so starved of matter that they become optically thin, and therefore unstable. Such a dynamical instability would naturally lead to rapid fluctuations in the emission from the inner portion of the disk, perhaps explaining some of the fluctuations observed in many X-ray binaries and active galactic nuclei.

It is easily seen that this instability could be avoided if the viscosity is a function of Σ (or r) rather than constant; in particular, if α falls at least as fast as Σ^{-1} then $\Sigma(r)$ rises rapidly near the black hole and the instability does not occur. In particular, Cunningham's (1973) suggested β-disk prescription,

$$t_{r\phi} = \beta P_g, \qquad (11)$$

is stable to this LE clumping. Such "β-disks" are reasonably similar to α-models, except that they have much higher values of Σ in their inner regions and thereby remain optically thick despite staying geometrically thin.

A more complete linear analysis was performed by SS76, who considered axisymmetric perturbations of wavelength Λ in a thin disk, such that $z_0 \ll \Lambda \ll r$; these perturbations are assumed to grow slowly compared to the dynamical timescale, which is locally given by Ω^{-1}. They found that the LE dynamical instability actually corresponded to the more slowly growing branch of a general dispersion relation. If one defines $P_r = \chi P$, the result of the SS76 analysis is that for $\chi < 0.6$ (i.e., gas pressure dominance), disk accretion is stable. For $\chi > 0.6$ there exists an unstable mode for $\Lambda/z_0 > f(\chi)$; for $\chi = 1, f(\chi) = 2$ (its minimum value), and this mode corresponds to rings moving across the disk with $2z_0 < \Lambda < 4z_0$. The growth rate, $\omega = 0$ at $\Lambda = 2z_0$ but is up to $\omega \simeq \alpha\Omega/10$ at $4z_0$. For longer wavelengths, the solution bifurcates into slower and faster growing branches. The lower (LE) branch corresponds to the situation where viscosity perturbation amplitudes $(\delta\eta/\eta)$ are much less than $\delta\Sigma/\Sigma$ or $\delta z_0/z_0$. Under these circumstances, the rate of heat loss can balance the rate of production, i.e., $Q^- = Q^+$, and the growth rate decreases as the wavelength of the perturbation increases,

$$\omega \simeq \alpha\Omega \frac{2(5-3\chi)}{3(5\chi-3)}\left(\frac{z_0}{\Lambda}\right)^2. \tag{12}$$

The faster growing thermal instability has $Q^+ \neq Q^-$, and in this situation $\delta\Sigma/\Sigma \ll (\delta z_0/z_0, \delta\eta/\eta)$; here the growth rate is given by

$$\omega = \alpha\Omega \frac{6(5\chi-3)}{A(\chi)}, \tag{13}$$

where $A(\chi) = P/(e\rho)$, with e the specific internal energy; for a $\gamma = 5/3$ gas, $A = 3(1+\chi)/2$. As $\chi \longrightarrow 1$ this perturbation growth rate is $\omega \simeq 0.21\alpha\Omega$ for $\Lambda > 12z_0$, and is therefore quite fast as long as α is not negligibly small.

If the viscosity only couples to the gas pressure, which is essentially equivalent to saying it is dominated by actual mass motions, and not magnetic fields, then the β prescription of $t_{r\phi} = \beta P_g$ can be analyzed in the same way. Such β disks are completely stable to these linear perturbations. Various authors have considered aspects of these β disks, among them Meyer & Meyer-Hofmeister (1982), Wandel & Petrosian (1988), and Mangalam & Wiita (1993). Key differences are that the β disks have somewhat larger central and surface temperatures and somewhat smaller thicknesses than do α disks for the same M and $\dot{M}$.

4. Possible Physical Mechanisms for Viscosity

In this section we briefly mention most of the mechanisms that have been suggested to provide viscosity in disks, but by no means do we attempt to provide a thorough review. Despite the crudity of the α model, it does provide a useful way of talking about the strength of viscosity, and we will summarize the viscosity mechanisms in terms of the values of the α parameter that they apparently produce.

4.1. HYDRODYNAMICAL MECHANISMS

One possibility is that the mere differential rotation of the disk will generate turbulence through shear instabilities. While two-dimensional laboratory experiments indicate that sheared cylindrical flows are indeed unstable, this possibility has remained very controversial when considered in the realm of accretion disks, since locally Keplerian flows are linearly stable (Stewart 1975). Non-linear HD instabilities may well be present (e.g., Dubrulle & Knobloch 1992), but only rather low (probably $< 10^{-2}$) values of α are anticipated from this possible mechanism.

A hydrodynamical modification on the basic SS73 idea of largest eddy turbulence asks if their limit, $L_{\text{turb}} \leq z_0$, is absolute. It is distinctly possible that the turbulent cells are elongated by the shear and accretion flows so that the turbulence is not isotropic. If the disk turbulence is sufficiently anisotropic, $L_{\text{turb}} \gg z_0$ is not inconceivable. This could produce, in principle, a very high value of $\alpha > 1$.

One way of guaranteeing the presence of turbulence in a disk is to have a need for convection to transport energy vertically through the disk; in analogy with stellar convection turbulence would then be clearly present, with a maximum eddy size corresponding the the thickness of the convection layer. Early estimates claimed that convection could be important (Bisnovatyi-Kogan & Blinnikov 1977), but later calculations (Meyer & Meyer-Hofmeister 1982) indicated that most disks are convectively stable. Even if convection is present and important for energy transport, it would limit the turbulence to subsonic speeds and probably smaller scales than z_0 in most parts of the disk. This implies low values of α if convection is the dominant source of viscosity.

Since galactic X-ray sources are in binary systems, it is well worth considering the influence of the binary companion on the accretion disk. Many authors have examined this possibility, and it is quite clear that a binary companion can drive nearly stationary spiral shocks through the disk (e.g., Sawada, Matsuda & Hachisu 1987; Różcyczka & Spruit 1993; Chakrabarti & Wiita 1993). In some cases these shocks provide very strong transport of angular momentum and allow adiabatic accretion with predicted values for $\alpha(r)$; these tend to be rather high, with $\bar{\alpha} \sim 0.1$.

Other interesting HD mechanisms that might be relevant to the generation of effective viscosity include inertial waves (e.g., Vishniac, Jin & Diamond 1990) and tidal forces driving small scale turbulence (e.g., Ryu & Goodman 1994), both of which might produce modest values of α.

4.2. MAGNETOHYDRODYNAMICAL MECHANISMS

There are several ways in which magnetic fields might dominate the viscosity and angular momentum transport. In this subsection we mention a few relatively speculative possibilities based on large scale fields; in the next section we discuss a mechanism that is now believed to be of key importance.

The possibility that a large-scale dynamo can be established in an accretion disk is worthy of consideration, and, while generally considered unlikely, it appears that this is possible under some circumstances, (e.g., Mangalam & Subramanian 1994). If so, large B fields could give rise to strong shear and reconnection. While such putative dynamo fields are likely to be important for producing heating and particle acceleration, there is no obvious way to derive the viscosity; nonetheless, the presence of an effective dynamo should yield a rather large value of α.

Many authors have suggested that a corona exists above and below accretion disks (e.g., Galeev, Rosner & Vaiana 1979; de Vries & Kuijpers 1992), though the lack of convection zones may mean that flux tubes cannot be anchored and may well escape without forming stellar-like coronæ (Chakrabarti & D'Silva 1994). If magnetically heated coronae do exist, then substantial dissipation of energy and angular momentum is possible through reconnection and flares in the corona. This is likely to produce moderate effective viscosities, with $\alpha \sim 0.1$.

Magnetic loops could stretch between annular rings at substantially different radii in an accretion disk, and the difference in velocity could shear and amplify the field. Locally driven winds as well as loops tying together different radii can very effectively transport angular momentum. In principle, this type of coronal torque could dominate, with (Burm & Kuperus 1988)

$$\frac{\text{coronal torque}}{\text{body torque}} = \Big(\frac{B_z B_\phi/8\pi}{\alpha P}\Big)\Big(\frac{r}{z_0}\Big) \approx \frac{r}{\alpha z_0} \gg 1. \qquad (14)$$

This could produce the equivalent of a medium to very large α; however, as in the case of global dynamos and coronæ, there is no guarantee that this type of MHD process plays a significant role in actual accretion disks.

5. Magnetic Instabilities Yield an Effective Viscosity

We finally turn to a form of magnetic viscosity that almost certainly is present in all accretion disks.

5.1. THEORETICAL ANALYSIS

A linear magnetic instability for magnetized Couette flow has been known for a long time (Velikhov 1959; Chandrasekhar 1961), but its key importance for accretion disks was stressed by Balbus & Hawley (1991, hereafter BH) and it is therefore usually called the Balbus-Hawley (B-H) instability. This instability only requires the presence of a very weak seed field, which could be advected in with the accreted material, anchored in the central mass (especially if it is a neutron star), or generated by a dynamo in the disk itself (e.g., Brandenburg et al. 1995).

The basic idea is simple (BH): consider a weak vertical field, B_z, in a disk undergoing differential rotation. Any small perturbation outward (in the r direction) grows rapidly (on the dynamical, Ω^{-1}, timescale). To see this, consider an outwardly displaced fluid element; the magnetic field simultaneously tries to: (i) enforce rigid rotation (by resisting shearing); (ii) return the element back to its original position (by resisting stretching). While the latter aspect is patently stabilizing, the former produces the instability, by forcing the element to rotate too fast for its displaced location. Since its velocity is too large, the inertial ("centrifugal") force pushes the element yet further outward. For short enough wavelengths the restorative force wins, but for long enough wavelengths the instability grows.

The B-H instability is clearly effective in transporting angular momentum outward from the faster inner region to the slower outer region. The linear analysis produces the following criterion for it to operate (in the limits that the disk is rotating supersonically or if the Brunt-Väisälä frequency is negligible; for the full expression, see BH):

$$k_z^2 v_{zA}^2 < \frac{d\Omega^2}{d \ln r}, \tag{15}$$

where k_z is the wavenumber in the vertical direction and v_{zA} is the vertical component of the Alfven speed, $v_{zA} = B_z/(4\pi\rho)^{1/2}$. The B-H instability also effectively transports magnetic flux inward, as can be shown using interchange stability arguments (Christodoulou, Contopoulos & Kazanas 1996). Although this instability in its simplest form depends upon the presence of a vertical field component, and the Keplerian and accretion flows in the disk tend to amplify the azimuthal and, to a lesser extent, the radial, components without affecting B_z, it is almost inconceivable that any realistic situation will not have at least *some* non-zero B_z. The most rapidly growing wave numbers have growth rates around 0.75Ω, implying that even minute fields will be amplified by factors of 10^6 in under 3 rotation periods. Another key result is that these growth rates are independent of the strength of the magnetic field, at least for the linear analytic theory (BH).

Additional work showed that a purely toroidal field also leads to instability (Balbus & Hawley 1992; Foglizzo & Tagger 1995); in this case the dominant restoring force involves pressure gradients. In both cases, the mechanism leading to the instability arises from the fact that when two fluid elements in the same Keplerian orbit are pulled together, they are torn apart in the radial direction due to the change in the centripetal acceleration. As noted by Brandenburg et al. (1996), this problem is equivalent to the difficulties encountered by spacecraft attempting docking maneuvers. Once the elements are at different radii, but coupled by either originally vertical or azimuthal fields, angular momentum exchange can occur and energy can be converted from the shear motion into heat through a turbulent cascade. In the limit where the fluid is locally incompressible, Goodman & Xu (1994) discovered an exact solution of the axisymmetric magnetic shearing instability that grows exponentially even in the non-linear stage.

5.2. NUMERICAL RESULTS

Clearly such analytical results demand numerical investigation to see if they survive non-axisymmetric perturbations and to see in what fashion(s) they saturate. A compressible two-dimensional MHD computation was immediately presented by Hawley & Balbus (1991), who showed that for weak fields, the numerical results were consistent with the linear perturbation theory. However, since the generation of magnetic fields is an inherently three-dimensional phenomenon, such restricted computations could not be convincing. Quite a few additional investigations have been performed and the general applicability of the instability under a wide range of conditions has now been demonstrated.

A very important step involved going to three-dimensional MHD simulations. One approach is that of Hawley, Gammie & Balbus (1995), who considered a "local" disk (shearing box) model: one where tidal and Coriolis forces are incorporated but background gradients in pressure and density are ignored, along with the vertical component of gravity; periodic boundary conditions were employed in the radial direction. They found that, regardless of whether the computations start with only vertical or only azimuthal fields (or with a combination of both), turbulence is initiated and sustained by the magnetic instability. The turbulence saturates and is anisotropic in the sense that angular momentum is transported outwards and the turbulent energy and angular momentum flux are concentrated on the largest scales. Stone et al. (1996) added vertical gravity and density stratification to this code and found quite similar results.

Three-dimensional simulations were also carried out by Matsumoto & Tajima (1995), who also used the shearing-box model but employed a different algorithm for the evolution of the magnetic field. They mainly studied initially purely toroidal fields and also saw the development of strong turbulence. Their results also indicate that the magnetic fluctuations saturate at late stages, and they explain this as arising from the enhanced resistivity produced by tangled field lines. Their relatively coarse simulations yield equivalent $\alpha_B \equiv -\langle B_x B_y \rangle/(4\pi P)$ values typically around 0.02 for initially toroidal fields, but they claim α_B can exceed 0.1 for initially vertical seed fields.

Yet another three-dimensional local simulation was produced by Brandenburg et al. (1995). Their model also included the effects of compressibility and stratification, but employed a significantly different algorithm, which is of great importance in checking such complex computations. Brandenburg et al. also dispensed with the periodic vertical boundary conditions used by Hawley et al. (1995) Stone et al. (1996) and Matsumoto & Tajima (1995), as these implied that the total magnetic flux through the box was always conserved at zero. Instead they assume that the magnetic field at the upper and lower faces of the box is vertical, but that the field lines can move freely across the surface, which allows for generation of a net magnetic field. They find that supersonic flows are first driven by the B-H instability; these flows regenerate a turbulent magnetic field which then reinforces the turbulence. Therefore, this system can be characterized as a dynamo, but one where the magnetic energy exceeds the turbulent kinetic energy, typically by factors of 3–10. Their original simulations yielded effective α values of between 0.001 and 0.005; these computations have the Maxwell (magnetic) stress corresponding to $\sim$5 times as much viscosity as the Reynolds (turbulent) stress.

More recent work by Brandenburg et al. (1996), improved their model by removing scale invariance and allowing finite accretion rates. This nonlocal model is more realistic and several different resolutions were used. They estimated the α parameter in three distinct ways: (i) via the mass accretion rate; (ii) using the heating rate; (iii) through horizontal components of the Maxwell and Reynolds stress tensors. All of these approaches were quite consistent, though now they found $\alpha_{\rm Maxw}/\alpha_{\rm Reyn} \approx 3$. When the numerical resolution is doubled, the effective values of α go up by factors of ~ 1.5 within the range they could investigate, and the highest resolution runs yielded $\alpha \approx 0.007$. The latest results from this program at slightly higher resolution (Brandenburg 1998) still yield $\alpha < 0.01$, though it remains to be seen if this value really saturates or if continued improvements in computational capabilities yield additional increases in the effective viscosity.

6. Conclusions

The only fairly direct observational evidence for the strength of disk viscosity comes from dwarf novae: binary systems where a white dwarf accretes through a disk. Substantial luminosity is released in a hot spot, where the accretion stream from the companion impinges on the outer portion of the disk, which complicates the analysis, but the spectrum can sometimes be confidently divided between the white dwarf, the companion, the disk and the hot-spot. These systems undergo strong, quasi-periodic outbursts. This large scale variability has long been convincingly understood in terms of thermal limit cycles behavior in the accretion disk when hydrogen recombination occurs (e.g., Meyer & Meyer-Hofmeister 1981; Cannizzo, Ghosh & Wheeler 1982; Mineshige & Osaki 1983). When these systems undergo eclipses, careful analyses of the light curves allow the temperature profile, $T(r)$ of the disk to be estimated. During quiescent phases, a flat brightness-temperature profile curve is found, while during outbursts, $T_{\text{eff}} \propto r^{-3/4}$ is more usual. Both these temperature variations and the relative lengths of the bursting and quiescent stages can be fit to the theory if $\alpha \sim 0.1$ in the bright phase and $\alpha \sim 0.02$ in the quiescent phase.

As mentioned in the introduction, no such observational evidence is yet available when black holes are the accreting masses, and the physical conditions in the disks around even stellar remnant black holes will be different from those around white dwarfs, not to mention the large differences for the parameters expected in disks around supermassive black holes. Nonetheless, such values of α are not unreasonable, and we have seen that a wide array of physical mechanisms are capable of producing effective viscosities of that order. Furthermore, it now appears that the local magnetic instabilities must yield some non-trivial amount of angular momentum transport. As has been seen from the discussion in §5.2, while there is general agreement on the viability and importance of the magnetic instabilities in accretion disks, there is some disagreement between groups as to the effective strength of the viscosity. So, although final estimates are not yet available, values of $\alpha \sim 0.01$ are certainly likely from this mechanism, and somewhat higher values are possible.

Therefore, even though there is still a substantial amount of ignorance concerning the complete nature of and strength of viscosity in accretion disks, we can now argue that taking $0.005 < \alpha < 0.3$ is a safe bet and that further constraining this parameter towards the geometrical middle of that range is not unreasonable. This is a major step forward, and it seems likely that a very good prescription for viscosity will be available long before the theory of accretion disks is 50 years old.

This work was supported in part by NASA grant NAG 5-3098 and Research Program Enhancement funds at Georgia State University. I am most grateful for the hospitality of the S.N. Bose Center for Basic Sciences.

References

Balbus, S.A. and Hawley, J.F. (1991) *ApJ*, **376**, 214
Balbus, S.A. and Hawley, J.F. (1992) *ApJ*, **400**, 610
Bisnovatyi-Kogan, G.S. and Blinnikov, S.I. (1977) *A&A*, **59**, 11
Brandenberg, A. (1998) in *Theory of Black Hole Accretion Disks*, eds. M. Abramowicz, G. Björnsson, and J. Pringle (Cambridge: Cambridge Univ. Press), in press
Brandenberg, A., Nordlund, Å., Stein, R.F. and Torkelson, U. (1995) *ApJ*, **446**, 741
Brandenberg, A., Nordlund, Å., Stein, R.F. and Torkelson, U. (1996) *ApJ*, **458**, L45
Burm, H. and Kuperus, M. (1988) *A&A*, **192**, 165
Cannizzo, J.K., Ghosh, P., and Wheeler, J.C. (1982) *ApJ*, **260**, L83
Chakrabarti, S.K. (1990) *Theory of Transonic Astrophysical Flows* (Singapore: World Scientific)
Chakrabarti, S.K. and D'Silva, S. (1994) *ApJ*, **424**, 138
Chakrabarti, S.K. and Wiita, P.J. (1993) *A&A*, **271**, 216
Chandrasekhar, S. (1961) *Hydrodynamic and Hydromagnetic Stability* (Oxford: Clarendon), 384
Christodoulou, D.M., Contopoulos, J. and Kazanas, D. (1996) *ApJ*, **462**, 865
Cunningham, C. (1973) Ph.D. dissertation, University of Washington
de Vries, M. and Kuijpers, J. (1992) *A&A*, **266**, 77
Dubrulle, B. and Knobloch, E. (1992) *A&A*, **256**, 673
Foglizzo. T. and Tagger, M. (1995), *A&A*, **287**, 297
Galeev, A.A., Rosner, R. and Vaiana, G.S. (1979) *ApJ*, **229**, 318
Goodman, J. and Xu, G. (1994) *ApJ*, **432**, 213
Hawley, J.F. and Balbus, S.A. (1991) *ApJ*, **376**, 233
Hawley, J.F., Gammie, C.F. and Balbus, S.A. (1995) *ApJ*, **440**, 742
Lightman, A.P. and Eardley, D.M. (1974) *ApJ*, **187**, L1 (LE)
Mangalam, A.V. and Subramanian, K. (1994) *ApJ*, **434**, 509
Mangalam, A.V. and Wiita, P.J. (1993) *ApJ*, **406**, 420
Matsumoto, R. and Tajima, T. (1995) *ApJ*, **445**, 767
Meyer, F. and Meyer-Hofmeister, E. (1981) *A&A*, **104**, L10
Meyer, F. and Meyer-Hofmeister, E. (1982) *A&A*, **106**, 34
Mineshige, S. and Osaki, Y. (1983) *Pub. Astr. Soc. Japan*, **35**, 377
Novikov, I. and Thorne, K.S. (1973) in *Black Holes*, ed. C. DeWitt & B.S. DeWitt (New York: Gordon & Breach), 343
Paczyński, B. and Wiita, P.J. (1980) *A&A*, **88**, 23
Różyczka, M. and Spruit, H.C. (1993) *ApJ*, **417**, 677
Ryu, D. and Goodman, J. (1994) *ApJ*, **422**, 269
Sawada, K., Matsuda, T. and Hachisu, I. (1987) *MNRAS*, **224**, 307
Shakura, N.I. and Sunyaev, R.A. (1973) *A&A*, **24**, 337 (SS73)
Shakura, N.I. and Sunyaev, R.A. (1976) *MNRAS*, **175**, 613 (SS76)
Stewart, J.M. (1975) *A&A*, **42**, 95
Stone, J.M., Hawley, J.F., Gammie, C.F. and Balbus, S.A. (1996), *ApJ*, **463**, 656
Velikhov, E.P. (1959) *Soviet Phys.-JETP*, **36**, 995
Vishniac, E.J., Jin, L. and Diamond, P. (1990) *ApJ*, **365**, 648
Wandel, A. and Petrosian, V. (1988) *ApJ*, **329**, L11

SHOCK FORMATION IN ADIABATIC ACCRETION FLOWS AROUND A KERR BLACK HOLE

JU-FU LU
Center for Astrophysics, University of Science and Technology of China, Hefei, Anhui 230026, China

Abstract. Shock formation in adiabatic accretion flows around a Kerr black hole is analytically proved for two cases, i.e. Rankine-Hugoniot shocks and isothermal shocks. In the former case it is shown that for given flow parameters (the specific total energy and the specific angular momentum) there can be no more than one stable standing shock, and the global flow solution can only be of α-x type. However, the solution topology in the latter case shows new remarkable characteristics: for given flow parameters there can be two stable standing shocks, and the global solution can be of three types, namely α-x, x-α, and α-α type. It is also significant that the effects of frame-dragging of Kerr geometry on the shock formation are found out. Finally, some comments are made on the reason why numerous authors who have studied advection dominated accretion flows have found no shocks, and on the observational significance of the shock study.

1. Introduction

Shock formation in black-hole accretion flows has been shown in the literature, both analytically (see Chakrabarti 1996a for references) and numerically (e.g. Molteni *et al.* 1996a). On the other hand, numerous authors who have studied advection dominated accretion flows (ADAF) have found no shocks (e.g. Chen *et al.* 1997; Narayan *et al.* 1997). In this paper I demonstrate analytically that shocks may indeed form for a large region of the flow parameter space, while shock-free solutions correspond to flow parameters outside the shock formation region. In order to relate to ADAF I concentrate on adiabatic flows, because such flows are, by definition, fully advection dominated. Two types of shocks are studied, namely Rankine-Hugoniot shocks which form when the radiative cooling is extremely in-

on each side of a shock.
The shock conditions can be generally written as follows:

$$E_- = \left(\frac{1+u_{r-}u_-^r}{-L}\right)^{1/2} /(1-nb_-^2) \tag{5}$$

$$E_+ = \left(\frac{1+u_{r+}u_+^r}{-L}\right)^{1/2} /(1-nb_+^2) \tag{6}$$

$$\dot{\mu}_- = \gamma^{-n} r_s^2 \left(\frac{b_-^2}{1-nb_-^2}\right)^n u_-^r \tag{7}$$

$$\dot{\mu}_+ = \gamma^{-n} r_s^2 \left(\frac{b_+^2}{1-nb_+^2}\right)^n u_+^r \tag{8}$$

$$\left(u_-^r + \frac{b_-^2}{\gamma u_{r-}}\right)/(1-nb_-^2) = \left(u_+^r + \frac{b_+^2}{\gamma u_{r+}}\right)/(1-nb_+^2) \tag{9}$$

where E_-, $\dot{\mu}_-$, E_+, $\dot{\mu}_+$ are the conserved energy and the conserved entropy accretion rate for the pre-shock flow and the post-shock flow respectively (both the pre-shock and the post-shock flows are adiabatic), u_-^r, b_-, u_+^r, b_+ are the radial four-velocity and the sound speed of the flow just before and just after the shock respectively, r_s is the location of the shock. The meaning of equations (5)-(8) is obvious. Equation (9) results from the conservation of momentum flux density across the shock, or, in other words, from the balance of total pressure [thermal pressure P plus ram pressure $(\varepsilon+P)u_r u^r$] on the two sides of the shock surface, and with the help of the polytropic equation of state and equation (2).

3. Shock Solutions

A shock-included global solution can be obtained in the following way.

(A) Rankine-Hugoniot shocks: $E_- = E_+$, $\dot{\mu}_- < \dot{\mu}_+$, $b_- < b_+$.

(a) Start with the two controlling parameters l and $E(=E_-=E_+)$, solve equations (1) and (4a, b) for r_c, u_c^r and b_c, then determine the value of $\dot{\mu}$ from equation (2) with known r_c, u_c^r and b_c. The values of l and E have to be suitably given, so that there exist two physical sonic points (i.e. saddle-type critical points) in the flow, i.e. r_c, u_c^r, b_c and accordingly $\dot{\mu}$ all have two different values. The smaller and the larger one of r_c are denoted as r_{in} and r_{out} respectively; the smaller one of $\dot{\mu}$ is just $\dot{\mu}_-$, and the larger one of $\dot{\mu}$ is just $\dot{\mu}_+$.

(b) Having l, E_- and $\dot{\mu}_-$, solve equations (1) and (2) for u^r and b at any r, i.e. obtain a shock-free global solution that passes through the outer

SHOCK FORMATION IN ADIABATIC ACCRETION FLOWS AROUND A KERR BLACK HOLE

JU-FU LU
Center for Astrophysics, University of Science and Technology of China, Hefei, Anhui 230026, China

Abstract. Shock formation in adiabatic accretion flows around a Kerr black hole is analytically proved for two cases, i.e. Rankine-Hugoniot shocks and isothermal shocks. In the former case it is shown that for given flow parameters (the specific total energy and the specific angular momentum) there can be no more than one stable standing shock, and the global flow solution can only be of α-x type. However, the solution topology in the latter case shows new remarkable characteristics: for given flow parameters there can be two stable standing shocks, and the global solution can be of three types, namely α-x, x-α, and α-α type. It is also significant that the effects of frame-dragging of Kerr geometry on the shock formation are found out. Finally, some comments are made on the reason why numerous authors who have studied advection dominated accretion flows have found no shocks, and on the observational significance of the shock study.

1. Introduction

Shock formation in black-hole accretion flows has been shown in the literature, both analytically (see Chakrabarti 1996a for references) and numerically (e.g. Molteni *et al.* 1996a). On the other hand, numerous authors who have studied advection dominated accretion flows (ADAF) have found no shocks (e.g. Chen *et al.* 1997; Narayan *et al.* 1997). In this paper I demonstrate analytically that shocks may indeed form for a large region of the flow parameter space, while shock-free solutions correspond to flow parameters outside the shock formation region. In order to relate to ADAF I concentrate on adiabatic flows, because such flows are, by definition, fully advection dominated. Two types of shocks are studied, namely Rankine-Hugoniot shocks which form when the radiative cooling is extremely inefficient, so that no energy is radiated away through the surface of the flow at the shock; and isothermal shocks which correspond to the opposite extreme of physical situations, i.e. the radiative cooling is very efficient, energy and entropy are lost from the surface of the flow at the shock to keep the post-shock temperature equal to its pre-shock value. Obviously, one needs knowledge of both cases, as the realistic situation must be between these two extremes. My analysis here is rigorous, in the sense that the full general relativistic equations of flow motion in Kerr geometry are solved. The content of this paper is described in more details in Lu et al. (1997) and Lu & Yuan (1998).

2. Basic Equations

I consider stationary, axisymmetric, geometrically thin, inviscid accretion fluid flows in the equatorial plane of a Kerr black hole. Wind flows can be accordingly understood as a motion in the opposite direction to accretion. A polytropic equation of state is adapted for the accretion matter: $P = K\rho^\gamma = K\rho^{1+1/n}$, $\varepsilon = \rho + nP$, where P, ρ and ε are the pressure, the rest mass density and the mass-energy density, respectively, γ and n are constant, and K is a measurement of the specific entropy, which is a constant in a shock-free flow, but can change across a shock. The flow motion is assumed to be adiabatic, i.e. (the Boyer-Lindquist coordinates are used together with $G = c = 1$ units and a $- + + +$ signature)

$$E = -hu_t = \left(\frac{1 + u_r u^r}{-L}\right)^{1/2} /(1 - nb^2) \tag{1}$$

is conserved. Here E is the specific total energy, h is the specific enthalpy [i.e. $h = (P + \varepsilon)/\rho$], u_μ are the four-velocity components obeying the normalization condition $u_\mu u^\mu = -1$, b is the sound speed defined as $b^2 = dP/d\varepsilon$, and $L(r, l) = g^{tt} - 2g^{t\phi}l + g^{\phi\phi}l^2$, with $g^{\mu\nu}$ being the metric tensor components, and l being the specific angular momentum defined as $l = -u_\phi/u_t$, which is a constant throughout the whole of the flow, because the flow here is axisymmetric and inviscid. Note that in the case of Rankine-Hugoniot shock E is conserved for the whole flow including the shock; while in the case of isothermal shock E is conserved for the pre-shock flow and for the post-shock flow respectively, the value of E changes across the shock.

The mass accretion rate $\dot{M} = r^2 \rho u^r$ is conserved for the whole flow, and it is convenient to define an 'entropy accretion rate' $\dot{\mu}$ as (Chakrabarti 1996b)

$$\dot{\mu} = K^n \dot{M} = r^2 \left[\frac{b^2}{\gamma(1 - nb^2)}\right]^n u^r \tag{2}$$

where ρ is expressed in terms of b by using the polytropic equation of state. Note also that for both the Rankine-Hugoniot shock and the isothermal shock case $\dot{\mu}$ is conserved for the pre-shock flow and for the post-shock flow respectively, but the value of $\dot{\mu}$ changes across the shock because of the change of K, i.e. the change of entropy; $\dot{\mu}$ is conserved for the whole flow either if the flow is shock-free, or if the shock is of isentropic type, which is not treated here. It is seen from equation (2) that the shape of the flow is taken to be conical. Such a shaped flow has all the properties characteristic of other types of thin flows, such as those of constant thickness and those in vertical equilibrium, as Chakrabarti (1996c) argued.

By combining the differential form of equations (1) and (2), the differential equation of radial motion is derived as

$$\frac{du^r}{dr} = \frac{N}{D} \tag{3}$$

with

$$D = \frac{b^2}{u^r} - \frac{u_r}{1+u_r u^r}$$

and

$$N = \frac{u^r u^r}{2(1+u_r u^r)} \frac{dg_{rr}}{dr} - \frac{2b^2}{r} - \frac{1}{2L}\frac{dL}{dr}$$

The critical (sonic) point condition of equation (3), i.e. $N = D = 0$, yields

$$b_c^2 = \left.\frac{u_r u^r}{1+u_r u^r}\right|_c \tag{4a} \tag{4}$$

and

$$b_c^2 \left(\frac{d\ln g_{rr}}{dr} - \frac{4}{r}\right)_c = \frac{1}{L_c}\left(\frac{dL}{dr}\right)_c \tag{4b}$$

where the subscript 'c' denotes the value of a quantity at the critical point. It is known that a critical point is a sonic point only for a corotating observer, because the radial three-velocity measured by such an observer is $v = u_r u^r/(1+u_r u^r)$, and is equal to the local sound speed at the critical point (Lu 1986). Equations (1) and (4a,b) enable one to solve for r_c, u_c^r and b_c from the given constants E and l, then the other constant $\dot{\mu}$ is accordingly determined from equation (2) with the values of r_c, u_c^r and b_c. It should be noticed that the formation of shock is based on the existence of at least two physical sonic points, because an accretion flow into a black hole is supposed to be initially subsonic and terminally supersonic (when crossing the black hole horizon), the flow must pass through a sonic point on each side of a shock.

The shock conditions can be generally written as follows:

$$E_- = \left(\frac{1 + u_{r-}u_-^r}{-L}\right)^{1/2} / (1 - nb_-^2) \tag{5}$$

$$E_+ = \left(\frac{1 + u_{r+}u_+^r}{-L}\right)^{1/2} / (1 - nb_+^2) \tag{6}$$

$$\dot{\mu}_- = \gamma^{-n} r_s^2 \left(\frac{b_-^2}{1 - nb_-^2}\right)^n u_-^r \tag{7}$$

$$\dot{\mu}_+ = \gamma^{-n} r_s^2 \left(\frac{b_+^2}{1 - nb_+^2}\right)^n u_+^r \tag{8}$$

$$\left(u_-^r + \frac{b_-^2}{\gamma u_{r-}}\right) / (1 - nb_-^2) = \left(u_+^r + \frac{b_+^2}{\gamma u_{r+}}\right) / (1 - nb_+^2) \tag{9}$$

where E_-, $\dot{\mu}_-$, E_+, $\dot{\mu}_+$ are the conserved energy and the conserved entropy accretion rate for the pre-shock flow and the post-shock flow respectively (both the pre-shock and the post-shock flows are adiabatic), u_-^r, b_-, u_+^r, b_+ are the radial four-velocity and the sound speed of the flow just before and just after the shock respectively, r_s is the location of the shock. The meaning of equations (5)-(8) is obvious. Equation (9) results from the conservation of momentum flux density across the shock, or, in other words, from the balance of total pressure [thermal pressure P plus ram pressure $(\varepsilon + P) u_r u^r$] on the two sides of the shock surface, and with the help of the polytropic equation of state and equation (2).

3. Shock Solutions

A shock-included global solution can be obtained in the following way.

(A) Rankine-Hugoniot shocks: $E_- = E_+$, $\dot{\mu}_- < \dot{\mu}_+$, $b_- < b_+$.

(a) Start with the two controlling parameters l and $E(= E_- = E_+)$, solve equations (1) and (4a, b) for r_c, u_c^r and b_c, then determine the value of $\dot{\mu}$ from equation (2) with known r_c, u_c^r and b_c. The values of l and E have to be suitably given, so that there exist two physical sonic points (i.e. saddle-type critical points) in the flow, i.e. r_c, u_c^r, b_c and accordingly $\dot{\mu}$ all have two different values. The smaller and the larger one of r_c are denoted as r_{in} and r_{out} respectively; the smaller one of $\dot{\mu}$ is just $\dot{\mu}_-$, and the larger one of $\dot{\mu}$ is just $\dot{\mu}_+$.

(b) Having l, E_- and $\dot{\mu}_-$, solve equations (1) and (2) for u^r and b at any r, i.e. obtain a shock-free global solution that passes through the outer sonic point r_{out}; similarly, a shock-free global solution that passes through the inner sonic point r_{in} can be obtained with known l, E_+ and $\dot{\mu}_+$.

(c) Having $l, E_-, E_+, \dot{\mu}_-$ and $\dot{\mu}_+$, solve the five equations (5)-(9) for the five unknowns: r_s, u^r_-, b_-, u^r_+, and b_+, i.e. determine the shock.

(d) Finally, obtain a shock-included global solution by connecting the two shock-free solutions of (b) at the shock of (c).

(B) Isothermal shocks: $b_- = b_+$, $E_- > E_+, \dot{\mu}_- > \dot{\mu}_+$.

(a) Start with l and E (which is just E_-), solve equations (1) and (4a,b) for r_c (it is the outer sonic point r_{out}), $u^r_c(r_{out})$ and $b_c(r_{out})$, then determine the value of $\dot{\mu}$ (it is just $\dot{\mu}_-$) from equation (2).

(b) Having l, E_- and $\dot{\mu}_-$, solve equations (1) and (2) for u^r and b at any r, i.e. obtain a shock-free global solution that passes through r_{out}.

(c) Having l, E_- and $\dot{\mu}_-$, solve the nine equations [i.e. equations (1), (2), (4a,b), (5)-(9)] for the nine unknowns: $E_+, \dot{\mu}_+, r_c$ (it is now the inner sonic point r_{in}), $u^r_c(r_{in})$, $b_c(r_{in})$, r_s, $b_-(= b_+), u^r_-$ and u^r_+, i.e. determine the inner sonic point and the shock simultaneously [note that E of equation (1) is now just E_+ of equation (6), and $\dot{\mu}$ of equation (2) is now just $\dot{\mu}_+$ of equation (8)].

(d) Having l, E_+ and $\dot{\mu}_+$, solve equations (1) and (2) for u^r and b at any r, i.e. obtain a shock-free global solution that passes through r_{in}.

(e) Finally, obtain a shock-included global solution by connecting the two shock-free solutions of (b) and (d) at the shock of (c).

It is clear that the global solution of the flow motion, with or without shocks, is completely determined by the two parameters of the inflow l and E (i.e. E_-).

In both the above two cases step (c) is the key one: as multiple formal solutions satisfying the shock conditions (5)-(9) for the same pair of l and E may appear, it is necessary to judge the nature of these formal shocks before going on to step (d). Those formal shocks located inside r_{in} or outside r_{out} can be ruled out simply by employing the boundary conditions, i.e. the flow must be initially subsonic and terminally supersonic. For shocks located between r_{in} and r_{out}, a stability analysis is required. By checking the variation of the total pressure of the perturbed post-shock flow, Chakrabarti & Molteni (1993) found for the first time that, of the two shocks located between r_{in} and r_{out} in an adiabatic accretion flow, the outer one (i.e. closer to r_{out}) is stable, but they did not give a definite conclusion about the inner one. Yang & Kafatos (1995) made a similar, but improved, stability analysis for shocks in isothermal accretion flows; we follow their method here. Across the shock, the momentum flux density, defined as [different from the total pressure $P + (\varepsilon + P)u_r u^r$ only by a constant factor]

$$F = \left(u^r + \frac{b^2}{\gamma u_r}\right) / (1 - nb^2) \qquad (10)$$

is conserved, resulting exactly in equation (9). If as a result of some perturbation the shock location is moved from r_s to $r_s + \delta r$, the momentum flux density may not keep its balance. The resulting difference across the shock is

$$\delta F = F_+ - F_- = \left(\left.\frac{dF}{dr}\right|_+ - \left.\frac{dF}{dr}\right|_-\right)\delta r \equiv \Delta\delta r \qquad (11)$$

where the subscripts $+$, $-$ denote the values of a quantity just after and just before the shock respectively. The stability of the shock depends on the sign of Δ. If $\Delta > 0$, when $\delta r > 0$ (or $\delta r < 0$), the momentum flux (i.e. the total pressure) just after the shock is larger (or smaller) than that just before the shock (remember that an accretion flow is moving along decreasing r), so the shock should be shifted towards further increasing (or further decreasing) r, the flow will never find a new location at which the shock condition of momentum conservation is satisfied, therefore the shock is unstable. In contrast, $\Delta < 0$ implies that the shock in accretion flows is stable, because the imbalance of the momentum flux resulting from δr would always cause the shock to move back towards its unperturbed location. It is seen from equation (10) that to evaluate $(dF/dr)_-$, the values of $(du^r/dr)_-$ and $(db/dr)_-$ are needed, which can be solved for by combining the differential form of equations (5) and (7), i.e. $dE_-/dr = 0$ and $d\dot{\mu}_-/dr = 0$. Similarly, the values of $(du^r/dr)_+$ and $(db/dr)_+$, needed to evaluate $(dF/dr)_+$, can be solved for by combining the differential form of equations (6) and (8), i.e. $dE_+/dr = 0$ and $d\dot{\mu}_+/dr = 0$; then the sign of Δ is determined.

4. Results

I present now the computational results, of which Figs 1(a)-(c) are an outline, where (a), (b) and (c) are for a Schwarzschild black hole (the specific angular momentum of the hole $a = 0$), a rapid Kerr black hole ($a = 0.99$) with prograde flows, and a rapid Kerr black hole with retrograde flows, respectively (in this paper a, l and r are all in units of the black hole mass). It is seen that in the parameter space spanned by l and E (i.e. E_-), there is a strictly defined region for each type of shock formation. For an adiabatic accretion flow with its parameters belonging to the region bounded by two dotted lines and the vertical line $E = 1$, there are two Rankine-Hugoniot

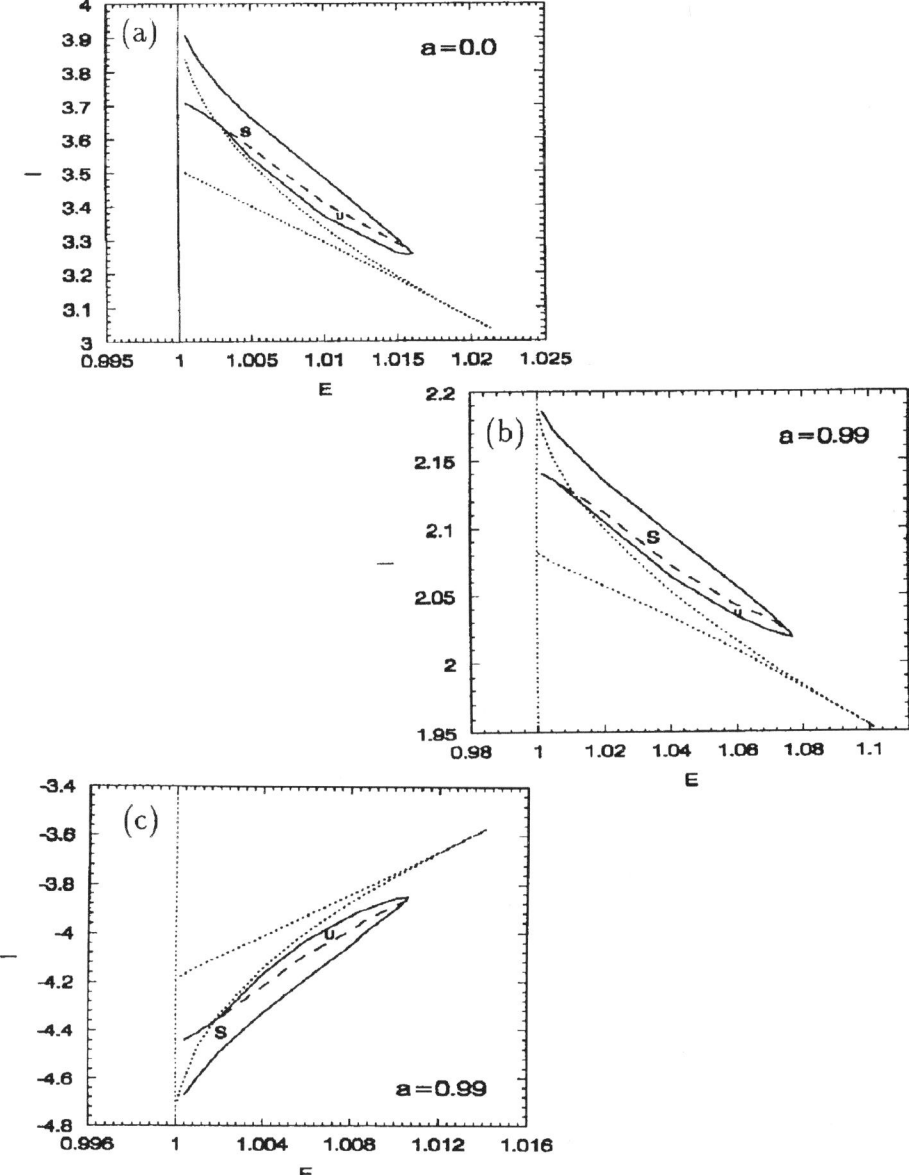

Fig. 1 : Energy-angular momentum parameter region of Rankine-Hugoniot shock formation in adiabatic black-hole accretion flows is bounded by two dotted lines and the vertical line $E = 1$. Region of isothermal shock formation is bounded by two solid lines and the vertical line $E = 1$, and is further divided by a dashed line into two parts: for the part marked by the letter U only unstable shocks can form, while for the part marked by the letter S both stable and unstable shocks can form. (a) is for a Schwarzschild black hole ($a = 0$), (b) and (c) are for a rapid Kerr black hole ($a = 0.99$) with prograde flows and with retrograde flows, respectively.

shocks located between r_{in} and r_{out}, the outer one (i.e. closer to r_{out}) is always stable, while the inner one is unstable except when the value of l is very close to the lower dotted line in (a) and (b), or very close to the upper dotted line in (c); flows with parameters located outside the region cannot develop Rankine-Hugoniot shocks at al. The region of isothermal shock formation in adiabatic accretion flows is bounded by two solid lines and the vertical line $E = 1$, it is further divided by a dashed line into two parts: for the part marked by the letter U only unstable shocks can form, while for the part marked by the letter S both stable and unsatble shocks can form; flows with parameters located outside the region cannot develop isothermal shocks at all.

Figs 2(a)-(d) are typical examples of shock-included global solutions of adiabatic accretion flows, in the form of the Mach number, defined as the ratio of the radial three-velocity measured by a corotating observer to the local sound speed, $M = v/b$, versus r. Following Abramowicz & Chakrabarti (1990) a shock-free global solution that is complete (i.e. is able to join the black hole horizon to the large distance where the flow starts) is called x type, and an incomplete shock-free global solution (i.e. either extending to the black hole horizon only, or extending to the large distance only) is called α type. In all four panels the shock-free global solution passing through the outer sonic point (where $M = 1$) is shown by solid lines; such a solution is of x type in Fig. 2(a), and of α type in the other three figures. In Fig. 2(a) there are two α-type shock-free global solutions passing through the inner sonic point (where $M = 1$), drawn as a dotted line and a dashed line respectively. Two shock locations are indicated by a dotted vertical line and a dashed vertical line respectively, of which only the outer shock is stable; thus the stable shock solution is obtained by following the arrows, i.e. the flow first passes through the outer sonic point to become supersonic, and forms a shock at the location of the dashed vertical line, then follows the dashed line to become supersonic again by passing through the inner sonic point; such a shock solution is of α-x type. In Fig. 2(b) there are three α-type solutions passing through the inner sonic point, drawn as a dot-dashed line, a dashed line and a dotted line respectively. Two of the three shock locations are indicated by a dot-dashed vertical line and a dashed vertical line respectively (the third one is further inward and is not drawn). Only the outer one (the dot-dashed vertical line) is stable, thus the stable shock solution is indicated by the arrows, and is of α-α type. In Fig. 2(c) there is an x-type and an α-type solution passing through the inner sonic point, drawn as dotted lines and a dashed line respectively. Two shocks are indicated by a dotted vertical line and a dashed vertical line respectively, and both of them are stable; thus there are two stable shock solutions as shown by the arrows, one of which is

SHOCK FORMATION

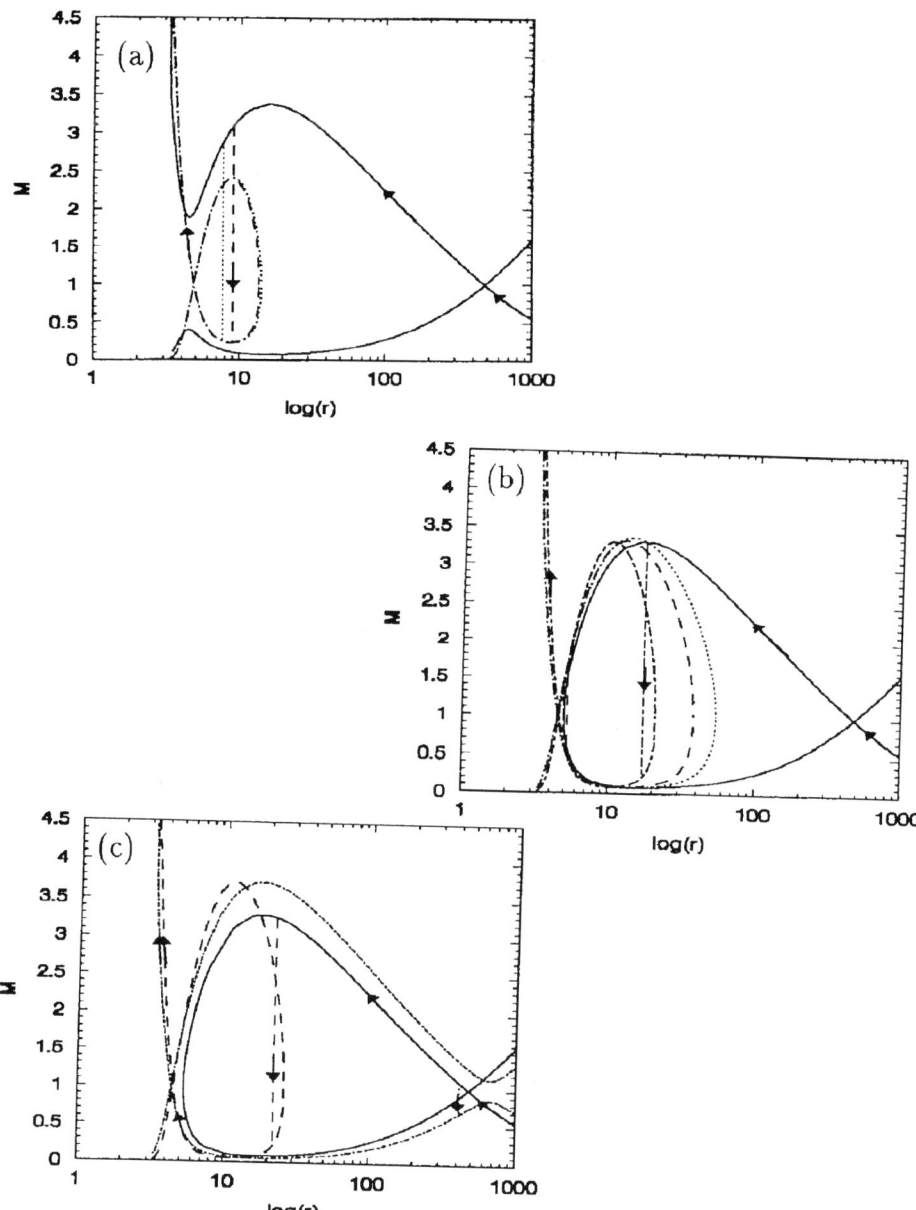

Fig. 2 (a-c): Typical examples of the global flow solution, i.e. the Mach number of the flow versus the radial coordinate. The global shock-free solution that passes through the outer sonic point is shown by solid lines, the solutions that pass through the inner sonic point are shown by other types of line, the shock location is shown by vertical lines, and the global solution with the stable shock is obtained by following the arrows.

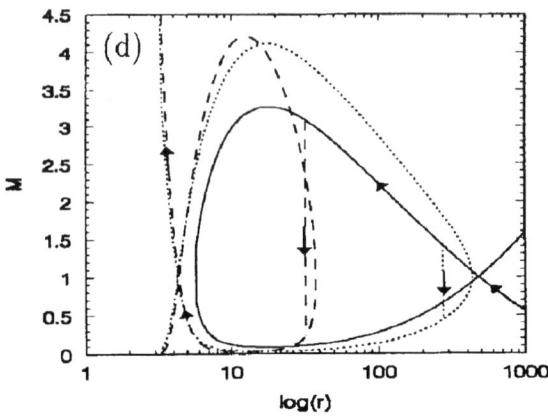

Fig. 2d

of x-α type, and the other of which is of α-α type. In Fig. 2(d) there are also two stable shocks and accordingly two stable shock-included global solutions. What is different from the case of Fig.2(c) is that both the solutions are of α-α type here. *Global solutions of adiabatic accretion flows with Rankine-Hugoniot shocks can only have the topology of Fig. 2(a), while solutions with isothermal shocks can have all the topologies in the four figures.* The possibility of three types of shock-included solutions, namely α-x, x-α and α-α type, was first pointed out by Abramowicz & Chakrabarti (1990), however only the α-x type solution had been actually obtained in previous studies (e.g. Chakrabarti 1996a). Here the existence of x-α type and α-α type solutions is proved; in addition, the 'bistability' of solution, i.e. the fact that for the same flow parameters there can be two stable shock-included global solutions, is found for the first time.

To sum up, the difference between the two types of shocks in adiabatic accretion flows is that (1) they occur for different regions in the flow parameter space; (2) for given flow parameters l and E, the stable Rankine-Hugoniot shock location is practically unique whenever it exists, while there can be as many as two stable isothermal shocks; and (3) the topology of global solution can only be of α-x type in the Rankine-Hugoniot shock case, while in the isothermal shock case the solution topology can be of α-x, x-α or α-α type.

It is also significant that the effects of frame-dragging of Kerr geometry on the formation and the properties of stable shocks are found out: (1) For retrograde flows around a rapid Kerr black hole (Fig. 1c), shocks form for higher values of l (the absolute value) and smaller values of E than in the Schwarzschild black hole case (Fig. 1a), reflecting the fact that in the former case some centrifugal force is 'spent' to fight against the frame-dragging of the black hole. For prograde flows around a rapid Kerr hole (Fig. 1b), on the other hand, the frame-dragging helps to strengthen the centrifugal barrier, so that only those flows with lower l and higher E are allowed to pass through the barrier, i.e. to form shocks. (2) Comparing with the Schwarzschild black hole case as well as the case of retrograde flows around a rapid Kerr black hole, the frame-dragging of a rapid Kerr hole causes shocks in prograde flows to be located closer to the hole, to be stronger (i.e. to have a larger ratio of Mach numbers just before to just after the shock), and (for isothermal shocks) to be able to release more energy by radiation.

5. Discussion

Numerous authors have studied transonic global solutions of advection dominated accretion flows (ADAF), but have found no shocks (e.g. Chen *et al.* 1997; Narayan *et al.* 1997). I am agreed that the shock-free solutions obtained by these authors are correct, but that is not all. The reason why they have found no shocks is probably a result of their philosophy on the boundary condition and the completeness of the solution. Namely, they all have required that a global solution should pass through the inner sonic point and keep being supersonic on one hand, and should connect to a Keplerian, highly subsonic solution on the other hand. Such a global solution is of x type. However, the most frequent topology of shock-included global solutions is α-x type (cf. Fig. 2a), as it may participate in both Rankine-Hugoniot shocks and isothermal shocks in adiabatic accretion flows (such flows are fully ADAF). Unfortunately, an α-type solution was excluded a priori by the above authors, because such a solution, although it passes through the inner sonic point, does not satisfy the Keplerian outer boundary condition, and is incomplete. An α-type solution passing through the inner sonic point, however, can be part of an α-x type, or part of α-α type (Fig. 2b), shock-included global solution, and such global solutions are complete and satisfy physical boundary conditions. Even for an x-type solution passing through the inner sonic point, it is possible that this solution connects with an α-type solution passing through the outer sonic point via a shock, forming an x-α type shock solution (Fig. 2c). The above authors seem to have stopped when such an x-type solution is obtained, however, without searching further for the possibility of an x-α type shock solution. In other words, they have concentrated their attention only on the global solution with a single sonic point. This means that they have chosen flow parameters outside the region of shock formation (cf. Fig. 1), because the shock formation requires the existence of at least two physical sonic points in the flow.

Finally, some very brief comments are in order concerning the observational significance of the shock study. Black-hole accretion models may need to include shocks, as then they can give a complete explanation of the observed steady, as well as time-dependent behaviors, such as UV spectra of active galactic nuclei (Chakrabarti & Wiita 1992), the transition of spectral states and quasi-periodic oscillations of black hole candidates (Chakrabarti & Titarchuk 1995; Yang & Kafatos 1995; Molteni et al. 1996b). The effects of frame-dragging of Kerr black holes on shocks may provide some clues to distinguish observationally different black-hole spin states, as Sponholz & Molteni (1994) suggested. Above all, however, it is well known that the key to identify stellar black holes lies in distinguishing them from neutron stars, when they both appear as X-ray sources in binaries. Indeed, all the results in this paper are in principle applicable to the case of neutron stars without strong magnetic fields. However there are some differences: first, in the black hole case the shock formation requires at least two physical sonic points in the flow, but this is not necessary for neutron star case, instead, a flow with only one sonic point can also develop a shock because the flow terminates subsonically; second, because of the same reason, when there are two sonic points in the flow, the formal shock located inside the inner sonic point is ruled out in the black hole case, but in the neutron star case this innermost shock is possible, and in fact was considered to be physically important (Fukue 1987). It is worthwhile to study the observational appearance of these differences.

References

Abramowicz, M.A. and Chakrabarti, S.K. (1990), *ApJ*, 350, 281.
Chakrabarti, S.K. (1996a), *ApJ*, 464, 664.
Chakrabarti, S.K. (1996b), *MNRAS*, 283, 325.
Chakrabarti, S.K. (1996c), *ApJ*, 471, 237.
Chakrabarti, S.K. and Molteni, D. (1993), *ApJ*, 417, 671.
Chakrabarti, S.K. and Titarchuk, L.G. (1995), *ApJ*, 455, 623.
Chakrabarti, S.K. and Wiita, P.J. (1992), *ApJ*, 387, L21.
Chen, X., Abramowicz, M.A. and Lasota J.P. (1997), *ApJ*, 476, 61.
Fukue, J. (1987), *PASJ*, 39, 309.
Lu, J.-F. (1986), *Gen. Relativ. Gravitation*, 18, 45.
Lu, J.-F., Yu, K.N., Yuan, F. and Young, E.C.M. (1997), *A&A*, 321, 665.
Lu, J.-F. and Yuan, F. (1998), *MNRAS*, in press.
Molteni, D., Ryu, D. and Chakrabarti, S.K. (1996a), *ApJ*, 470, 460.
Molteni, D., Sponholz, H. and Chakrabarti, S.K. (1996b), *ApJ*, 457, 805.
Narayan, R., Kato, S. and Honma, F. (1997), *ApJ*, 476, 49.
Sponholz, H. and Molteni, D. (1994), *MNRAS*, 271, 233.
Yang, R. and Kafatos, M. (1995), *A&A*, 295, 238.

ROTATING ACCRETION FLOWS NEAR A BLACK HOLE: A NUMERICAL STUDY

DONGSU RYU

Dept. of Astro. & Space Science, Chungnam Nat. Univ., Korea
email: ryu@canopus.chungnam.ac.kr

Abstract.

The characteristics of thin, axisymmetric, supersonic accretion flows near a black hole is presented. Such flows of inviscid, cold adiabatic gas are characterized by the specific angular momentum and the specific energy. Using two-dimensional numerical simulations in cylindrical geometry, we show that there are various regimes in which the accretion flows behave distinctly differently. We compare such numerical solutions with analytical solutions. We confirm that for a wide range of above parameters a stable standing shock wave with a vortex inside it forms close to the black hole. Apart from steady state solutions, we show the existence of non-steady solutions for thin accretion flows where the accretion shock is destroyed and re-generated periodically. The period is roughly equal to $4 - 6 \times 10^3 R_g/c$ depending on the angular momentum of the flow. The unstable behavior should be caused by dynamically induced instabilities, since inviscid, adiabatic gas is considered. We discuss possible relevance of the periodic behavior on quasi-periodic oscillations (QPOs) observed in galactic and extragalactic black hole candidates.

1. Introduction

The gravitational interaction between a compact object and the surrounding gas has been the topic of many studies. Bondi (1952) first considered the accretion of adiabatic gas into a gravitating point mass, M_{bh}, with the assumptions of non-relativistic (but Newtonian) treatment and spherical symmetry, and found a shock-free solution with a steady state accretion rate given by

$$\dot{M}_{acc} = 4\pi\alpha\frac{(GM_{bh})^2 \rho}{c_s^3} \quad (1)$$

where ρ and c_s are the density and sound speed of the gas at infinity. Here, α is a numerical constant of order unity.

However, more realistic accretion involves a finite amount of angular momentum forming rotating accretion flows. Previous studies of such non-spherical accretion demonstrated the variety and complexity of structures forming around the central object. Rotating accretion flows are important ingredients in many astrophysical systems containing a black hole, which involve mass transfer from one object to another (such as in a binary system) or from set of objects to another (such as in a galactic center). The standard disk model of such accretion flows by Shakura & Sunyaev (1973) assumes Keplerian distribution of accreting matter. There, the inner edge of the disk is chosen to coincide with the marginally stable orbit located at three Schwarzschild radii, $r_i = 3R_g$, where the Schwarzschild radius

$$R_g = \frac{2GM_{BH}}{c^2} \quad (2)$$

is the horizon of a black hole of mass M_{BH}. This disk model is clearly incomplete, since the inner boundary condition on the horizon was not taken care of. As an accretion flow approaches the horizon, its radial velocity reaches the velocity of light. Therefore, a black hole accretion flow is necessarily supersonic and must pass through a sonic point where the flow has to be sub-Keplerian. Thus, independent of heating and cooling processes, a black hole accretion has to deviate from a standard Shakura-Sunyaev type Keplerian disk. The disk with realistic accretion flows is called the *advective disk* (Chakrabarti, 1996).

Here, we present the results of numerical study of accretion flows near a black hole by assuming they are thin, axisymmetric, and inviscid.

2. Analytic Consideration

We choose cylindrical coordinates (r, θ, z) and place a black hole at the center. We assume that the gravitational field of the black hole can be described in terms of the potential introduced by Paczyński & Wiita (1980)

$$\phi(r, z) = -\frac{GM_{bh}}{R - R_g} \quad (3)$$

where $R = \sqrt{r^2 + z^2}$. The accreting matter is assumed to be adiabatic gas without cooling and dissipation and described with a polytropic equation of state, $P = K\rho^\gamma$, where γ is the adiabatic index which is considered to be constant with $4/3$ throughout the flow. K is related to the specific entropy of the flow, s, and varies only at shocks, if present. Since we consider only weak viscosity limit, the specific angular momentum of the accretion flow

$$\lambda = rv_\theta \quad (4)$$

is assumed to be conserved. Thus, unlike a Bondi flow, which is described by a single parameter (say, specific energy), the one-dimensional accretion flows are described by two parameters which are the specific energy

$$\mathcal{E} = \frac{v_r^2}{2} + \frac{c_s^2}{\gamma - 1} + \frac{\lambda^2}{2r^2} + \phi = 0 \quad (5)$$

and the specific angular momentum, λ. Here, c_s is the gas sound speed.

Fig. 1 shows the classification in the parameter space (Chakrabarti, 1989):
N: No sonic points. Shock only if supersonic injection.
O: Outer sonic point only as in a Bondi solution. No shock.
I: Inner sonic point only. Shock only if supersonic injection.
O*: Outer and center sonic points. Solution does not extend to the horizon.
I*: Inner and center sonic points. Solution does not extend to large distance.
SA: Two (outer and inner) sonic points. Shock in accretion solutions but not in wind solutions.
SW: Two sonic points. Shock in wind solutions but not in accretion solutions.
NSA: Two sonic points. No shock condition satisfied in accretion solutions.
NSW: Two sonic points. No shock condition satisfied in wind solutions.

Although these solutions are strictly valid only for inviscid flows, even when viscosity is high, given that the viscous time-scale is likely to be much larger compared to the infall time-scale, the inviscid solutions are likely to remain important.

3. One-dimensional Numerical Solutions

Fig. 2 shows an example numerical solution from the 'SA' region. We superpose analytical solution (solid lines) and numerical solutions with a grid-based code, the TVD code (dashed lines) and with a particle-based code, the SPH code (dotted lines) (Molteni et al., 1996a). The TVD calculation was done with 512 grids and the SPH calculation was done with ~ 560 particles of size $h = 0.3R_g$. Matter was injected at the outer boundary located at $r = 50R_g$. An absorption condition was used to mimic the black hole horizon at the inner boundary which is located either at $r = 1.5R_g$ for the

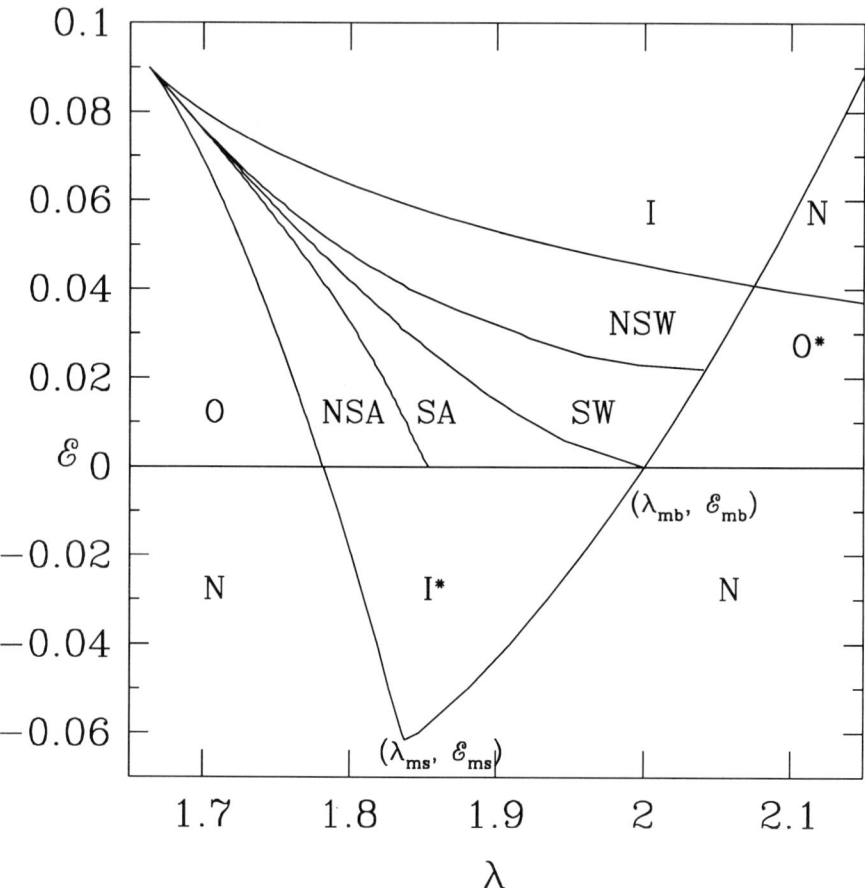

Figure 1. Classification of one-dimensional accretion flows in the parameter space of specific energy, $\mathcal{E}$, and the specific angular momentum, λ. $\mathcal{E}$ is given in unit of c^2 and λ in unit of $R_g c$.

TVD calculation or at $r = 1.25 R_g$ for the SPH calculation. The adiabatic index $\gamma = 4/3$ was used.

The figure shows an excellent agreement between the analytic and numerical solutions. Here, the flow starts out subsonically, presumably from a Keplerian disk. Then, it enters through the outer sonic point (located at $r = 27.97 R_g$), passes through the shock (located at $r = 7.98 R_g$), and subsequently passes through the inner sonic point (located at $r = 2.57 R_g$) before entering the black hole.

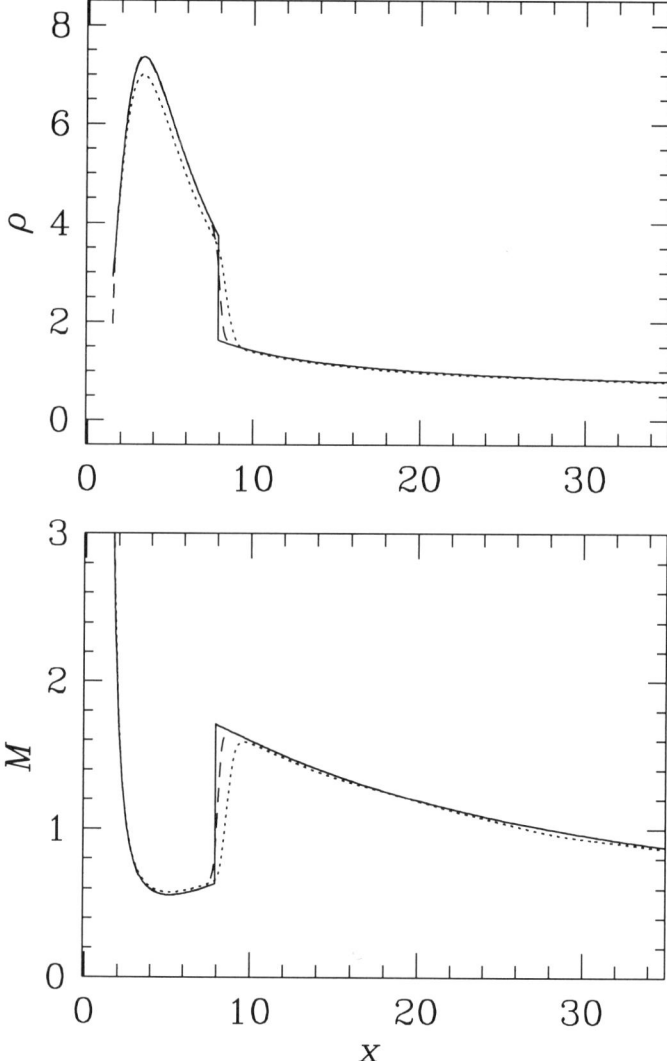

Figure 2. One-dimensional accretion flow with a standing shock in the region of 'SA'. $\mathcal{E} = 0.036$ and $\lambda = 1.80$ are used. The solid line is the analytical solution, and the long and short dashed lines are the solutions of the TVD and SPH simulations, respectively. Upper panel is the mass density in arbitrary units and the lower panel is the Mach number of the flow.

4. Two-dimensional Numerical Solutions

In multi-dimensional accretion flows, non-steady solutions, as well as steady solutions, exist. A realistic accretion disk is expected to be three-dimensional, of course. Assuming axisymmetry, the problem is reduced to two-dimensions. It is difficult to solve analytically the problem in full generality even in two-

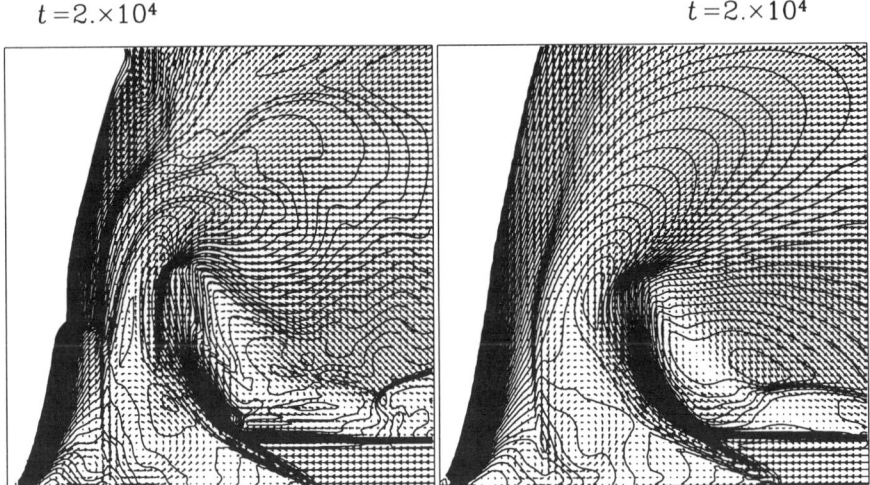

Figure 3. Two-dimensional accretion flow showing a stable behavior in the region of 'O*'. $\mathcal{E} = 0$ and $\lambda = 2$ are used. Contours are for density and arrows are for velocity vector. Numerical solutions in two different resolutions are shown, high resolution with 256×256 cells (left) and low resolution with 128×128 cells (right).

dimensions. Here, we discuss some examples of two-dimensional accretion flows with zero specific energy ($\mathcal{E} = 0$). More extensive discussion on accretion flows with $\mathcal{E} = 0$ was reported in Ryu, Chakrabarti & Molteni (1997). Discussion on extensive calculations for accretion flows with non-zero specific energy ($\mathcal{E} \neq 0$) will be reported in a future paper.

In these simulations, we inject supersonic matter (with a radial Mach number $M = v_r/c_s = 10$) at the outer boundary, $r_b = 50R_g$. The inflow at the outer boundary is assumed to have a small thickness, h_{in}, or a small arc angle, $\theta_{in} = \arctan(h_{in}/r_b) \ll 1$. If zero-energy accretion flows belong the region of 'O', most of the material is accreted into the black hole forming a stable quasi-spherical flow or a simple disk-like structure around it (just as a Bondi flow). If accretion flows belong the region of 'SA' and 'O*', the incoming material produces a stable standing shock with one or more vortices behind it and some of it is deflected away at the shock as a conical outgoing wind of higher entropy. An Example is give in Fig. 3, where the numerical solution of an accretion flow with $\lambda = 2R_g c$ in the region 'O*' is shown (with two different numerical resolutions).

Accretion flows with parameters in the region 'NSA' with $1.782R_g c < \lambda < 1.854R_g c$ show an unstable behavior. Figs. 4 and 5 shows the numerical solution of a flow with $\lambda = 1.85R_g c$. In this case, the structure with an accretion shock and a generally subsonic high density disk is established

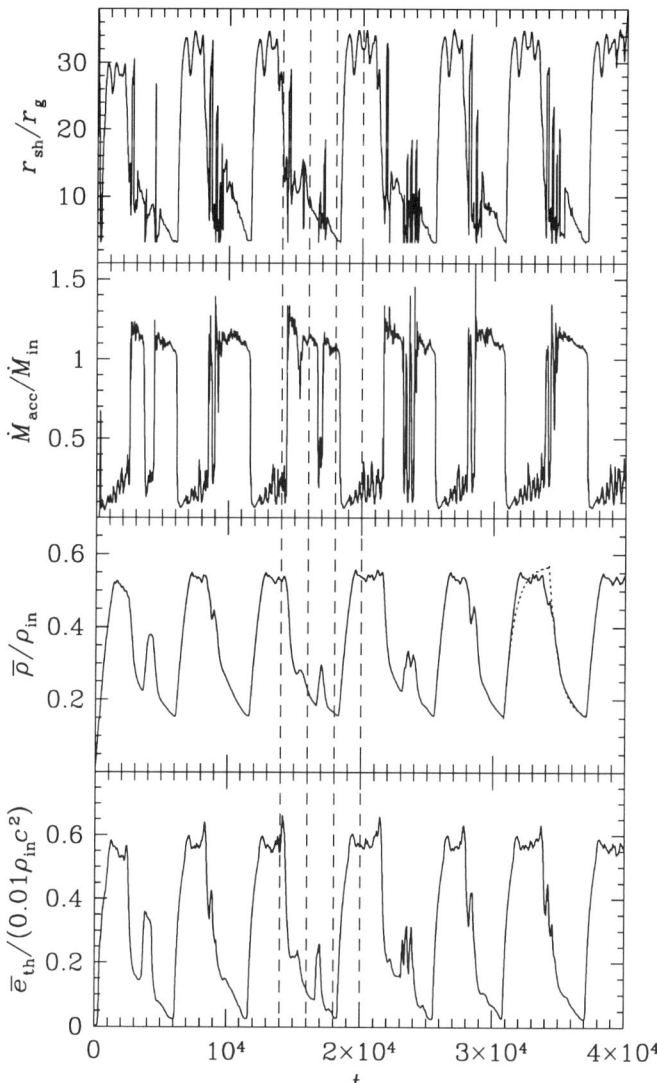

Figure 4. Two-dimensional accretion flow showing an unstable behavior in the region of 'NSA'. $\mathcal{E} = 0$ and $\lambda = 1.85$ are used. The temporal evolution of the position of the accretion shock along the equatorial plane, r_{sh}, the mass accretion rate into the black hole, $\dot{M}_{acc}$, and the mean density, $\bar{\rho}$, and the mean thermal energy, $\bar{e}_{th}$, of disk matter inside the computational domain are shown.

around the black hole. However, the structure is not stable. At the accretion shock, the incoming flow is deflected. But some of the post-shock flow, which is further accelerated by the pressure gradient behind the shock, goes through a second shock, where the flow is deflected once more downwards.

$t = 1.4 \times 10^4$ $t = 1.6 \times 10^4$

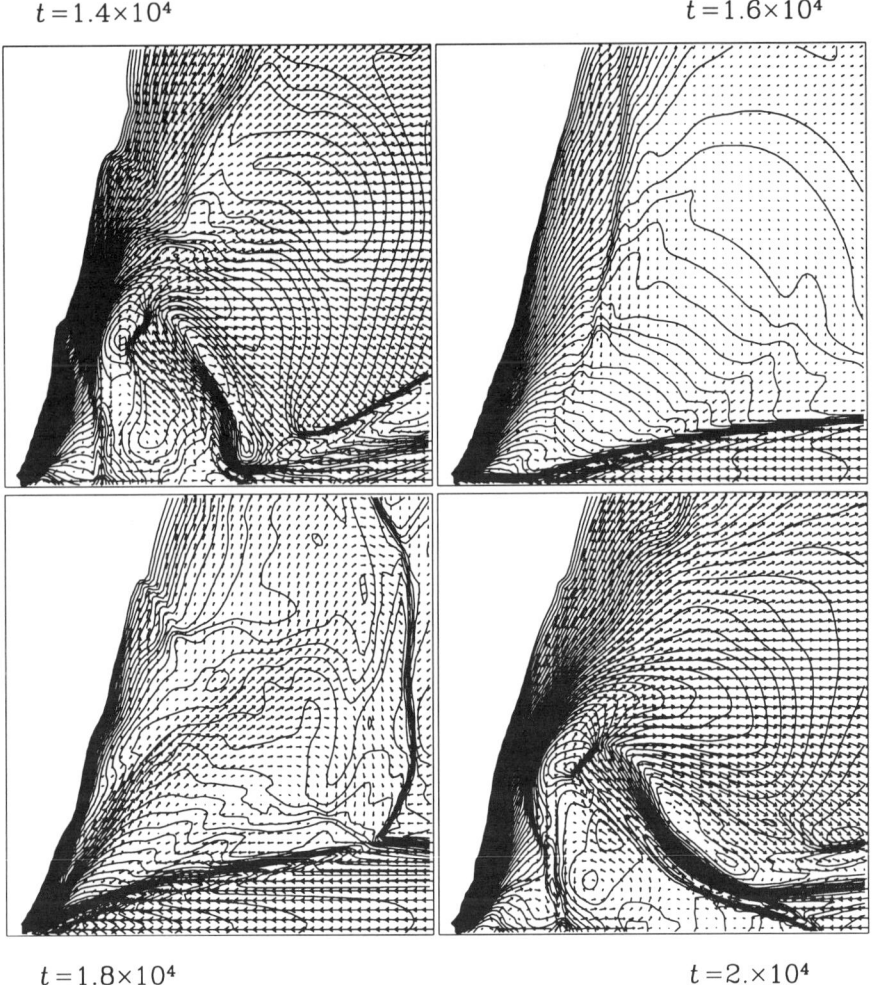

$t = 1.8 \times 10^4$ $t = 2. \times 10^4$

Figure 5. Two-dimensional accretion flow showing an unstable behavior in the region of 'NSA'. $\mathcal{E} = 0$ and $\lambda = 1.85$ are used. Contours are for density and arrows are for velocity vector. Time is given in unit of R_g/c.

The downward flow squeezes the incoming material, and the accretion shock starts collapsing ($t = 1.4 \times 10^4 R_g/c$). In the process of the collapse, some of the post-shock material escapes as wind but most is absorbed into the black hole. After the collapse, the re-building of the accretion shock starts with the incoming material bouncing back from the centrifugal barrier. The subsonic post-shock region becomes a reservoir of material, so the material is accumulated behind the shock. With the accumulated material a giant vortex is formed, which in turn supports the accretion shock ($t =$

$2 \times 10^4 R_g/c$). This continues until the incoming flow is squeezed enough so the accretion shock collapsed, and the cycle continues.

5. Discussion

The time scale of the periodicity of unstable flows is interesting. It is in the range of
$$\tau \approx 4 - 6 \times 10^3 \frac{R_g}{c} = 4 - 6 \times 10^{-2} \left(\frac{M_{bh}}{M_\odot}\right) \text{s}. \quad (6)$$

The modulation of amplitude is also very significant, and could be as much as a hundred percent depending on detailed processes.

Oscillations with these characteristics have been observed in black hole candidates and are called the QPOs. For instance, in the QPOs from the low mass x-ray binaries, the oscillation frequency has been found to lie typically between 5 and 60 Hz (Van der Klis, 1989). Thus, compact objects with mass $M \approx 0.3 - 5 M_\odot$ could generate oscillations of right frequencies due to the instability discussed in this paper. Similar oscillations of period on the order of a few hours to a few days are expected in soft X-rays and UV emissions from galactic centers.

However, by considering simplified physics which we have assumed here, it may be premature to assume that the presented mechanism would explain all the QPOs observed. In some of the cases, the oscillation may be due to the dynamic instability considered in Ryu *et al.* (1995), or it may be due to resonance of the cooling time scale (bremsstrahlung or Comptonization, whatever the case may be) and the infall time scale in the enhanced density region near the centrifugal barrier as shown by Molteni, Sponholz and Chakrabarti (1996b). In detailed works, radiative processes as well as viscosity should be included in the accretion calculations to examine the observational consequences of the present instability.

References

Bondi, H., 1952, *Monthly Notices of the R. A. S.*, 112, 195.
Chakrabarti, S.K., 1989, *The Astrophysical Journal*, 347, 365.
Chakrabarti, S.K., 1996, *Physics Reports*, 266, 229.
Molteni, D., Ryu, D., & Chakrabarti, S. K., 1996a, *The Astrophysical Journal*, 470, 460.
Molteni, D., Sponholz, H. & Chakrabarti, S. K., 1996b, *The Astrophysical Journal*, 457, 805.
Paczyński, B., & Wiita, P. J., 1980, *Astronomy and Astrophysics*, 88, 23.
Ryu, D., Brown, G. L., Ostriker, J. P., & Loeb, A., 1995, *The Astrophysical Journal*, 452, 364.
Ryu, D., Chakrabarti, S. K., & Molteni, D., 1997, *The Astrophysical Journal*, 474, 378.
Shakura, N. I., & Sunyaev, R. A., 1973, *Astronomy and Astrophysics*, 24, 337.
Van der Klis, M., 1989, *Annual Review of Astronomy and Astrophysics*, 27, 517.

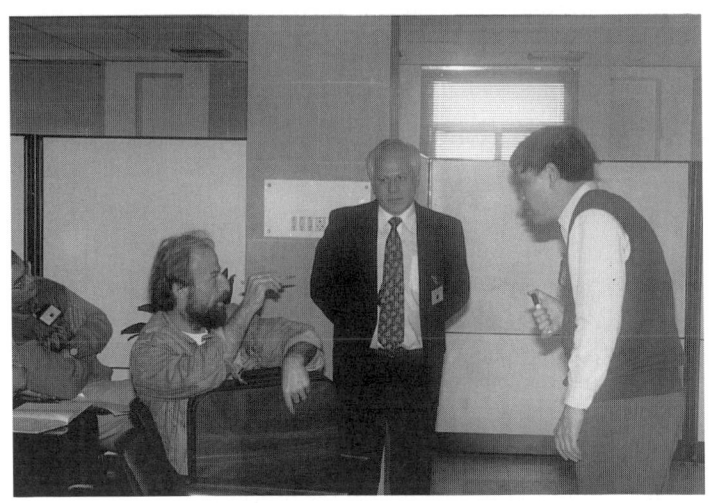

Ju-Fu Lu (right) listens to T. Zannias (left)

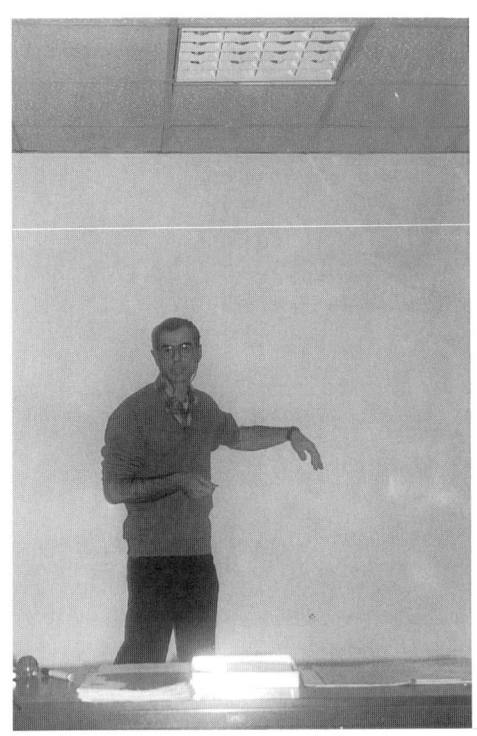

D. Molteni

SIMULATIONS OF SHOCKS WITH SMOOTHED PARTICLES HYDRODYNAMICS METHOD

D. MOLTENI, G. GERARDI AND M.A. VALENZA
Department of Physical and Astronomical Sciences
Via Archirafi 36, 90123,Palermo, Italy

AND

G. LANZAFAME
Institute of Astronomy
University of Catania, Viale A. Doria 6, 95125, Catania, Italy

1. Introduction

Shock waves in accretion flows onto compact objects are very important to transform the gravitational potential energy into radiation. The possibility that they exist also around black holes has been suggested years ago by Hawley, Smarr and Wilson (1984a, 1984b). Recently it has been shown that there is a wide range of parameters (essentially angular momentum and specific energy [for adiabatic flows] or temperature [for isothermal flows]) for which steady shock configurations are possible (Chakrabarti 1990, Chakrabarti & Molteni, 1993). Recent analytical calculations have shown the stability for isothermal and adiabatic shocks (Lu 1997; Nakayama, 1994). This fact has great astrophysical relevance since, in this way, the emission due to the shock is not a transient episode but can be a permanent mechanism responsible for the radiated energy. Here we review the relevant results obtained in the last few years through numerical simulations by a novel technique named Smoothed Particles Hydrodynamics (SPH). We concentrate on the axisymmetric problems which are done using cylindrical coordinates. The plan of this contribution is the following : in Section 2 we describe the SPH numerical method in cylindrical coordinates for axisymmetric problems, then, in Section 3 and its subsections, we give the basic ingredients, analytical equations, of the physical problem of accretion flows onto Black Holes. In Section 4 we present some results of the simulations and comment on them.

2. SPH in Cylindrical Coordinates

SPH is a Lagrangean method. As is well known, Lagrangean methods are not widely used in more than one spatial dimension due to complexities in deriving the mesh distortion in general two and three dimensional motions. Since SPH is essentially grid free, use of such method is possible because spatial derivatives at each point are computed by interpolation over neighbouring points, instead of by usual difference techniques. The interpolation points, usually called "particles", are moved following the fluid dynamics equations. It has been shown that SPH is equivalent to a finite elements method with a moving grid (Dilts, 1996). The SPH method is extensively described in various fundamental papers (Monaghan, 1985; Benz, 1990). We resume this article with its basic aspects and comment on some peculiarities involved in writing the axisymmetric version of SPH.

2.1. BASIC FLUID DYNAMICS EQUATIONS

Let us start with the basic equations we integrate. They are the classical equations for compressible viscous fluid valid for axisymmetric configurations. Let us denote the cylindrical coordinates by r (radial distance from the z-axis) z, ϕ and assume axial symmetry i.e. $\frac{d}{d\phi} = 0$.

We have for the mass conservation,

$$\frac{D\varrho}{Dt} = -\varrho \left[\frac{1}{r} \frac{\partial(rv_r)}{\partial r} + \frac{\partial v_z}{\partial z} \right]. \tag{1}$$

with the Lagrangian derivative,

$$\frac{D}{Dt} = \frac{\partial}{\partial t} + \mathbf{v} \cdot \nabla. \tag{2}$$

Since our major goal is to simulate astrophysical accretion disks, in which the viscosity is essentially derived from the tangential component of the stress tensor, we will assume that the only nonvanishing components of the stress tensor are the cross terms of the azimuthal with the radial (i.e., $\tau_{r\phi}$) and the vertical (i.e., $\tau_{z\phi}$) component, respectively:

$$\tau_{r\phi} = \mu r \frac{\partial \Omega}{\partial r} \quad , \quad \tau_{z\phi} = \mu \frac{\partial v_\phi}{\partial z} \quad , \tag{3}$$

where we note for the angular velocity $\Omega = v_\phi/r$ and for the dynamical viscosity $\mu = \nu\varrho$ with the kinetic viscosity ν. The conservation of ϕ-momentum is given by:

$$\frac{Dv_\phi}{Dt} = -\frac{v_\phi v_r}{r} + \frac{1}{r^2\varrho} \left[\frac{\partial}{\partial r} \left(\mu r^3 \frac{\partial \Omega}{\partial r} \right) + \frac{\partial}{\partial z} \left(\mu r^3 \frac{\partial \Omega}{\partial z} \right) \right] \tag{4}$$

For the radial and the vertical momentum we have

$$\frac{Dv_r}{Dt} = -\frac{v_\phi^2}{r} - \frac{1}{\varrho}\frac{\partial p}{\partial r} + g_r \quad , \quad \frac{Dv_z}{Dt} = -\frac{1}{\varrho}\frac{\partial p}{\partial z} + g_z \tag{5}$$

respectively. It is worthwhile to note that it is possible, for any Lagrangean method, to integrate in a very accurate way the angular momentum equation in the case of zero viscosity and this is achieved very trivially. Indeed the angular momentum equation (with zero viscosity) $\frac{Dv_\phi}{Dt} = -\frac{v_\phi v_r}{r}$ can be written $\frac{Dv_\phi}{v_\phi} = -\frac{v_r}{r}Dt$ and therefore discretized over finite time steps as

$$(v_\phi)_i^{n+1} = (v_\phi)_i^n \, e^{-\left(\frac{v_r}{r}\right)_i^n \Delta t}$$

where the index 'n' is the time level and index 'i' refers to the spatial evaluation point.

For the energy equation we could use

$$\frac{D\epsilon}{Dt} = -\frac{P}{\rho}\nabla\mathbf{v} + \frac{\Phi}{\rho}$$

with Φ the dissipation function. In case of zero viscosity this equation provides an interpolation formula for SPH which is sufficiently good. But for problems which include viscosity, we prefer to start from the following analytical equation (Batchelor, 1981). This uses the total kinetic and thermal energy and has better accuracy in the SPH formulation (it conserves exactly the total energy, as it can be put in a symmetric form so that pair of particles exchange equal amounts of energy, cf. Monaghan 1985):

$$\frac{D}{Dt}\left(\epsilon + \frac{1}{2}\mathbf{v}^2\right) = -\frac{P}{\rho}\nabla\mathbf{v} + \mathbf{v}\cdot\left(\frac{D\mathbf{v}}{Dt}\right) + \frac{1}{\rho}\nabla\left(\overleftrightarrow{\tau} : \mathbf{v}\right) \tag{6}$$

$$\left(\frac{D\mathbf{v}}{Dt}\right) = -\frac{1}{\rho}\nabla P + \mathbf{g}$$

where γ is the usual ratio of the specific heats, $\mathbf{g}$ is the gravitational acceleration. $\overleftrightarrow{\tau} : \mathbf{v}$ is the vector resulting from the contraction of the stress tensor with the velocity vector.

We use the equation of state for an ideal gas,

$$p = (\gamma - 1)\varrho\epsilon. \tag{7}$$

2.2. INTERPOLATION METHOD

As already mentioned, the greatest difficulty for a Lagrangean method is the evaluation of the spatial derivatives in an arbitrarily deforming mesh. This difficulty is overcome in SPH using an otherwise well known interpolation criterion. The value of any function can be computed from the values at neighbouring points (Shoenberg, 1973):

$$f(\mathbf{P}_i) = \int_V f(\mathbf{P'})W(h, \mathbf{P}_i - \mathbf{P'})dV' \simeq \sum_{j=1}^{N} f(\mathbf{P}_j)\frac{m_j}{\varrho_j}W_{ij}, \qquad (8)$$

where W_{ij} stands for $W(\mathbf{P}_i - \mathbf{P}_j)$ and is the interpolating function or 'kernel'. This function must have the properties of the Dirac delta function, h is its amplitude. The properties of this function should be such that optimal interpolation are produced as discussed for example in Balsara (1995). In general, it has been shown that the kernel and its first derivative must be continuous and have a compact support.

Our basic approach in cylindrical geometry is the same as one uses in Cartesian coordinates and therefore we simply assume an usual kernel function, depending directly on the radial (r) and vertical (z) variables, and therefore retaining the usual normalization factor and width. Now pseudo-particles are small tori of mass $dm_k = 2\pi \varrho_k r_k dr_k dz_k$. In this way, we may use the same grid in the (r, z) domain and the same procedure for searching the nearest neighbours of each particle as in Cartesian coordinates. Therefore, applying the usual procedure for the evaluation of any smooth function in the point (r_i, z_i) we have:

$$f(\mathbf{R}_i) = \int_V f(\mathbf{R'})W(h, \mathbf{R}_i - \mathbf{R'})\frac{2\pi r\varrho}{2\pi r\varrho}d\mathbf{R'} \simeq \sum_{j=1}^{N} f(\mathbf{R}_j)\frac{m_j}{2\pi r_j\varrho_j}W_{ij}, \qquad (9)$$

where $\mathbf{R}_k = (r_k, z_k)$. Since the 2π is a common factor in all the formulae, we include it into the kernel definition.

We remind the readers the procedure to make spatial derivatives in SPH in cylindrical coordinates:

$$\nabla f(\mathbf{P}_i) = \sum_{j=1}^{N} \frac{m_j}{r_j \rho_j} f(\mathbf{P}_j) \nabla_i W_{ij}.$$

This is valid if the function $f(\mathbf{P})$ goes to zero at the boundaries of the domain. More complex procedures have been developed in case it is required to take into account the contribution of boundary conditions (Campbell, 1988).

For the density we have the simple expression that identically satisfy the continuity equation in the cylindrical form:

$$\varrho(\mathbf{R}_i) \simeq \sum_{j=1}^{N} \frac{m_j}{r_j} W_{ij}. \tag{10}$$

Rewriting the fundamental equations in the formulation in a manner more suitable for the SPH evaluation (Monaghan, 1985), and applying the previous criteria we have the following expression;
For the radial (r) momentum we obtain from (5)

$$\left(\frac{Dv_r}{Dt}\right)_i = -\frac{v_{\phi i}^2}{r_i} - \sum_{j=1}^{N} \frac{m_j}{r_j} \left(\frac{p_i}{\varrho_i^2} + \frac{p_j}{\varrho_j^2} + \Pi_{ij}\right) \frac{\partial W_{ij}}{\partial r_i} \quad. \tag{11}$$

The vertical (z) momentum satisfies:

$$\left(\frac{Dv_z}{Dt}\right)_i = -\sum_{j=1}^{N} \frac{m_j}{r_j} \left(\frac{p_i}{\varrho_i^2} + \frac{p_j}{\varrho_j^2} + \Pi_{ij}\right) \frac{\partial W_{ij}}{\partial z_i} \quad. \tag{12}$$

For the thermal energy per unit mass we obtain:

$$\left(\frac{D\epsilon}{Dt}\right)_i = -\frac{1}{2}\sum_{j=1}^{N} \frac{m_j}{r_j} \left(\frac{p_i}{\varrho_i^2} + \frac{p_j}{\varrho_j^2} + \Pi_{ij}\right) \frac{r_i \mathbf{v}_i - r_j \mathbf{v}_j}{r_i} \cdot \nabla_i W_{ij} + \frac{\Phi_i}{\varrho_i} \quad. \tag{13}$$

The dissipation function may be included in the simplest form:

$$\Phi_i = \mu_i \left(r\frac{\partial \Omega}{\partial r}\right)_i^2 + \mu_i \left(r\frac{\partial \Omega}{\partial z}\right)_i^2 \quad. \tag{14}$$

However this form doesn't exactly conserve the total energy, as can been shown analytically by summing the contribution of the thermal and momentum equations, and suffer large errors due to the bad approximation involved in squaring of the Ω derivative.

The discretization of the energy equation in its total energy form can be easily done exploiting the antisymmetric SPH version of any divergence of a vector field $\mathbf{S}$:

$$\frac{1}{\rho}\nabla \mathbf{S} = \sum_{j=1}^{N} m_j \left(\frac{1}{\rho_i^2} + \frac{1}{\rho_j^2}\right) \frac{r_i \mathbf{S}_i - r_j \mathbf{S}_j}{r_i r_j} \cdot \nabla_i W_{ij}.$$

We note that for gas simulations the contribution due to internal tension, the 'hoop' effect (Petschek and Libersky, 1993), is usually small. A first order approximation of this effect can however easily set up adding for each particle an acceleration term : $P_i/(2\pi \varrho_i)$.

Following the standard prescriptions (Monaghan, 1985), the artificial viscous pressure Π_{ij} is formulated as

$$\Pi_{ij} = \frac{\alpha \tilde{\mu}_{ij} \bar{c}_{ij} + \beta \tilde{\mu}_{ij}^2}{\bar{\rho}_{ij}} \quad , \tag{15}$$

with the averaged quantities

$$\bar{c}_{ij} = \frac{c_i + c_j}{2} \quad , \quad \bar{\rho}_{ij} = \frac{\rho_i + \rho_j}{2} \quad , \quad \tilde{\mu}_{ij} = \frac{r_i v_{ri} - r_j v_{rj}}{r_i(R_{ij}^2 + \eta^2)} + \frac{(v_{zi} - v_{zj})(z_i - z_j)}{(R_{ij}^2 + \eta^2)} \quad ,$$

$$R_{ij}^2 = (r_i - r_j)^2 + (z_i - z_j)^2 \quad , \quad \eta = 0.1h \quad ,$$

α and β are artificial viscosity coefficients used to damp out oscillations in shock transitions, c denotes the sound speed. We found, by numerical test experiments, that the quality of shock treatment is greatly improved if the geometric terms $r_i r_j$ terms are appropriately inserted.

Since our aim is to simulate accretion disks, it is essential make a correct treatment of the tangential velocity and its diffusion due to the viscosity. We integrate explicitly the viscous diffusion term; the cylindrical SPH-version of the diffusion term of equation (4) is given following the criteria by Brookshaw (1985):

$$\left(\frac{\partial \Omega}{\partial t}\right)_i = \sum_{j=1}^N \frac{m_j}{r_j} \left(\frac{\Omega_i - \Omega_j}{\varrho_i \varrho_j}\right) D_{ij} \frac{\mathbf{R}_{ij}}{R_{ij}^2} \cdot \nabla_i W_{ij} \quad , \tag{16}$$

where

$$D_{ij} = \frac{\mu_i r_i^3 + \mu_j r_j^3}{r_i^3} \quad , \quad \mathbf{R}_{ij} = (r_i - r_j, z_i - z_j). \tag{17}$$

In the solutions we present here, the cooling terms, such as bremsstrahlung, are just pure analytical expressions without any derivative and therefore they don't require evaluation of any special interpolation. They are computed for each particle straightaway by evaluating the values of temperature, density at each particle position. More complex physics can be incorporated into SPH formalism with difficulties not more than with any other discretization procedure. As an example we examine the case of a two component gas. We consider the possibility of different temperatures of ions and electrons in a plasma, due to different sources of heating and cooling for each species and slow relaxation between them, when only binary collisions are included.

We assume no mass or charge segregation, the total pressure will be given by the sum of the partial contributions and no relative drift speed between electrons and ions. So the mass conservation and momentum equations have the usual single species structure. The abundances of the species may change with time, but since they usually depend on local properties like the temperature or the density, the related equations can be easily integrated explicitly by a fractional step method. Therefore in this case all new physical properties are taken into account by the energy equations, typically they can be formulated as in Wolff et al. (1989):

$$\frac{D\epsilon_e}{Dt} = -\frac{P_e}{\rho}\nabla\mathbf{v} + \nabla\mathbf{q}_e - \Lambda_e + \Lambda_x - \Lambda_C,$$

where, $\mathbf{q}_e$ is the electron conductive flux given by:

$$\mathbf{q}_e = \frac{1.85 \times 10^{-5}}{\ln(\Lambda)} T_e^{\frac{5}{2}} \nabla T_e,$$

where, $\ln(\Lambda) = \frac{1.11 \times 10^{-5} T_e}{\rho^{\frac{1}{2}}}$ is the Coulomb logarithm, Λ_e is the energy loss rate due to bremsstrahlung given by:

$$\Lambda_e = 4.76 \times 10^{20} \rho^2 T^{\frac{1}{2}} \left(1 + 2.594 \cdot \Theta_e + 0.809 \cdot \Theta_e^2 + 0.175 \cdot \Theta_e^3 - 0.1892 \cdot \Theta_e^4\right)$$

where $\Theta_e = kT_e/m_e c^2$, Λ_x is the electron-ion energy exchange rate give by:

$$\Lambda_x = -2.04 \times 10^{29} \frac{\rho^2 \ln(\Lambda)}{T_e^{\frac{3}{2}}} (T_e - T_i)$$

and Λ_C is the Compton cooling loss rate given by:

$$\Lambda_C = 6.84 \rho T_e U_\nu \left(1 + 2.5 \cdot \Theta_e + 1.876 \cdot \Theta_e^2 + 1.891 \cdot \Theta_e^3 + 2.0048 \cdot \Theta_e^4\right)$$

All quantities are in c.g.s. units.
The total energy equation becomes

$$\frac{D}{Dt}\left(\epsilon_{Tot} + \frac{1}{2}\mathbf{v}^2\right) = -\frac{1}{\rho}\nabla\left(\overleftrightarrow{\sigma} : \mathbf{v}\right) + \nabla\mathbf{q}_e - \Lambda_e - \Lambda_C.$$

where, $\epsilon_{Tot} = \epsilon_e + \epsilon_i$, $E = \epsilon_{Tot} + \frac{1}{2}\mathbf{v}^2$, ϵ is the thermal energy per unit mass, $\overleftrightarrow{\sigma} = P_{Tot}\overleftrightarrow{\mathbf{I}} + \overleftrightarrow{\tau}$ is the full stress tensor, and $\overleftrightarrow{\tau}$ is the pure shear component.

As already said this is better suited for SPH, due to its conservation properties and, because shock treatment can be done with the usual artificial viscosity term, affecting only one energy equation. With a fractional step procedure we compute separately the contribution to the total energy $E = V^2/2 + \epsilon_{Tot}$ coming from the viscosity.

Since the electron-ion coupling and cooling terms depend on the thermal content of each species we have to recover their value from the total energy. We proceed in this way: to obtain the thermal energy per unit mass of the electrons we subtract from E^{n+1} the kinetic energy (coming from the momentum equation) and assume the reasonable fact that viscous heating rate of electron and ions is given by the ratio $\xi = Q_i/Q_e = \sqrt{(m_i/m_e)}$, giving $\Delta\epsilon_e = \Delta\epsilon_{Tot}/(1+\xi)$. Each separate energy species term is then modified by cooling, energy exchange, diffusion, integrating the appropriate equations in an independent fractional step.

For the Compton cooling we need to have the radiation field energy density U_ν. This term should be computed by the radiation energy transport equation, but it can be evaluated by some *ad hoc* procedure for specific physical context. If radiation comes from a geometrically well defined object, star surface or some disk surface, the problem is trivial. With U_ν the radiation energy density, for radiation emitted by a spherical star $U_\nu = \frac{L_\nu}{4\pi r^2 c}$.

If a turbulent equipartition B field is assumed to be present, then the total pressure has to contain also this term, and it affects both the momentum equation and the total energy equation. The total energy is now:

$$E = V^2/2 + \epsilon_{Tot} + \mathbf{B}_{turb}^2/(8\pi\rho)$$

The equipartition hypothesis fixes, at every time step, the **B** field density

$$\mathbf{B}_{turb}^2/(8\pi\rho) = \epsilon_{Tot}$$

so we may obtain the thermal energies in the same way as described above.. Furthermore, the energy balance requires magnetic field annihilation, but we may even disregard the specific physical process producing such an effect, provided that the magnetic field energy lost by a gas parcel in the time step is attributed to the particles energies. Following Bisnovatyi-Kogan and Lovelace (1997) we may choose to attribute it mainly to the electron energy by a constant fraction g and by a constant fraction $1-g$ to the ion energy. If the magnetic energy is increased somewhat we have to provide an *ad hoc* prescription (e.g., decrease in the kinetic energy associated with the shear motion) unless we want to integrate in a self consistent way the induction equation. This is a far more difficult problem (see, however, Monaghan, 1985).

As we observed here, the formulae for cylindrical geometry are similar to the Cartesian ones and the most relevant changes are:

(i) the mass of a particle which appears in the equation is divided by its distance from the z-axis
(ii) mutual velocity difference $v_j - v_i$ between two particles must be replaced by the more sophisticated term $(r_i \mathbf{v}_i - r_j \mathbf{v}_j)/r_i$.

The force $F_{r_{ji}}$ differs from $F_{r_{ij}}$ while $F_{z_{ij}} = F_{z_{ji}}$ for particles at the same radial coordinate. This difference in the force is due to the geometry. It has been noted (Erant and Benz, 1992) that in the cylindrical geometry, the collisions between particles don't conserve the radial momentum. However, we remind the readers that the linear momentum of a toroidal particle is zero. Thus even when the radial velocity is different from zero, the quantity $m_i v_{r_i}$ is not a conserved quantity. Instead, the angular momentum is exactly conserved in the nonviscous cases, and its total amount is also exactly conserved since the discretized version of the diffusion equation is antisymmetric with respect to the exchange of the particle index.

We note that the artificial viscosity acts only on the radial and vertical motion, so no numerical viscosity is operating on the angular speed, resulting in an exact conservation of angular momentum.

The quantity representing differential mass $dm_k = 2\pi \varrho_k r_k dr_k dz_k$ is needed for the derivation of the formulae irrespective of whether the particles in the simulations have the same mass or not. Obviously, the density is no longer directly proportional to the number of particles per unit volume as in Cartesian coordinate even when the particles have the same mass.

3. Physical Models

We numerically solve the fluid dynamic equations describing the time dependent rotational motion of gas falling into a Schwarzschild black hole. We assume that the initial angular momentum is sub-Keplerian at large distances. We take into account the general relativistic effects using the Paczyński and Wiita pseudo-Newtonian potential (1980).

$$\phi(r, \varphi) = -\frac{GM}{r - r_g},$$

This approach is frequently used when finer relativistic details are not required; since we are studying the general physical properties of the phenomenon, we consider this approximation to be sufficiently accurate. Exact analytical general relativistic treatment for similar, although more idealized configurations, have shown that the basic properties of the flow are essentially unaltered as far as the existence of steady shock are concerned (Chakrabarti, 1990; Nakayama, 1994; Lu and Yuan, 1997). In general we will use non-dimensional units, with the reference quantities : the light

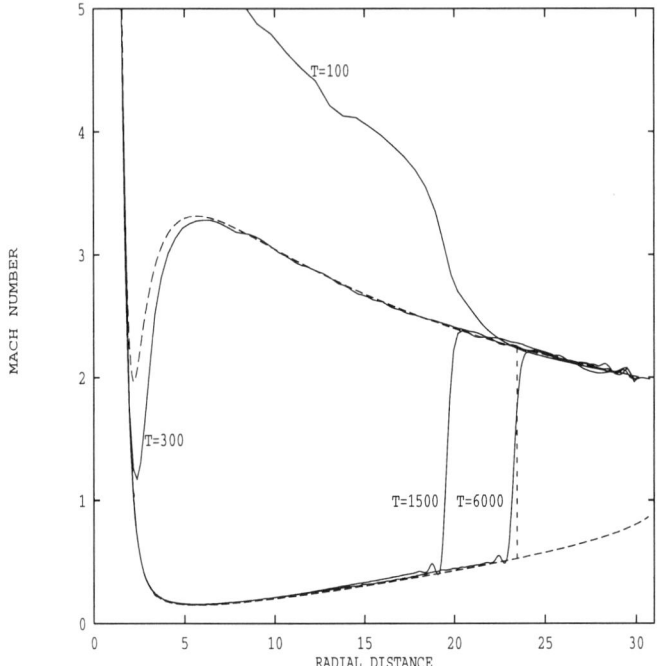

Fig. 1 : Comparison of analytical and numerical results in a one dimensional accretion flow. The dashed lines show the theoretical supersonic, subsonic and shock solutions while the solid curves are drawn at various times. The numerical (solid) shock location merges with theoretical location when steady state is reached.

speed c and the Schwarzschild radius r_g. So the time scale is in units of $r_g/ = 2GM/c^3$.

3.1. NONVISCOUS IDEAL GAS WITH $\gamma = 4/3$

We first study a gas with polytropic index $\gamma = C_p/C_v = 4/3$. In this case, the relation between pressure and thermal energy per unit mass is the same $P \propto \epsilon^4$ as the one for a gas with radiation in local thermal equilibrium (LTE) and radiation pressure is dominant. Therefore, we foresee that the solutions have the same general qualitative behaviour of the gas with radiation pressure dominant and in LTE.

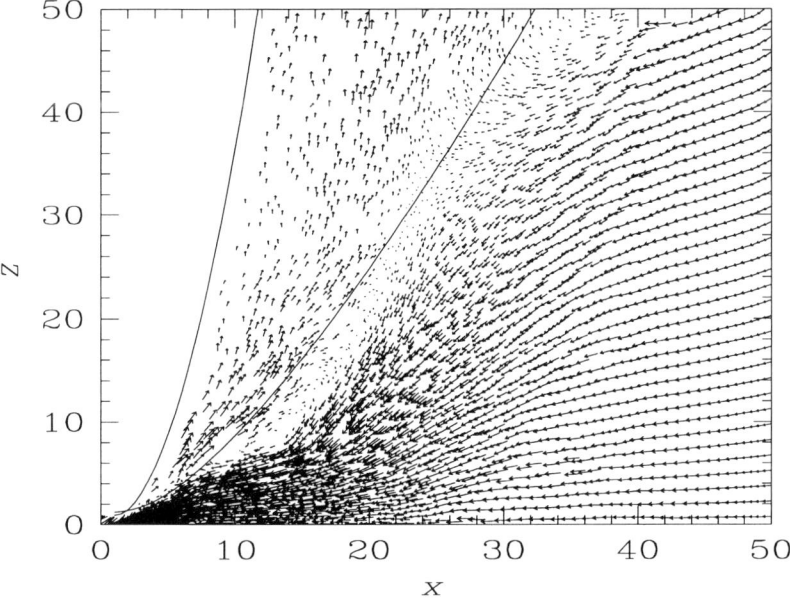

Fig. 2: 2D simulation result: Particles positions and their velocity field

3.1.1. *1D Simulations*

One dimensional simulations are important to establish the accuracy of the code. In this case, exact analytical solutions can be derived. The shock position is therefore determined exactly and the agreement between the analytical results and the numerical code can be verified. Since the solution is very sensitive to the angular momentum value, it is important that the code conserves angular momentum very accurately. We have verified that even few percent of non conservation may change drastically the solutions. In this respect the SPH in cylindrical coordinates is very powerful. As already shown in previous Section, it is indeed possible to integrate very accurately the tangential momentum equation. Fig.1 shows the variation of the Mach number of the radial motion at different times, as shock is forming (Chakrabarti and Molteni, 1993). Asymptotically the solution matches with the theoretical curve showing that the SPH codes are sufficiently accurate.

3.1.2. *2D Simulations*

Two dimensional simulations have been performed for different ranges of the energy and angular parameters.

Many interesting new features appeared in the 2D simulations. In order to obtain analytical solutions the vertical equilibrium assumption was made. But by simulations it appears that it is not quite satisfied. This is obvious since close to the black hole, the z component of the gravitational forces increases greatly. One can compare these simulation results with those obtained from the assumption of vertical equilibrium (Chakrabarti 1990) as they are guides to find the appropriate range of (E,λ) to have steady shocks.

In general we observe that:

1) the *actual* (E,λ) range that produces steady shock solutions in a two dimensional flow is much wider than the analytical range obtained using vertical equilibrium model (Molteni, Lanzafame and Chakrabarti, 1994).

2) The shock position is usually at a larger radial distance than the analytically predicted location, since turbulence is also present.

3) The shock structure may be very complex, exhibiting bifurcations, oblique fronts in regions away from the equatorial plane.

4) The shocked gas may be heated so much that it is ejected at high speed, forming a supersonic wind expanding in an hollow cone.

5) The post-shock region is usually turbulent, with matter forming vortices in between the shock and the centrifugal barrier.

Fig.2 shows a typical distribution of the particles in the r-z plane.

We also find that when the energy parameter E and the angular momentum parameter λ are such that a steady shock cannot form (despite having two saddle type sonic points) then oscillations of the shock position are set up. The basic ideas underlying this phenomenon have been discussed in Ryu et al. (1996) and Ryu and Chakrabarti (1998)

All these simulations are rather challenging for conventional codes since high density regions, close to the BH, have to be treated together with very low density ones that occur in the funnel near the $r = 0$ axis. SPH proves to be very useful since, as is usual for its approach, particles don't go in empty regions and therefore no stability problem arises at the matter boundary.

3.2. NONVISCOUS FLOW WITH BREMSSTRAHLUNG LIKE COOLING

We now present results of accretion flows with cooling so that more realistic accretion disks may be tackled. We use adiabatic index $\gamma = 5/3$ to mimic the accreting gas with low radiation pressure and include a power-law cooling process $\Lambda \approx \rho^2 T^\beta$. When $\beta = 0.5$, the bremsstrahlung cooling is obtained. The main idea here is to check if the centrifugal barrier could behave similar to a rigid surface of a star. If so, then the gas cooled by bremsstrahlung could show oscillations like the ones obtained in the accretion column in white dwarfs (Langer, Chanmugam and Shaviv, 1981). The

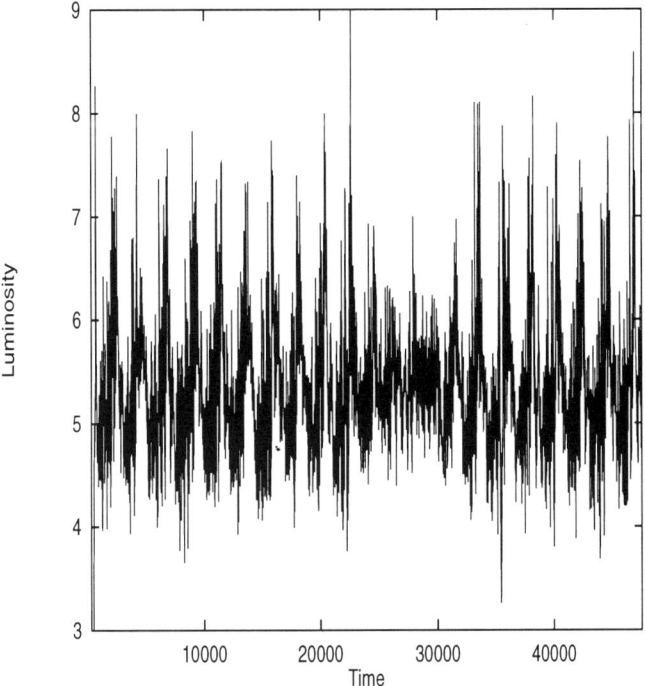

Fig. 3 : Luminosity versus time in non-dimensional units

results of our simulations show that this indeed occurs. We have shown that the shock wave undergoes a steady quasi-periodic oscillation in the radial direction with an amplitude roughly $\sim 5\% - 10\%$ of the steady state shock distance and a time period close to the cooling time. We find that the oscillation is actually of "resonance" type, namely it occurs when the cooling timescale roughly matches with the infall timescale.

Both 1D and 2D simulations show periodicities of the order of 1500 non-dimensional time units for shocks around $30r_g$.

Fig.3, Fig.4a and Fig.4b show the luminosity versus time and a sequence of the isodensity contours of the disk at different phases of the oscillations. The post-shock region forms a hot corona which could be the site of hard X-rays and γ-ray production in galactic and extra-galactic black hole candidates. A detailed discussion is given in Molteni, Sponholz, Chakrabarti (1996).

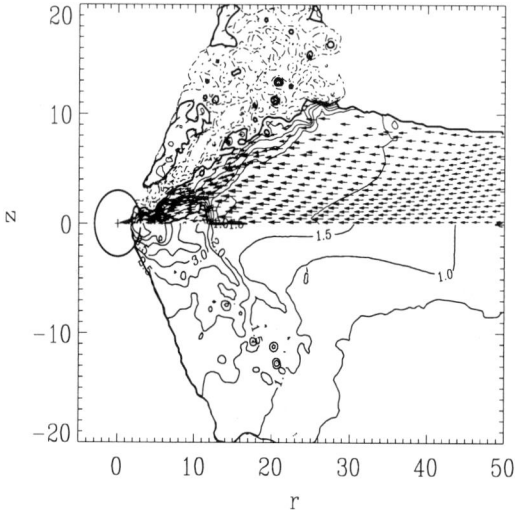

Fig. 4a: Contours of constant density (lower half) and constant Mach number (lower half) of the flow particles along with the particle velocity vectors at the maximum of the oscillations.

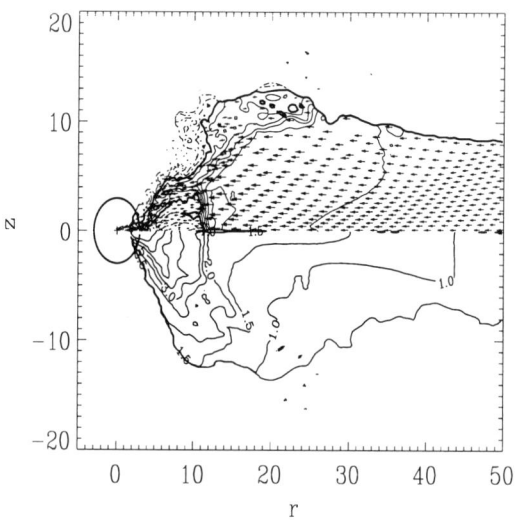

Fig. 4b: Same as Fig. 4a at the minimum of the oscillations.

3.3. VISCOUS AND ISOTHERMAL

A comparison with analytical results can still be done if viscosity is included when the flow is assumed to be one dimensional and isothermal (Chakrabarti, 1990; Chakrabarti & Molteni, 1995). Here also, the analytical and numerical results agree very well.

In the numerical simulations we also studied how different viscosity prescriptions affect the flow structure (Chakrabarti & Molteni, 1995). We tested Shakura-Sunyaev viscosity (α_s), mass flux balanced viscosity (α_m), pressure balanced viscosity (α_p). In general we see that the standard Shakura-Sunyaev viscosity prescription brings to a discontinuity in the angular momentum therefore leading to instabilities due to the sharp decrease in the angular momentum with increasing r. This is simply because the thermal pressure that is used in standard Shakura-Sunyaev viscosity is not continuous across the discontinuities. We thus recommend that the net pressure (thermal plus ram) be used instead (Chakrabarti & Molteni, 1995). Fig. 5 shows the comparison of the angular momentum distribution of the disk (y axis) as a function of the radial distance (x axis). Solid ragged curve is obtained for pressure balance prescription, with $\alpha_p = 0.1$ everywhere in the flow. Long dashed curve is obtained with Shakura-Sunyaev prescription ($\alpha_s = 0.01$) and the dotted curve is drawn for mass flux prescription ($\alpha_m = 0.01$), The angular momentum distribution is non-monotonic but is smooth and continuous apart from some numerical noise. The jump in angular momentum is smaller for pressure and mass flux limited prescription. Smoothed solid curve marked 'Keplerian' represents the Keplerian distribution for reference purpose.

A salient point that was observed is that when viscosity is very low, a stable and an unstable shock may, in principle, form with the stable shock gradually becoming weaker as viscosity is increased. When viscosity crosses a critical value, the stable shock disappears altogether (Chakrabarti, 1990).

3.4. VISCOUS AND ADIABATIC FLOW WITH $\gamma = 4/3$

We examined also the behavior of the more general polytropic thick, viscous, accretion disks.

In Fig. 6(a-d), we show the effect of the introduction of viscosity in the flow (Lanzafame, Molteni & Chakrabarti, 1998). Higher viscosity causes higher differential angular momentum transport between the pre- and the post-shock solutions and as a result the shock is drifted away in the radial direction until the momentum balance is reached. The viscosity parameter α is 5×10^{-4} in (b), 10^{-3} in (c) and 1.5×10^{-3} in (d) respectively. In the increasing order of viscosity, the shock locations are $X_s = 6.0$, 7.5, and 10.5 respectively. Another point of interest: as viscosity is raised, the amount

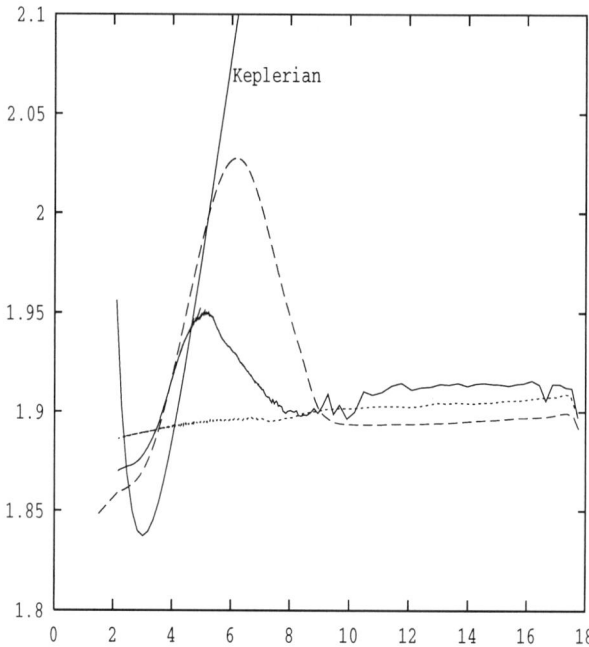

Fig. 5: Comparison of angular momentum distributions (y axis) as a function of the radial distance (x axis) in disks with various viscosity prescriptions: with $\alpha_s = 0.01$ (long dashed curve), with $\alpha_m = 0.01$ (dotted curve) and with $\alpha_p = 0.1$ (solid, ragged). Smooth solid curve is the Keplerian distribution for comparison.

of outflowing matter in the wind is decreased (to less than a percentage of inflow matter, e.g, Eggum, Coroniti & Katz, 1985). This is because of weakening of the centrifugal barrier of the in-going matter. Lower viscosity causes matter to bounce from the barrier and escape as winds. In higher viscosity, higher turbulence are also seen to form in the post-shock flow. This is because more matter from higher elevation falls on the equatorial plane and convert their potential energy to turbulent energy.

Occasionally, observed radiations show quasi-periodic oscillations. Molteni, Sponholz & Chakrabarti (1996), and Ryu, Chakrabarti & Molteni (1997) suggested that the so-called QPOs observed in the compact objects could be due to the oscillations of the 'boundary layers', i.e., the centrifugally supported denser region which may or may not include shocks. The oscillations seen in Molteni et al (1996) occur even when steady shocks are expected analytically. The oscillations seen in Ryu et al (1997) occur for those flow parameters where the theoretical analysis does not predict the formation of

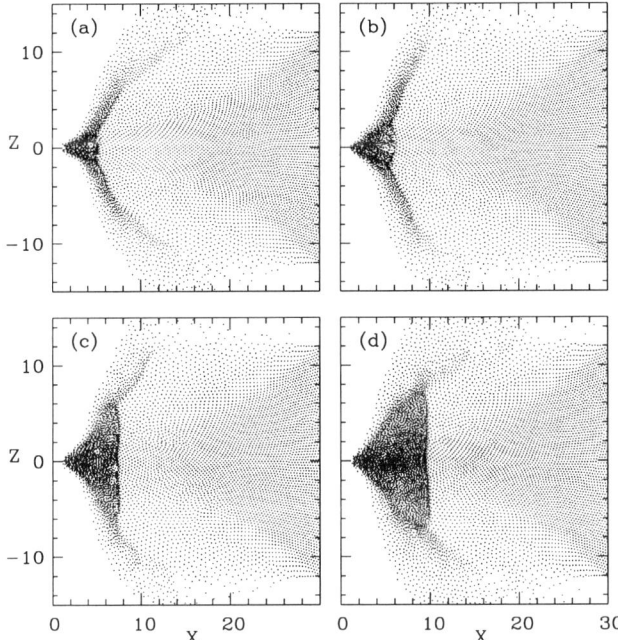

Fig. 6 (a-d): Outward drift of the shock location with viscosity. See text for details.

any steady shocks, even though two sonic points exist. Lanzafame, Molteni & Chakrabarti, 1998 found the oscillations for viscous flows as well. In 7(a-c), roughly half of the cycle is shown where the shock location decreases monotonically. Total number of particles in this simulation is on an average around 10,000. In Fig. 7d, the shock drifted again outward. Note that apart from the axisymmetric shock oscillation, a new, corrugated instability, is also apparent in Fig. 6(d). The time (in units of $2GM/c^3$) at which the simulations are shown are $t = 4280$, $t = 5580$, $t = 6210$ and $t = 6790$ respectively. Thus, in a stellar black hole of mass $10 M_\odot$, the period would be around 0.4s, whereas in a supermassive black hole of mass $10^7 M_\odot$, the period would be 10^6s or a few days. These time scales are comparable with the time scale of observed QPOs in black hole candidates. With a different disk input parameters (e.g., for different accretion rate or viscosity) the periods will vary, but order of magnitude does not change.

Fig. 8 shows the variation of the number of simulation particles as a function of time for the case shown in Fig. 7(a-d). Here the total number N and the number of sub-sonic particles N_{sub} (presumably, participating in the Compton reprocessing of the soft-photons from the Keplerian disk; see, Chakrabarti & Titarchuk, 1995) are shown. The amplitude modulation

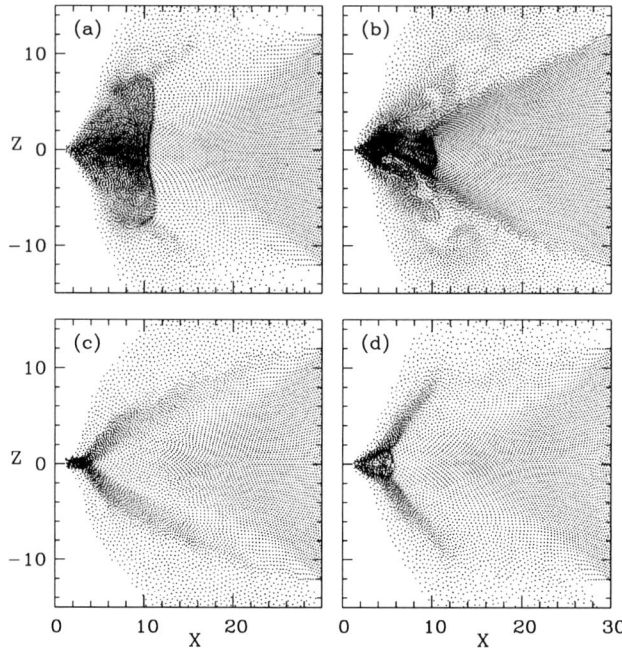

Fig.7 (a-d): Oscillation of the shocks with time when a steady shock is not predicted, yet two sonic points exist in a viscous transonic flow. These solutions provide reasonable explanation to the quasi-periodic modulation of X-rays coming from the black hole candidates.

is significant: 50 percent variation in total particle number, and more than 700 percent variation in the sub-sonic particle number. Note in Fig. 7(a-d) that the vertical height of the shock increases proportionately with the shock location. Thus it is capable of intercepting more soft photons from the Keplerian disk. We believe that this is important: the observed significant (10 − 100 percent) variations in QPO cannot be explained away by any means other than such a dynamical variation of the X-ray emitting region.

So we may say that above a critical viscosity, when the steady shock is not expected, the flow could still form an unsteady shock which periodically evacuates the disk. Beyond another critical viscosity (keeping other parameters unchanged), where the flow passes only through the inner sonic point, the shock disappears and only the smooth sub-Keplerian disk (originating from a Keplerian disk) remains. Oscillation in non-viscous two-dimensional flow is discussed in Ryu (this volume).

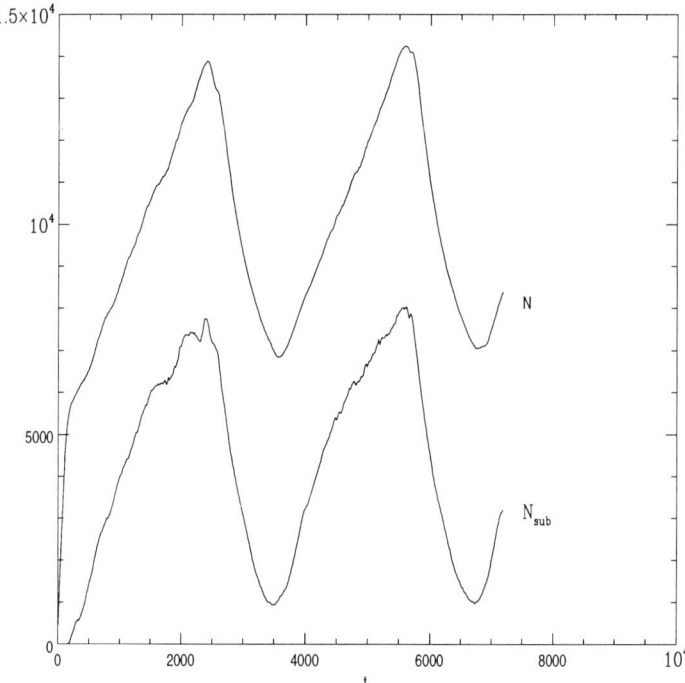

Fig. 8: Oscillation of the total amount of matter in the accretion disk with a shock wave. Total particle number N and subsonic particle number N_{sub} are plotted as functions of time

4. Non-axisymmetric Perturbations and Non-axisymmetric Shocks by SPH

Recently, it has been shown that the axisymmteric shocks described so far are stable for non-axisymmtric perturbations as well. Fig. 9 shows the isodensity contour lines for the simulation of a circular shock initially perturbed by a density blob at four different times t. This result was obtained using a Total Variation Diminishing code. For details see Molteni, Tóth & Kuznetsov (1998).

Earlier, Molteni, Gerardi and Chakrabarti (1994) in the context of studying star-disk interaction found, for the first time, the formation of non-axisymetric spiral shocks in a black hole accretion disk (in 2D) using SPH simulations. Such results have been found since then, using 3D SPH simulations (Yukawa, Boffin & Matsuda, 1997).

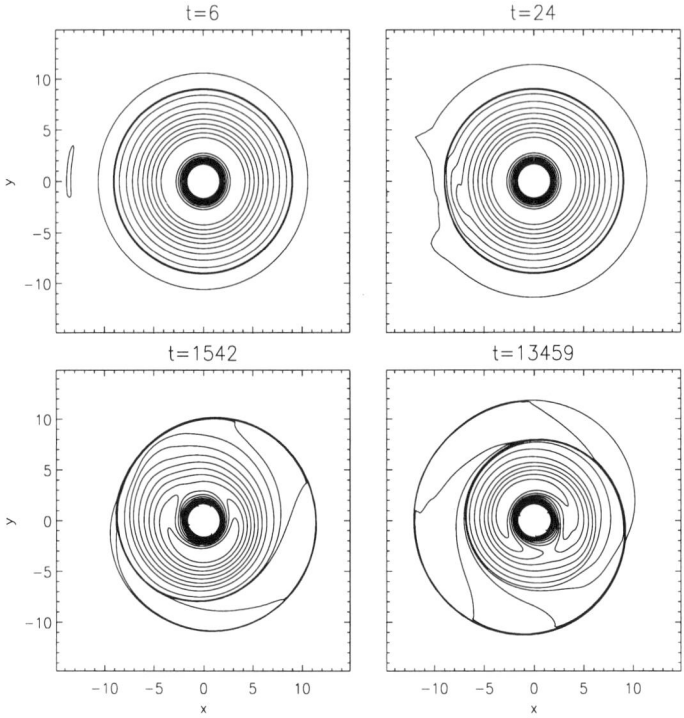

Fig. 9: Contours of density are shown at four different times t. The simulation is done in r, φ coordinates, but the picture is shown in the x, y plane for easier interpretation. The initial perturbation has not reached the axisymmetric shock in the upper left figure, and it just crosses the shock front at $t = 24$. A spiral spiral shock appears a long time ($t \approx 1000$) after this blob crosses the axisymmetric shock and it remains stable indefinitely thereafter.

5. Conclusions

From the *numerical* point of view we may say that the Smoothed Particle Hydrodynamics method is a reliable one. It has to be used with adequate care, in particular the number of overlapping particles has to be large enough to ensure that the physical variables and their derivatives are interpolated correctly. Typically a minimum number of nine neighbours per particle is required. It is therefore computationally expensive if the integration domain has no large empty regions. Fortunately, in our shock simulations the hollow funnel region along the z axis is totally empty! Despite its moderate shock resolution the simulation is able to capture the good qualities, at least in its cylindrical coordinates version: the numerical shear viscosity in the ϕ direction is practically zero, that is important to capture steady shock solutions.

From the point of view of the *black hole physics* we may say that numerical simulations prove that:

- black holes also have centrifugally driven boundary layer. It is largely proved that steady shock configurations exist for a large range of initial temperature and for initially sub-Keplerian angular momentum. The shock positions range from hundreds down to a few Schwarzschild radii, implying that extremely high energy emission is possible, from this region, especially if the accreting gas has moderate optical thickness.

- the centrifugal layer is, in general, stable under axisymmetric and non-axisymmetric perturbations.

- the role of the physical viscosity is debated, but some points are clear: i) - for moderate viscosities steady shock configurations still exist; ii) - a sub-Keplerian accreting flow may form inside the nearly Keplerian disk and a Keplerian disk may form inside a sub-Keplerian disk (this may be important, since even in binary systems when accretion occurs by wind capture, carrying small angular momentum, a Keplerian disk close to the compact object can form). iii) - the centrifugal barrier goes away when the viscosity is sufficiently large.

- this boundary layer can emit steady and non-steady X-rays and cause QPOs. Simulations prove that there is a great variety of mechanisms (both for viscous and nonviscous cases) that trigger oscillations of the shock around its equilibrium position: resonance between cooling time and in-falling time, pure gas dynamical instabilities, viscous accumulation and depletion of inner Keplerian zone are some of the processes.

We may conclude that the exploration of the physics of black hole accretion with numerical methods revealed very fruitful, rich and extremely relevant physical effects apart from the intrinsic value as a fluid dynamical problem. The phenomena is clearly still to be fully understood to explain the observational results.

References

Balsara D. S. (1995), *J. Comput. Phys.* **121**, 357,
Batchelor C.K. (1981), *An introduction to fluid dynamics*, Cambridge University Press
Benz W. (1990) in *Numerical Modeling of Nonlinear Stellar Pulsation: Problems and Prospects*, ed. J.R. Buchler (Dordrecht: Kluwer) 269
Bisnovatyi-Kogan, G.S. and Lovelace, R.V.L. (1997) *Astrophys. J.* **486**, L43
Brookshaw L. (1985) *Proc. Astron. Soc. Aust.* **6** 207
Campbell P. M. (1988) *Technical report, Defense Nuclear Agency*, Washington
Chakrabarti, S. K. (1989), *Astrophys. J.* **347**, 365
Chakrabarti, S. K. (1990) *Theory of Transonic Astrophysical Flows* , (Singapore: World Scientific)
Chakrabarti, S. K., & Molteni, D. (1993) *Astrophys. J.* **417**, 671
Chakrabarti, S. K., & Molteni, D. (1995) *M.N.R.A.S* **272**, 80

Chakrabarti, S. K., & Titarchuk, L. G. (1995) *Astrophys. J.*, **455**, 623
Dilts G. A., Los Alamos N.R.L., LA-UR 96-134, (1996) *Equivalence of the SPH Method and a Space-Time Galerkin Moving Particle Method*
Erant M. and Benz W. (1992), *Astrophys. J.* **387**, 294
Harten, A. (1983), *J. Comp. Phys.* **49**, 357
Hawley, J. F., Smarr, L. L., & Wilson, J. R. (1984a), *Astrophys. J.* **277**, 296.
Hawley, J. F., Smarr, L. L., & Wilson, J. R. (1984b), *Astrophys. J. Suppl.*, **55**, 211
Langer S. H., Chanmugam G., & Shaviv G. (1981) *Astrophys. J.* **245**, L23
Lanzafame, G., Molteni, D. & Chakrabarti, S.K. (1998) *M.N.R.A.S.* (in press)
Lu J. and Yuan F. (1997), *PASJ* **49**, 525
Makishima, K., in Physics of Neutron Stars and Black Holes (1988) ed. Y. Tamaka, (Tokyo: Universal Academic Press), 177
McHardy, I. M., & Pounds, K. A., (1988) in *Physics of Neutron Stars and Black Holes* ed. Y. Tamaka, (Tokyo: Universal Academic Press), 288
Molteni, D., Gerardi, G., & Chakrabarti, S. K. (1994), *Astrophys. J.* **436**, 249
Molteni, D., Lanzafame, G., & Chakrabarti, S. K. (1994) *Astrophys. J.* **425**, 161
Molteni, D., Sponholz, H., & Chakrabarti, S. K. (1996) *Astrophys. J.* **457**, 805
Molteni, D., Ryu, D., & Chakrabarti, S. K. (1996), *Astrophys. J.* **470**, 460
Molteni, D., Tóth G., & Kuznetsov O. (1998) *Submitted*
J.J. Monaghan and J.C. Lattanzio. (1985), *A & A* **149**, 135
J.J. Monaghan (1985), *Computer Physics Reports* **3**, 71
Nakayama, K. (1994), *MNRAS* **270**, 871
Paczyński, B., & Wiita, P. J. (1980), *A & A* **88**, 23
J.C.B. Papaloizou and J.E. Pringle (1985), *MNRAS* **213**, 799
Petschek A. G., Libersky L.D. (1993), *J. Comput. Phys.* **109**, 76
J.E. Pringle (1981), *Ann. Rev. Astron. Astrophys.* **19**, 137
Ryu, D., Brown, G. L., Ostriker, J. P., & Loeb, A. (1995), *Astrophys. J.* **452**, 364.
Ryu, D., Chakrabarti S., Molteni D. (1997), *Astrophys. J.* **474**, 378
Ryu, D. (1998) (this volume)
Schüssler M. and Schmitt D. (1981), *A & A* **97**, 373
Shoenberg I. J. (1973) *Cardinal Spline Interpolation, SIAM*, Philadelphia
Wolff M.T.,Gardner J.H. and Wood K.S. (1989), *Astrophys. J.* **346**, 833
Yukawa, H., Boffin, H.M.J. & Matsuda, T. (1997), *MNRAS* **292**, 321

NUCLEOSYNTHESIS IN ADVECTIVE ACCRETION DISKS AROUND GALACTIC AND EXTRA-GALACTIC BLACK HOLES

B. MUKHOPADHYAY
S. N. Bose National Centre For Basic Sciences JD Block, Salt Lake, Sector-III, Calcutta-700091, India

1. Introduction

Many of the observational evidences for black hole rely on the fact that the incoming gas has the potential to become as hot as its virial temperature $T_{virial} \sim 10^{13}\ ^oK$ (Rees, 1984). This flow is usually cooled down through bremsstrahlung and Comptonization effects and hard and soft states are produced depending on the degree by which this cooling takes place (Chakrabarti & Titarchuk, 1995). The generally sub-Keplerian, advective flow after deviating from a Keplerian disk, especially in the hard states, remains sufficiently hot to cause a significant amount of nuclear reactions around a black hole before plunging in it. The energy generated could be high enough to destabilize the flow and the modified composition may be dispersed through winds to change the metalicity of the galaxy (Chakrabarti, Jin & Arnett, 1987 [CJA]; Jin, Arnett & Chakrabarti, 1988; Chakrabarti, 1988; Mukhopadhyay & Chakrabarti, 1998). Earlier works have been done in cooler thick accretion disks only. Below, we present a few examples of nuclear reactions in advective flows and discuss the implications. Results of more detailed study could be seen in Mukhopadhyay & Chakrabarti (1998) [MC98].

2. Physical Systems Under Considerations

Black hole accretion is by definition advective, i.e., matter must have *radial* motion, and transonic, i.e., matter must be supersonic (Chakrabarti 1996 [C96] and references therein). The supersonic flow must be sub-Keplerian and therefore deviate from the Keplerian disk away from the black hole. The study of viscous, transonic flows was initiated by Paczyński & Bisnovatyi-Kogan (1981).

By and large, we follow C96 for thermodynamical parameters along a flow and Chakrabarti & Titarchuk (1995) [CT95] and Chakrabarti (1997a) [C97a] to compute the temperature of the Comptonized flow in the advective region which may or may not have shocks. According to these solutions, a black hole accretion may be thought to be similar to a sandwich whose sub-Keplerian flow rate ($\dot{m}_h$) in the 'bread' part progressively increases and that ($\dot{m}_d$) in the 'meat' part progressively decreases as flow moves in towards the black hole. Finally at $x = x_K$, the equatorial flow also deviates from a Keplerian disk and for $x < x_K$ the entire flow is sub-Keplerian. Among the major reactions which are taking place inside the disk, we note that, due to hotter nature of the advective disks, especially when the accretion rate is low and Compton cooling is negligible, the major process of hydrogen burning is the rapid proton capture process (which operates at $T \gtrsim 5 \times 10^8 K$) as opposed to the PP chain (which operates at much lower temperature $T \sim 0.01 - 0.2 \times 10^9 K$) and CNO (which operates at $T \sim 0.02 - 0.5 \times 10^9 K$). The present paper being exploratory in nature, we do not include nuclear heating and cooling in determining the structure and stability of the accretion flow. We do not assume here heating due to magnetic dissipation (see, Shapiro, 1973 and Bisnovatyi-Kogan, 1998).

For simplicity, we take the solar abundance as the abundance of the Keplerian disk. Furthermore, Keplerian disk being cooler, no composition change is assumed inside it. In other words, our computation starts only from the time when matter is launched from the Keplerian disk ($x = x_K$). Most of the cases were repeated with initial abundance same as the output of big-bang nucleosynthesis (hereafter referred to as 'big-bang abundance').

According to CT95, and C97a, for two component accretion flows, for $\dot{m}_d \lesssim 0.1$ and $\dot{m}_h \lesssim 1$ the black hole remains in hard states. Lower rate in Keplerian disks *generally* implies a lower viscosity and a larger x_K ($x_K \sim 30 - 1000$; see, C96 and C97a). In this parameter range the protons remain hot, typically, $T_p \sim 1 - 10 \times 10^9$ degrees or so. This is because the efficiency of emission is lower ($f = 1 - Q^-/Q^+ \sim 0.1$, where, Q^+ and Q^- are the heat generation [due to viscous processes] and heat loss rates respectively. Also see, Rees [1984], where it is argued that $\dot{m}/\alpha^2$ is a good indication of the cooling efficiency of the hot flow.). We studied a large region of parameter space in details where $0.0001 \lesssim \alpha \lesssim 1$, $0.001 \lesssim \dot{m} \lesssim 100$, $0.01 \lesssim F_{Compt} \lesssim 0.95$, $4/3 \lesssim \gamma \lesssim 5/3$ are chosen. Here, F_{Compt} is the factor by which the proton temperature is reduced due to bremsstrahlung and Comptonization effects. Results with several sets of initial conditions are in MC98. Since shocks can form in advective disks for a large region of parameter space (C96 and references therein) we use a case with a standing shock in this paper.

In selecting the reaction network we kept in mind the fact that hotter flows may produce heavier elements through triple-α and rapid proton and α capture processes. Furthermore due to photo-dissociation significant neutrons may be produced and there is a possibility of production of neutron rich isotopes. Thus, we consider sufficient number of isotopes on either side of the stability line. The network thus contains protons, neutrons, till ^{72}Ge – altogether 255 nuclear species. The standard reaction rates were taken [MC98].

3. Results

We present now with a typical case which contained a shock wave in the advective region. We use the mass of the black hole $M/M_\odot = 10$, Π-stress viscosity parameter $\alpha_\Pi = 0.07$, the location of the inner sonic point $x_{in} = 2.9115$ and the value of the specific angular momentum at that point $\lambda_{in} = 1.6$, the polytropic index $\gamma = 4/3$ as free parameters. The net accretion rate $\dot{m} = 1$, which is the sum of (very low) Keplerian component and the sub-Keplerian component. Results of CT95 and C97a for $\dot{m}_d \sim 0.1$ and $\dot{m}_h \sim 0.9$, fix $F_{Compt} = 0.03$, $x_K = 401$. This factor is used to convert the temperature distribution of solutions of C96 (which does not explicitly uses Comptonization) to temperature distribution *with* Comptonization. The proton temperature and velocity distribution computed in this manner are shown in Figs. 1(a-b). (velocity is measured in units of 10^{10} cm sec^{-1}).

In Fig. 1c, we show the composition change close to the black hole both for the shock-free branch (dotted curves) and the shocked branch of the solution (solid curves). Only prominent elements are plotted. The difference between the shocked and the shock-free cases is that in the shock case the similar burning takes place farther away from the black hole because of much higher temperature in the post-shock region. A significant amount of the neutron (with a final abundance of $Y_n \sim 10^{-3}$) is produced due to photo-dissociation process. Note that closer to the black hole, ^{12}C, ^{16}O, ^{24}Mg and ^{28}Si are all destroyed completely, even though at around $r = 3$ or so, the abundance of some of them went up first before going down. Among the new species which are formed closer to the black hole are ^{30}Si, ^{46}Ti, ^{50}Cr. Note that the final abundance of ^{20}Ne is significantly higher than the initial value. Thus a significant metallicity could be supplied by winds from the centrifugal barrier. In Fig. 1d, we show all the energy release/absorption components for the shocked flow. The viscous energy generation (Q^+) and the loss of energy (Q_-) from the disk (short dashed) are shown. These quantities, had the advective regime had Keplerian distribution, are also plotted (dotted). Solid curve represents the nuclear energy release/absorption for the shocked flow and the long dashed curve is that for the shock-free flow. Dot-dashed curve represents the nuclear energy release/absorption for big-

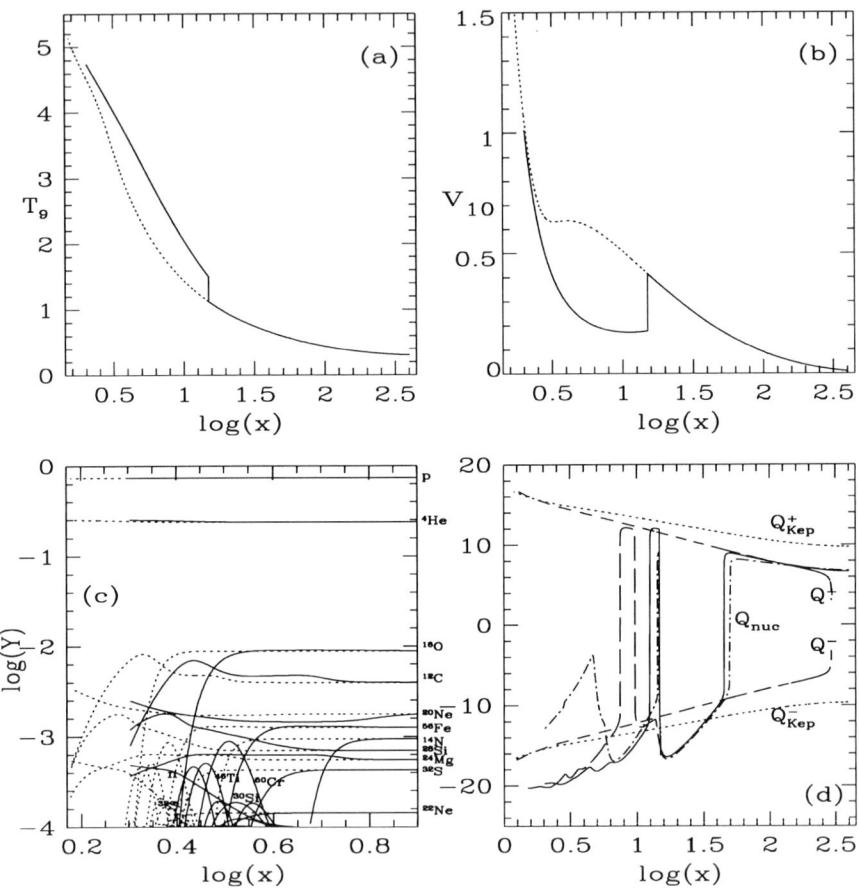

Fig. 1 : Variation of (a) proton temperature (T_9), (b) radial velocity v_{10}, (c) matter abundance Y_i in logarithmic scale and (d) various forms of specific energy release and absorption rates as functions of logarithmic radial distance (x in units of Schwarzschild radius). See text for parameters. Solutions in the stable branch with shocks are solid curves and those without the shock are dotted in (a-c). Curves in (d) are described in the text. At the shock temperature and density rise significantly and cause a significant change in abundance even farther out. Shock induced winds may cause substantial contamination of the galactic composition when parameters are chosen from these regions.

bang abundance. As matter leaves the Keplerian flow, the rapid proton capture (rp-) processes (such as, $p+^{18}O \rightarrow ^{15}N+^{4}He$ etc.) burn hydrogen and releases energy to the disk. At around $x = 45$, $D \rightarrow n + p$ dissociates D and the endothermic reaction causes the nuclear energy release to become 'negative', i.e., a huge amount of energy is absorbed from the disk. At around $x = 14$ the energy release is again dominated by the original rp-processes. Excessive temperature at around $x = 12.6$ breaks ^3He down into deuterium This type of reactions absorb a significant amount of energy from the flow. When big-bang abundance is chosen to be the initial abundance, the net composition does not change very much, but the dominating reactions themselves are somewhat different because the initial compositions are different. For instance, in place of rapid proton capture reactions as above, the fusion of deuterium into ^4He plays dominant role via $D + D \to\, ^3He + n$, $D + p \to\, ^3He$, $D + D \to p + T$, $^3He + D \to p +\, ^4He$. This is because no heavy elements were present to begin with. Endothermic reactions at around $x = 20-40$ are dominated by deuterium dissociation as before. However, after the complete destruction of deuterium, the exothermic reaction is momentarily dominated by neutron capture processes (due to the same neutrons which are produced earlier via $D \to n + p$) such as $n +\, ^3He \to p + T$ which produces the spike at around $x = 14.5$. Following this, ^3He and T are destroyed as in solar abundance case and reaches the minimum in the energy release curve at around $x = 6$. The tendency of going back to the exothermic region is stopped due to the photo-dissociation of ^4He via $^4He \to p + T$ and $^4He \to n +\, ^3He$. At the end of the big-bang abundance calculation, a significant amount of neutrons are produced. It is interesting to note that the radial dependence as well as the magnitude of the energy release due to rp-process and that due to viscous dissipation (Q^+) are *very* similar (save the region where endothermic reactions dominate). This suggests that even with nuclear reactions, at least some part of the advective disk may be perfectly stable.

We now present another interesting case where lower accretion rate ($\dot{m} = 0.01$) but higher viscosity (0.2) were used and the efficiency of emission is intermediate ($f = 0.2$). That means that the temperature of the flow is high ($F_{Compt} = 0.1$, maximum temperature $T_9^{max} = 13$). $x_K = 8.4$ in this case, if the high viscosity is due to stochastic magnetic field, protons would be drifted towards the black hole due to magnetic viscosity, but the neutrons will not be drifted (Rees et al., 1982) till they decay. This principle has been used to do the simulation in this case. The modified composition in one sweep is allowed to interact with freshly accreting matter with the understanding that the accumulated neutrons do not drift radially. After few iterations or sweeps the steady distribution of the composition may be achieved. Figure 2 shows the neutron distributions in iteration numbers 1, 7, 14 & 21 respectively (from bottom to top curves) in the advective region. The formation of a 'neutron torus' (Hogan & Applegate, 1987) is very apparent in this result and generally in all the hot advective flows. Details

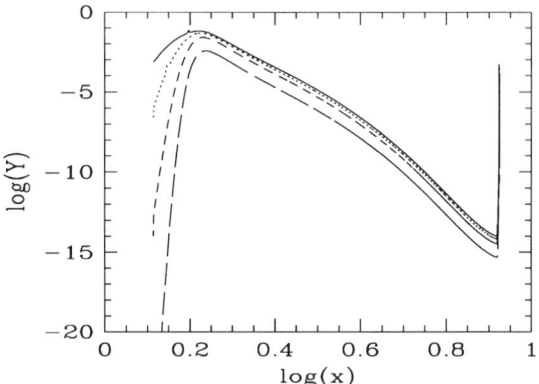

Fig. 2 : The convergence of the neutron abundance through successive iterations in a very hot advective disk. From bottom to top curves 1, 7, 14 and 21 iteration results are shown. A neutron torus with a significant abundance is formed in this case.

are in Chakrabarti & Mokhopadhyay (1998).

4. Discussions and Conclusions

In this paper, we have explored the possibility of nuclear reactions in advective accretion flows around black holes. Although this region is not fully self-consistently computed yet, particularly near the region where the advective disk joins with a standard Keplerian disk, we have used the best model that is available in the literature so far (C96). Temperature in this region is controlled by the efficiencies of bremsstrahlung and Comptonization processes (CT96, C97a) and possible heating and cooling due to magnetic fields (Shapiro, 1973; Bisnovatyi-Kogan, 1998). For a higher Keplerian rate and higher viscosity, the inner edge of the Keplerian component comes closer to the black hole and the advective region becomes cooler (CT95). However, as the viscosity is decreased, the inner edge of the Keplerian component moves away and the Compton cooling becomes less efficient.

The composition changes especially in the centrifugal pressure supported denser region, where matter is hotter and slowly moving. Since centrifugal pressure supported region can be treated as an effective surface of the black hole which may generate winds and outflows in the same way as the stellar surface, one could envisage that the winds produced in this region would carry away modified composition (Chakrabarti, 1997b; Das & Chakrabarti 1998; Das, 1998). In very hot disks, a significant amount of free neutrons are produced which, while coming out through winds may recombine with outflowing protons at a cooler environment to possibly form deuteriums a process originally suggested by Ramadurai & Rees (1985) in the context of ion tori around black holes. A few related questions have been asked lately: Can lithium in the universe be produced in black hole accretion (Jin 1990; Yi & Narayan, 1997)? We believe that this is not possible. The spallation reactions may produce such elements when only He-He reactions are considered. But when the full network is used we find that the hotter disks where spallation would have been important also photo-dissociate heliums to deuteriums and then to protons and neutrons before any significant lithiums could be produced. Another question is: Could the metalicity of the galaxy be explained, at least partially, by nuclear reactions? We believe that this is quite possible. Details are in MC98.

An interesting possibility of formation of the neutron torus was also discussed by Hogan & Applegate (1987): Can a neutron torus be formed around a black hole? We find that in the case of hot inflows, such formation of neutron tori is a very distinct possibility (Chakrabarti & Mukhopadhyay, 1998). Presence of a neutron torus around a black hole would help the formation of neutron rich species as well, a process hitherto attributed to the supernovae explosions only.

The advective disks as we know today do not perfectly match with a Keplerian disk. The shear, i.e., $d\Omega/dx$ is always very small in the advective flow compared to that of a Keplerian disk near the outer boundary of the advective region. We believe that such behavior is unphysical and had the viscosity α parameter or the cooling function were allowed to be changed continuously, such deviation would not have occurred. Thus some improvements of the disk model at the transition region is needed, but since major reactions are closer to the black hole, we believe that such modifications of the model would not change our conclusions. The neutrino luminosity is generally very large compared to the photon luminosity in case of hot disk (Mukhopadhyay & Chakrabarti 1998). In the first Case that we discussed above, neutrinos typically carry an energy of around 10^{30} ergs sec^{-1} gm^{-1}. Assuming that a typical neutrino is of energy ~ 1 MeV, and appreciable neutrinos are emitted only from a region of a radial extent of the order of a Schwarzschild radius where the disk is also around a Schwarzschild radius thick and the density is around 10^{-9} gm sec^{-1}. In presence of hot advective disks, the number of neutrinos that should be detected per square cm area on the surface of earth would be at least a few per second provided the source is a $10 M_\odot$ black hole at a distance of 10kpc. On the other hand, neutrino luminosity from a cool advective disk is low (around 10^{15} ergs

$\sec^{-1}$ gm^{-1}) and no appreciable number of neutrino are expected. Thus, probably one way to check if hot, and stable advective disks exist is to look for neutrinos from the suspected black hole candidates, especially in the hard states.

In all the cases, even when the nuclear composition changes are not very significant, we note that the nuclear energy release due to exothermic reactions or absorption of energy due to endothermic reactions is of the same order as actual radiation from the disk. Unlike the gravitational energy release due to viscous processes, nuclear energy release strongly depends on temperatures. Thus, the additional energy source or sink may destabilize the flow. This aspect has not been studied in this work yet. A realistic way to do this is to include the nuclear energy also in time dependent studies of the black hole accretion (e.g., Molteni, Lanzafame & Chakrabarti, 1994; Molteni, Ryu & Chakrabarti, 1997). Such works are in progress and the results would be reported elsewhere.

References

Bisnovatyi-Kogan, G., 1998, this volume.
Chakrabarti, S. K., Jin, L. & Arnett, W.D. *ApJ*, **313**, 674 [CJA], 1987
Chakrabarti, S. K., *ApJ*, **464**, 623 [C96] 1996
Chakrabarti, S. K., *ApJ*, **324**, 391 1988
Chakrabarti, S. K., *ApJ*, **484**, 313 [C97a], 1997a
Chakrabarti, S. K., *ApJ* (submitted), 1997b
Chakrabarti, S.K., & Mukhopadhyay, B. (submitted), 1998
Chakrabarti, S.K., & Titarchuk, *ApJ*, **455**, 623 [CT95], 1995
Das, T.K. & Chakrabarti, S.K., *ApJ* (submitted), 1998
Das, T.K., 1998, this volume
Hawley, J.W., Smarr, L.L. & Wilson, J.R., *Astrophys. J.*, **277**, 296, 1984
Hawley, J.F., Smarr, L.L., & Wilson, J.R., *ApJS*, **55**, 211, 1985
Hogan, C.J. & Applegate, J.H. *Nat*, **390**, 236, 1987
Jin, L., Arnett, W.D. & Chakrabarti, S.K. *ApJ*, **336**, 572, 1989
Jin, L., *ApJ*, **356**, 501, 1990
Molteni, D., Lanzafame, G., & Chakrabarti, S.K., *ApJ* , **425**, 161,1994
Molteni, D., Ryu, D., & Chakrabarti, S. K., *ApJ*, **470**, 460, 1997
Mukhopadhyay, B. & Chakrabarti, S.K., *ApJ*, 1998, (submitted) [MC98]
Paczyński, B. & Bisnovatyi-Kogan, G.,*Acta Astron.*, **31**, 283, 1981
Ramadurai, S. & Rees, M.J., *MNRAS*, **215**, 53P-56P, 1985
Rees, M.J., *Ann. Rev. Astron. Astrophys.*, **22**, 471, 1984
Shapiro, S.L., *ApJ*, **180**, 531, 1973
Yi, I. & Narayan, R., *Astrophys. J.*, **486** 383, 1997

COMPUTATION OF MASS-OUTFLOW RATES FROM ADVECTIVE ACCRETION DISKS AROUND BLACK HOLES

TAPAS K. DAS
S. N. Bose National Centre For Basic Sciences
JD Block, Salt Lake, Sector-III, Calcutta-700091,
India
e-mail:tdas@boson.bose.res.in

1. Introduction

The existing models which study the origin, acceleration and collimation of mass outflow in the form of jets from AGNs and Quasars are roughly of three types. The first type of solutions confine themselves to the jet properties only, completely decoupled from the internal properties of accretion disks (see, e.g., Begelman, Blandford & Rees, 1984). In the second type, efforts are made to correlate the internal disk structure with that of the outflow using both hydrodynamic (e.g., Chakrabarti 1986) and magnetohydrodynamic considerations (Königl 1989; Chakrabarti & Bhaskaran 1992). In the third type, numerical simulations are carried out to actually see how matter is deflected from the equatorial plane towards the axis (e.g., Hawley, Smarr & Wilson, 1984, 1985; Eggum, Katz & Coroniti, 1985; Molteni, Lanzafame & Chakrabarti, 1994, hereafter MLC94; Molteni, Ryu & Chakrabarti, 1996, hereafter MRC96; Ryu, Chakrabarti & Molteni, 1997; Nobuta & Hanawa, 1998). ¿From the analytical front, although the wind type solutions and accretion type solutions come out of the same set of governing equations (Chakrabarti 1990), there are few, and mostly qualitative attempt to find connections among them (Chakrabarti, 1997a). As a result, the estimation of the outflow rate from the inflow rate has been difficult. Our work, *for the first time*, quantitatively connects the topologies of the inflow and the outflow. The simplicity of black holes and neutron stars lie in the fact that they do not have atmospheres. But the disks surrounding them have, and similar method as employed in stellar atmospheres should be applicable to the disks. Our approach in this paper is precisely this. We first determine the properties of the rotating inflow and outflow and identify solutions to connect them. In this manner, we self-consistently determine the mass outflow rates.

Before we present our results, we describe basic properties of the rotating inflow and outflow. As is well known (MLC94, MRC96 and references therein), in the centrifugal pressure supported boundary layer (CENBOL) the flow becomes hotter and denser and for all practical purposes behaves as the stellar atmosphere so far as the formation of outflows are concerned. In case where the shock does not form, regions around pressure maximum achieved just outside the inner sonic point would also drive the flow outwards. We calculate the mass outflow rates ($R_{\dot{m}}$) rate as a function of the inflow parameters, such as specific energy and angular momentum, accretion rate, polytropic index etc. We explore both the polytropic and the isothermal outflows. A detailed report on this type of outflows is presented elsewhere (Das & Chakrabarti, 1998, hereafter DC98).

The plan of this paper is the following: In the next Section, we describe our model and present the governing equations for the inflow and outflow. We also provide the solution procedure of those equations. In §3, we present results of our computations. Finally, in §4, we draw our conclusions.

2. Model Description, Governing Equations and the solution Procedure

We consider thin, axisymmetric polytropic inflows in vertical equilibrium (otherwise known as 1.5 dimensional flow). We ignore the self-gravity of the flow and viscosity is assumed to be significant only at the shock so that entropy is generated. We do the calculations using Paczyński-Wiita (1980) potential which mimics surroundings of the Schwarzschild black hole. The equations (in dimensionless units) governing the inflow are:

$$\mathcal{E} = \frac{u_e^2}{2} + na_e^2 + \frac{\lambda^2}{2r^2} - \frac{1}{2(r-1)}. \tag{1}$$

$$\dot{M}_{in} = u_e \rho_e r h_e(r), \tag{2}$$

(For detail, see, Chakrabarti, 1989 hereafter C89). The equations governing the polytropic outflow are

$$\mathcal{E} = \frac{\vartheta^2}{2} + n' a_e^2 + \frac{\lambda^2}{2r_m^2(r)} - \frac{1}{2(r-1)} \tag{3}$$

And

$$\dot{M}_{out} = \rho \vartheta \mathcal{A}(r). \tag{4}$$

Where r_m is the mean axial distance of the flow and $\mathcal{A}(r)$ is the cross sectional area through which mass is flowing out. (For detail, see, DC98.)

γ of the outflow was taken to be smaller than that of the inflow because of momentum deposition effects. The outflow angular momentum λ is chosen to be the same as in the inflow, i.e., no viscous dissipation is assumed to be present in the inner region of the flow close to a black hole. Considering that viscous time scales are longer compared to the inflow time scale, it may be a good assumption in the disk, but it may not be a very good assumption for the outflows which are slow prior to the acceleration and are therefore, prone to viscous transport of angular momentum. Detailed study of the outflow rates in presence of viscosity and magnetic field is in progress and would be presented elsewhere. The Isothermal outflow is governed by the following equations

$$\frac{\vartheta_{iso}^2}{2} + C_s^2 \ln\rho + \frac{\lambda^2}{2r_m(r)^2} - \frac{1}{2(r-1)} = \text{Constant} \qquad (5)$$

And
$$\dot{M}_{out} = \rho \vartheta_{iso} \mathcal{A}(r). \qquad (6)$$

Here, the area function remains the same above. A subscript *iso* of velocity ϑ is kept to distinguish from the velocity in the polytropic case. This is to indicate the velocities are measured here using completely different assumptions. For details, see DC98.

In both the models of the outflow, we assume that the flow is primarily radial. Thus the θ-component of the velocity is ignored ($\vartheta_\theta << \vartheta$).

2.1. PROCEDURE TO SOLVE FOR DISKS AND OUTFLOWS SIMULTANEOUSLY

For polytropic outflows, we solve equations (1-4) simultaneously using numerical techniques (for detail, see, DC98). In this case the specific energy $\mathcal{E}$ is assumed to remain fixed throughout the flow trajectory as it moves from the disk to the jet. At the shock, entropy is generated and hence the outflow is of higher entropy for the same specific energy.

A supply of parameters $\mathcal{E}$, λ, γ and γ_o makes a self-consistent computation of $R_{\dot{m}}$ possible when the shock is present. In the case where the shocks do not form, the procedure is a bit different. It is assumed that the maximum amount of matter comes out from the place of the disk where the thermal pressure of the inflow attains its maximum and the outflow is assumed to have the same quasi-conical shape with annular cross-section $\mathcal{A}(r)$ between the funnel wall and the centrifugal barrier as already defined. For this case, the compression ratio of the gas at the pressure maximum between the inflow and outflow R_{comp} is supplied as a free parameter, since it may be otherwise very difficult to compute satisfactorily. In the presence of shocks, such problems do not arise as the compression ratio is obtained self-consistently. For isothermal outflow, it is assumed that the outflow has exactly the *same* temperature as that of the post-shock flow, but the energy is not conserved as matter goes from disk to the wind. The polytropic index of the inflow can vary but that of the outflow is always unity. The other assumptions and logical steps are exactly same as those of the case where the outflow is polytropic. Here we solve equations (1-2) and (5-6) simultaneously using numerical technique to get results. (For details, see, DC98).

3. Results

3.1. POLYTROPIC OUTFLOW COMING FROM THE POST-SHOCK ACCRETION DISK

Figure 1 shows a typical solution which combines the accretion and the outflow. The input parameters are $\mathcal{E} = 0.00689$, $\lambda = 1.65$ and $\gamma = 4/3$ corresponding to relativistic inflow (see Fig. 5 and 6 of C89). The solid curve with an arrow represents the pre-shock region of the inflow and the long-dashed curve represents the post-shock inflow which enters the black hole after passing through the inner sonic point (I). The solid vertical line at X_{s3} with double arrow represents the shock transition. Three dotted curves represent three outflow solutions for the parameters $\gamma_o = 1.3$ (top), 1.1 (middle) and 1.03 (bottom). The outflow branches shown pass through the corresponding sonic points. It is evident from the figure that the outflow moves along solution curves which are completely different from that of the 'wind solution' of the inflow which passes through the outer sonic point 'O'. The mass loss ratio $R_{\dot{m}}$ in these cases are 0.47, 0.22 and 0.06 respectively. Figure 2 shows the ratio $R_{\dot{m}}$ as γ_o is varied. Only the range of γ_o for which the shock-solution is present is shown here. In Fig. 3a we show the variation of the ratio $R_{\dot{m}}$ of the mass outflow rate inflow rate as a function of the shock-strength (solid) M_-/M_+ (Here, M_- and M_+ are the Mach numbers of the pre- and post-shock flows respectively.), the compression ratio (dotted) Σ_+/Σ_- (Here, Σ_- and Σ_+ are the vertically integrated matter densities in the pre- and post- shock flows respectively), and the stable shock location (dashed) X_{s3} (in the notation of C89). Other parameters are $\lambda = 1.75$ and $\gamma_o = 1.1$. Note that the ratio $R_{\dot{m}}$ does not peak near the strongest shocks! Shocks are stronger when they are located closer to the black hole, i.e., for smaller energies. In Fig. 3b where $R_{\dot{m}}$ is plotted as a function of the specific energy $\mathcal{E}$ (along x-axis) and γ_o (marked on each curve). Specific angular momentum is chosen to be $\lambda = 1.75$ as before. The peak in $R_{\dot{m}}$ is observed (see also, Chakrabarti, 1997a, and Chakrabarti, this volume). To have a better insight of the behavior of the outflow we plot in Fig. 4 $R_{\dot{m}}$ as a function of the polytropic index of the incoming flow γ. The range

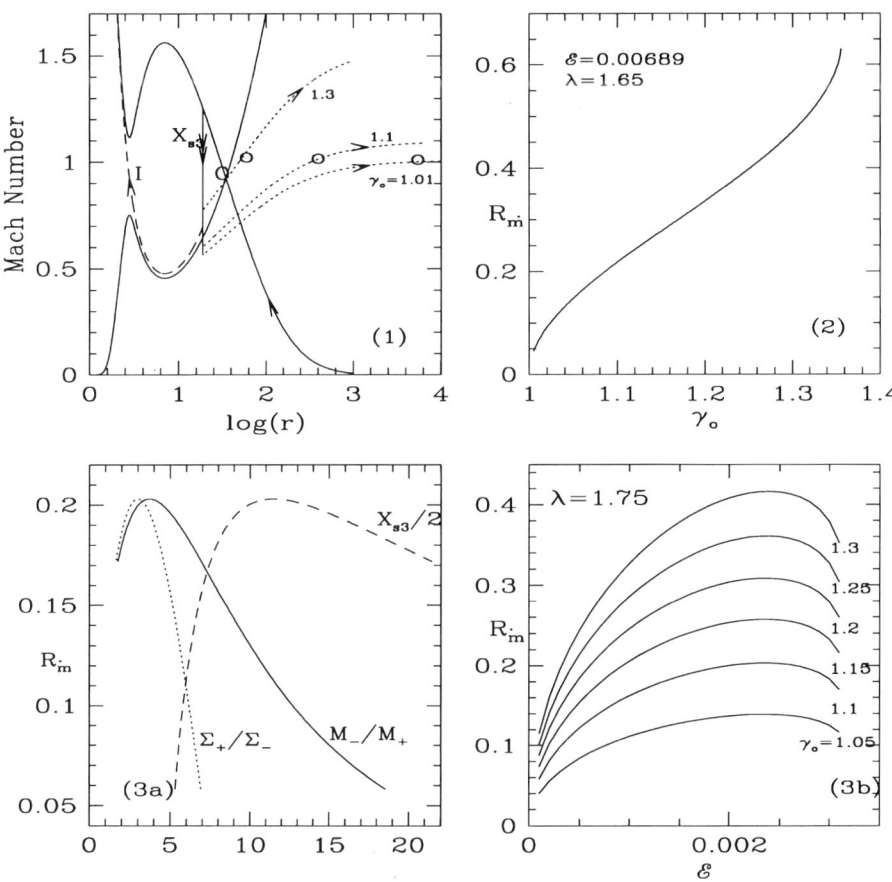

Fig. 1-3: Mach number of the flow is plotted against logarithmic radial distance both for the inflow and outflow (Fig. 1). The ratio of mass outflow rate and mass inflow rate is plotted against the polytropic index of the outflow (Fig. 2). The same ratio is plotted against the shock strength, shock location and the ratio of the integrated density (Fig. 3a) and specific energy $\mathcal{E}$ and polytropic index of the outgoing flow γ_o (Fig. 3b). See text for details.

of γ shown is the range for which shock forms in the flow. We also plot the variation of velocity ϑ_o, density ρ_o and area $\mathcal{A}(r)$ of the outflow at the location where the outflow leaves the disk. These quantities are scaled from the corresponding dimensionless units as $\vartheta_o \to 2 \times 10^4 \vartheta_o - 558$, $\rho_o \to 10^{22} \rho_o$ and $\mathcal{A} \to 0.0005 \mathcal{A}$ respectively in order to bring them in the same scale.

The non-monotonic nature of the variation of $R_{\dot{m}}$ with γ is observed.

3.2. POLYTROPIC OUTFLOW COMING FROM THE REGION OF THE MAXIMUM PRESSURE

In this case, the inflow parameters are chosen from region I (see C89) so that the shocks do not form. Here, the inflow passes through the inner sonic point only. The outflow is assumed to be coming out from the regions where the polytropic inflow has maximum pressure. Figure 5a shows a typical solution. The arrowed solid curve shows the inflow and the dotted arrowed curves show the outflows for $\gamma_o = 1.3$ (top), 1.1 (middle) and 1.01 (bottom). The ratio $R_{\dot{m}}$ in these cases is given by 0.66, 0.30 and 0.09 respectively. The specific energy and angular momentum are chosen to be $\mathcal{E} = 0.00584$ and $\lambda = 1.8145$ respectively. The pressure maximum occurs outside the inner sonic point at r_m when the flow is still subsonic. Figure 5b shows the variation of thermal pressure of the flow with radial distance. The peak is clearly visible. Figure 6 shows the ratio $R_{\dot{m}}$ as a function of γ_o for various choices of the compression ratio R_{comp} of the outflowing gas at the pressure maximum: $R_{comp} = 2$ for the bottom curve and 7 for the top curve. Note that flows with highest compression ratios produce highest outflow rates, evacuating the disk which is responsible for the quiescent states in X-ray Novae systems and also in some systems with massive black holes (e.g., our own galactic centre?).

The location of maximum pressure being close to the black hole, it may be very difficult to generate the outflow from this region. Thus, it is expected that the ratio $R_{\dot{m}}$ would be larger when the maximum pressure is located farther out. This is exactly what we see in Fig. 7, where we plot $R_{\dot{m}}$ against the location of the pressure maximum (solid curve). Secondly, if our guess that the outflow rate could be related to the pressure is correct, then the rate should increase as the pressure at the maximum rises. That's also what we observe in Fig. 7. We plot $R_{\dot{m}}$ as a function of the actual pressure at the pressure maximum (dotted curve). The mass loss is found to be a strongly correlated with the thermal pressure. Here we have multiplied non-dimensional thermal pressure by 1.5×10^{24} in order to bring them in the same scale.

3.3. ISOTHERMAL OUTFLOW COMING FROM THE POST-SHOCK ACCRETION DISK

Here the temperature of the outflow is obtained from the proton temperature of the advective region of the disk. The proton temperature is obtained using the Comptonization, bremsstrahlung, inverse bremsstrahlung and Coulomb processes (Chakrabarti, 1997b and references therein). Fig-

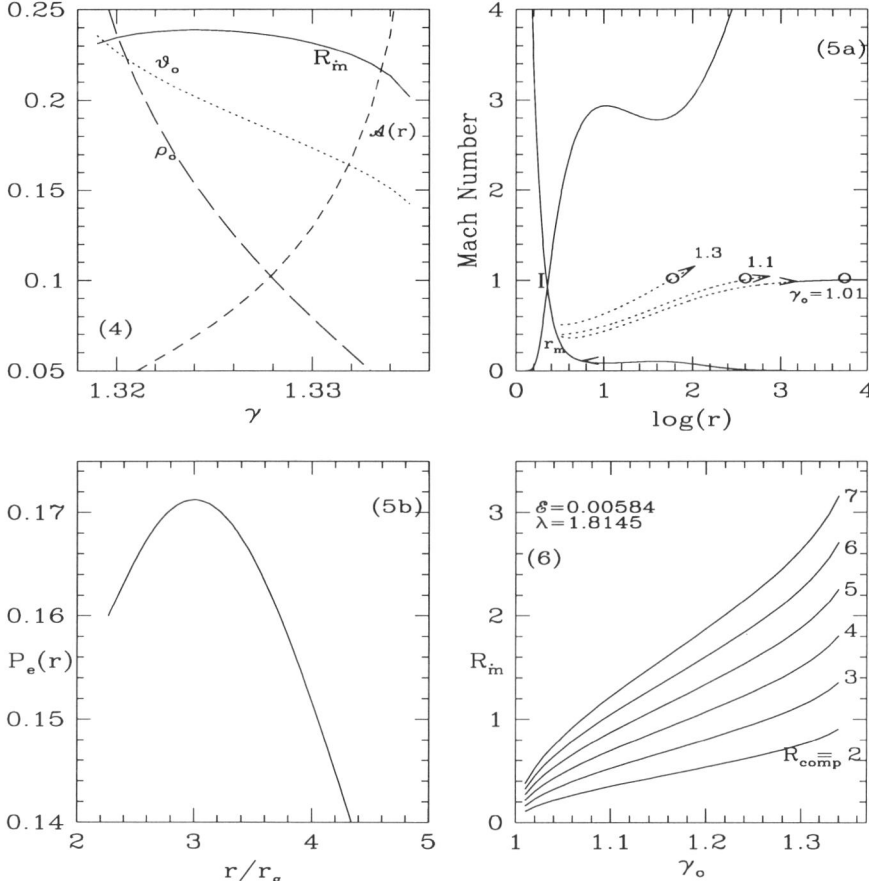

Fig. 4-6: Variation of velocity, density, cross sectional area and the rate ratio as a function of the polytropic index of the inflow (Fig. 4). Variation of the Mach number for inflow and outflow when shocks are not present (Fig. 5a). Thermal pressure variation as a function of the radial distance r/r_g showing a distinct maximum (Fig. 5b). Variation of $R_{\dot{m}}$ when both the compression ratio at the pressure maxima and polytropic index of the outflow are changed (Fig. 6). See text for details.

ure 8 shows the effective proton temperature and the electron temperature of the post-shock advective region as a function of the accretion rate (in logarithmic scale) of the Keplerian component of the disk. In Fig. 9a, we show the ratio $R_{\dot{m}}$ as a function of the Eddington rate of the incoming flow

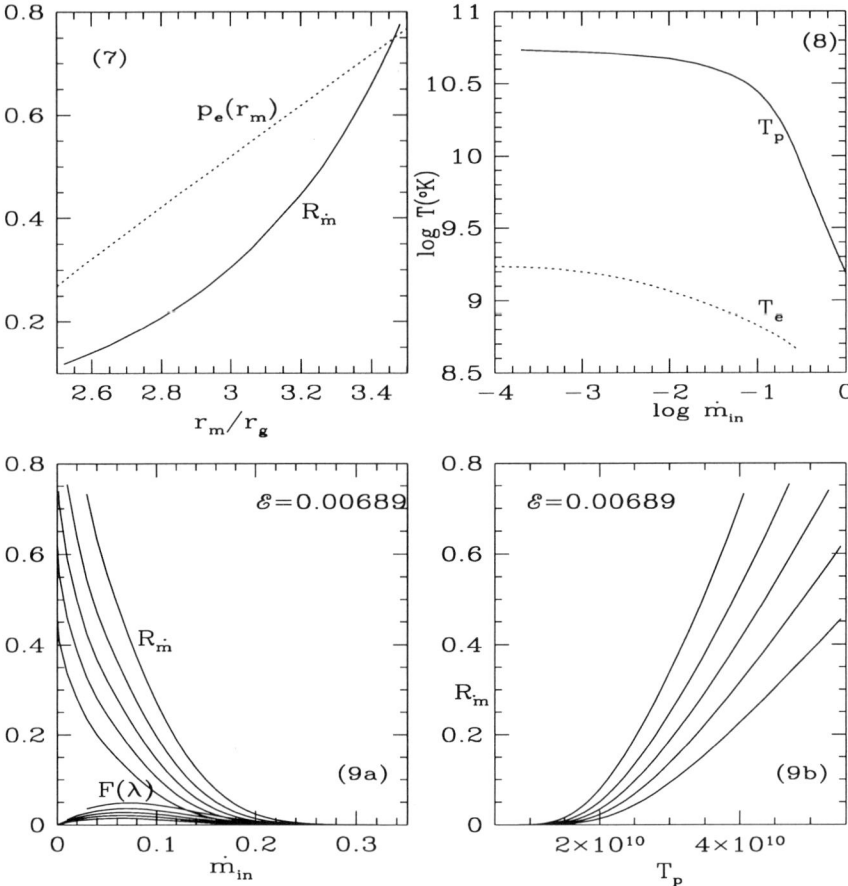

Fig. 7-9: Variation of the maximum pressure and $R_{\dot{m}}$ with the location where the pressure maxima occur (Fig. 7). Proton and electron temperatures in the advective region as a function of the inflow disk accretion rate $\dot{m}_{in}$ (Fig. 8). Variation of the $R_{\dot{m}}$ and angular momentum flux $F(\lambda)$ as a function of the accretion rate of the inflow (Fig. 9a). Variation of $R_{\dot{m}}$ with proton temperature T_p (Fig. 9b). See text for details.

for a range of the specific angular momentum. In the low luminosity objects the ratio is larger. Angular momentum is varied from $\lambda = 1.63$ (top curve) to 1.65 (bottom curve). An interval of $\lambda = 0.005$ was used. The ratio is very sensitive to the angular momentum since it changes the shock location rapidly and therefore changes the post-shock temperature very much.

We also plot the outflux of angular momentum $F(\lambda) = \lambda \dot{m}_{in} R_{\dot{m}}$ which has a maximum at intermediate accretion rates. In dimensional units, these quantities represent significant fractions of angular momentum of the entire disk and therefore the rotating outflow can help accretion processes. Curves are drawn for different λ as above. In Fig. 9b, we plot the variation of the ratio directly with the proton temperature of the advecting region. The outflow is clearly thermally driven. Hotter flow produces more winds as is expected. The angular momentum associated with each curve is same as before.

3.4. ISOTHERMAL OUTFLOW COMING FROM THE REGION OF THE MAXIMUM PRESSURE

This case produces very similar result as in the above case, except that like Section 3.2 the outflow rate becomes more than a hundred percent of the inflow rate when the proton temperature is very high. This phenomenon may be responsible for producing quiescent states in some black hole candidates.

4. Conclusions

The basic conclusions of this paper are the followings:

a) It is possible that most of the outflows are coming from the centrifugally supported boundary layer (CENBOL) of the accretion disks.
b) The outflow rate generally increases with the proton temperature of CENBOL. In other words, winds are, at least partially, thermally driven. This is reflected more strongly when the outflow is isothermal.
c) Even though specific angular momentum of the flow increases the size of the CENBOL, and one would have expected a higher mass flux in the wind, we find that the rate of the outflow is actually anti-correlated with the λ of the inflow. This is because the average proton temperature of CENBOL goes down with λ.
(d) Presence of significant viscosity in CENBOL may reduce angular momentum of the outflow. When this is taken into account, we find that the rate of the outflow is correlated with λ of the outflow. This suggests that the outflow is partially centrifugally driven as well.
e) The ratio $R_{\dot{m}}$ is generally anti-correlated with the inflow accretion rate. That is, disks of lower luminosity would produce higher ratio $R_{\dot{m}}$.
f) Generally speaking, supersonic region of the inflow do not have pressure maxima. Thus, outflows emerge from the subsonic region of the inflow, whether the shock actually forms or not.

An interesting situation arises when the polytropic index of the outflow is large and the compression ratio of the flow is also very high. In this case, the flow virtually bounces back as the winds and the outflow rate can be temporarily larger compared with the inflow rate, thereby evacuating the disk. In this range of parameters, most, if not all, of our assumptions breakdown completely because the situation becomes inherently time-dependent. It is possible that some of the black hole systems, including that in our own galactic centre, may have undergone such evacuation phase in the past and gone into quiescent phase.

So far, we made the computations around a Schwarzschild black hole. The mass outflow rates for Kerr black holes are being studied and the results would be reported elsewhere (Das, 1998a). For the quasi-spherical Bondi-type accretion onto black holes, the accretion disk does not form, the freely falling matter may produce a standing collisionless shock due to the plasma instabilities and the nonlinearity introduced in the flow due to even a small density perturbation. Outflow from these cases is also being studied and would be reported elsewhere (Das, 1998b).

We made a few assumptions, some of which may be questionable. Nevertheless, we believe that our calculation is sufficiently illustrative and gives a direction which can be followed in the future.

References

Begelman, M.C., Blandford, R.D. & Rees, M.J. 1984, Rev. Mod. Phys., 56, 255
Chakrabarti, S. K. 1986, Astrophys. J., 303, 582
Chakrabarti, S. K. 1989, Astrophys. J., 347, 365
Chakrabarti, S. K. 1990, Theory of Transonic Astrophysical Flows (Singapore: World Sci.) (C90)
Chakrabarti, S.K., 1997a, Astrophys. J., submitted
Chakrabarti, S. K. 1997b, Astrophys. J., 484, 313
Chakrabarti, S. K. & Bhaskaran, P. 1992, MNRAS, 255, 255
Das, T.K. & Chakrabarti, S.K. 1998, *Astrophys. J.* (submitted)
Das, T.K. 1998a, in preparation
Das, T.K. 1998b, in preparation
Königl, A. 1989, Astrophys. J., 342, 208
Hawley, J.W., Smarr, L. & Wilson, J. 1984, Astrophys. J., 297, 296.
Hawley, J.F., Smarr, L.L., & Wilson, J.R. 1985,Astrophys. J., 55, 211
Eggum, G. E., Coroniti, F. V., Katz, J. I. 1985, Astrophys. J., 298, L41
Molteni, D., Lanzafame, G. & Chakrabarti, S. K. 1994 Astrophys. J., 425, 161
Molteni, D., Ryu, D. & Chakrabarti, S.K. 1996 *Astrophys. J.* **470** 460
Paczyński, B. & Wiita, P.J. 1980 *Astron. Astrophys.*, **88**, 23
Ryu, D., Chakrabarti, S. K., & Molteni, D. 1997, Astrophys. J., 474, 378

BLACK HOLE SOLUTIONS OF EINSTEIN'S EQUATIONS AN OVERVIEW

T. ZANNIAS
Instituto de Física y Matemáticas
Universidad Michoacana de SNH
MORELIA MICH. MEXICO
e-mail:zannias@ginette.ifm.umich.mx

Abstract We review the currently known exact non vacuum black hole solutions of the Einstein's equations and access their astrophysical significance. The role of Penrose Cosmic Censorship hypothesis as the most essential principle underlying black hole astrophysics is also discussed. It is argued that from an astrophysicist's point of view the two parameter family of Kerr black holes is for the moment sufficient for accretion modeling.

1. Introduction

A forum involving theoretical astrophysicists and observers analyzing the latest observations supporting the existence of black hole offers the ideal environment for an assessment of the principles underlying the field of black hole astrophysics. It has been a rule in this meeting and is rule of every theoretical astrophysicist, that whenever in an astrophysical setting there are evidences of black holes always they assumed to be Schwarzschild ones, or members of the Kerr family. How well justified are those assumptions? and what are the equations of General Relativity predict? We shall take the opportunity to discussed those issues and at the same time indicate which of the principles underlying the field of black hole astrophysics stand on firm basis. To do so, first we shall briefly review how the black hole concept emerged within Einstein's theory of relativity. In that way it is my hope to demonstrate that the most essential principle underlying black hole astrophysics is Penroses Cosmic Censorship hypothesis (Penrose 1969). Once the Cosmic Censorship hypothesis conjecture is accepted as a valid principle of nature, it follows then that the implicit assumptions made by astrophysicists to work exclusively with a Schwarzschild or a Kerr black hole are rather well justified. Due to the significance of the Cosmic Censorship hypothesis conjecture, we shall also briefly discuss the theoretical evidences in its favor.

2. WHY COSMIC CENSORSHIP HYPOTHESIS ?

For an astrophysicist now days a black hole stands for an object carrying a strong gravitational field so that light originated within a limited region of space, cannot escape to infinite distances. Furthermore a black hole , and in contradistinction to other compact objects is lucking a hard surface, and for all purposes it acts as a perfect sink of any kind of matter and radiation. It is perhaps of interest to recall that the black hole concept is not really new, but the idea and goes long way back. Carter in his 1986 Cargese (Carter 1986) lectures documents that the clergyman J.Michell (and not Laplace) appeared to be the first person that examined the possibility of light-trapping by strong (Newtonian) gravitational fields. Mitcell came to the affirmative conclusion and thus 1784 (the year that J.Michell published his research) marks the birthday of black holes. The same problem also reconsider by Laplace later on and in the book of Hawking and Ellis Appendix A (Hawking and Elis 1973), the reader can find a translation of Laplace essay on the issue. Of course this primitive idea of black holes soon faded away, partially because of absence of any observations and partially because with the advance of the wave theory of light, scientist of the pre-Einstein era thought that gravity should not have any effect upon light propagation.

With the arrival of Einstein's general theory of relativity, gravity is not any longer described by a scalar potential but instead manifests itself as the geometry of the spacetime. Even though the relativistic description of gravity is radically different than its Newtonian counterpart, black holes (a nick name coined by J. Wheeler in (1967)) resurfaced again. However now black holes are dynamical objects and are not as simple and passive as pictured in the Newtonian theory. The celebrated Hawking's black hole evaporation phenomenon, suggests a deep and presently not entirely understood, connection between black holes, thermodynamics and quantum theory. Leaving asides those considerations to theoretical physicists, black holes also play an extremely importance role in astrophysical settings as well. They do provide a safe net protecting the environment from the almost infinite temperatures reached during the gravitation collapse.

Chandrasekhar in his profound investigation of Newtonian equilibrium configurations of cold stars, came to the conclusion that electron degeneracy pressure cannot withheld an infinity amount of gravitational squeezing. There is a limited maximum mass, the famous Chandrasekhar mass, above which there exist no Newtonian equilibrium configurations supported by electron degeneracy pressure. Even after the discovery of the neutrons, and the theoretical introduction of Neutron stars by Baade and Zwicky, again calculations show that the degeneracy pressure of neutrons can not withheld a infinite amount of gravitational squeezing, as well. Still there is a maximum mass which although above the Chandrasekhar mass, not by much. The issue as to what are the end states of cold stars whose masses at the end of their normal activity exceeds those maximum masses, has been answered by Oppenheimer in 1940s and by Wheeler and his collaborators later in the 1960s (Harrison et.al 1965). Oppenheimer in his pioneering work on the end states of spherical neutron stars models come to an astonishing conclusion. Within the framework of Einstein's equations he concluded that the final end-states are singular. The entire star is squeezed to pointlike dimensions and via Einstein's equations a spacetime singularity results. Wheeler and his collaborators repeated the calculations of Oppenheimer and even though more realistic equations of states for neutron star matter had been employed, Oppenheimer conclusions remained intact. Those unpleasant singular end states of gravitational collapse originally though as been the byproduct of the spherical assumptions employed in the calculations, and perhaps whenever rotation or and magnetic fields would be taken into account such singular end-states would be eliminated. All those expectations however shaterred in 1965, when Penrose (1965) in a landmark paper showed that the space time singularities seen by Oppenheimer and Wheeler are not the artifact of the spherical symmetry but rather due to the attractive nature of gravity. More precisely Penrose by employing entirely new techniques, showed that under physically reasonable assumptions, for instance no presence of matter with negative densities, and irrespectively of the symmetries involved, the gravitational collapse of a star would lead to a space time singularity once the collapse has proceed far enough so that a trapped surface forms. Subsequent work by Penrose, Hawking's, Geroch and others (see for example Hawking and Ellis 1973) strengthened further the original conclusion of Penrose and the presence of spacetime singularities has been indeed establish beyond any doubts. Of course one could argue that those spacetime singularities are not really present in nature but are artifacts of Einstein's field equations and consequently should be considered as a defect of the theory. However such reasoning is entirely unjustifiable. Einstein's theory of general relativity besides its beauty and simplicity has passed with flying colors to any test that has been so far subjected. From the classical solar type tests to the stringent test posed by the famous Taylor-Hulse pulsar. Its predictions are in perfect agreement with the observations and this agreement has been accomplished without any adjustments of free parameters. The theory, besides the cosmological constant does not contain other adjustable parameters. To blame the existence of spacetime singularities to Einstein's theory and look for alternative options is unwisfull thinking.

On the contrary, Einstein's theory predicts a graceful exit of the deadlock posed by the spacetime singularities. Even before 1965 relativist were well aware of the Schwarzschild and Kerr solutions. In those exact solutions of the theory the spacetime singularities are hidden within event horizons. By the defining property of the event horizon, no influences originated within it can affect outside observers. Penrose Cosmic Censorship hypothesis conjecture (Penrose 1969) essentially ensures that those properties of the Schwarzschild and Kerr solutions are not really the exemption but rather the absolute rule. More concretely, Penrose postulated that all spacetime singularities, which are unavoidable outcomes of the complete gravitational collapse of a bounded system, are hidden with event horizons. This postulate is the content of the Cosmic Censorship hypothesis . In astrophysical terms the principle ensures that the gravitational collapse of stars produces only black holes and never naked singularities (ie singularities not clothed by horizons). Before we continue, let us consider what would happen in the case that the principle fails. In such event, one of the possible outcomes of gravitational collapse could be naked singularities. In the presence of naked singularities the Einstein's theory looses completely its predictive power. Perhaps the interest reader may contemplate for instance, to construct accretion flows on a negative mass Schwarzschild solution instead of a Schwarzschild black hole solution. There exist an infinite set of solutions each one obeying differennt boundary conditions on the naked singularity and even worst there is not a physical principle that would dictate a preferable set of boundary conditions. The same situation would occur if one for instance consider electromagnetic, gravitational or other kind of perturbations on the presence of a naked singularity. Even though locally, the equations obeyed by the perturbations are unaffected by the presence of the singularity, the construction of global solutions needs boundary conditions on the location of the singularity. On the contrary if the principle is correct then at a stroke restores the predictive power of Einstein's theory. Typically one imposes regularity conditions on the event horizon and that is sufficient to specify unambigously the physics outside the horizon. In more physical terms, physics outside the event horizon is entirely unaware and at the same time unaffected, by the existence of the space time singularity.

To this moment and despite considerable efforts the Cosmic Censorship hypothesis is the major unresolved problem of classical general relativity. In principle the problem is physically well posed. One needs to specify the initial state of the precollapsing star, cast the Einstein's equations into an initial value formulation, and then using the dynamics of general relativity propagate the state to the infinite future. This process would unambiguously pick up the spacetime geometry and is capable to verify or refute the conjecture. Unfortunately there exist unsurmable mathematical difficulties in executing the required steps. Currently with the available mathematical machinery we are unable to predict analytically or even numerically, the behavior of solutions of the highly non linear Einstein's equations to the far future of the initial slice. Thus for the time being, we cannot pass a verdict on the conjecture. There are however evidences in its favor. Attempts by Penrose and others to reach contradictions with the content of the conjecture have failed. Numerical simulations, as far as they can go, do not contradict it. Even though there exist some counter-examples ie exact solutions of the theory where the singularity appears naked, the general consensus is that those counterexamples are posing no serious threat. The encountered shell crossing singularities (Yodzis et al. 1973) or shell focusing singularities (Christodoulou 1984, Eardley et al. 1979, Ori et al. 1987, Lake et al. 1990, 1991) are essentially due to the breakdown of the phenomenological description of matter, rather resulting from the dynamics of the theory. In fact many of them are reproducible in hydrodynamical flows propagating on flat spacetime. On the other hand if the conjecture fails, and one assumes that gravitational collapse is a common event in the universe, the supernova 1987A verifies that, then some evidences of the enormous amount of heat produced during the collapse ought to have been detected. Further the participants of the present meeting report strong evidences of black holes in close X-ray binary systems and at the centers of galaxies. In an indirect way, those evidences are also supporting the Cosmic Censorship hypothesis . Accepting the Cosmic Censorship hypothesis conjecture as a working principle of nature, we shall now argue that the theory of black hole astrophysics lies on strong foundations.

3. BLACK HOLE UNIQUENESS THEOREMS

The post Cosmic Censorship hypothesis era, meaning after 1965, was dominated by enormous efforts by relativist trying to get a grip on the possible set of solutions of Einstein's equations describing black holes. In an astrophysisict language the relativists during that era were trying to settle the following problem: Suppose a star is undergoing complete gravitational collapse. Originally the star may be characterized by a multipole set of attributes such as angular momentum, multipole moments, magnetic fields etc. If the Cosmic Censorship hypothesis is correct, then the final spacetime geometry definitely would be a black hole. What is the spacetime geometry of this final black hole state? and how the initial state of the star will influence the final spacetime geometry? I would like to cite here a passage from Carter's 1972 Les Houches lectures (Carter 1973) describing clearly the feelings and thoughts of the early era of black hole physics:

.... I think that we had all thought of a black hole even in the stationary limit, as potentially a fairly complicated object. I know that I had always imagined it should have many degrees of freedom representing the vestigial multipole structure of the star or other object from whose collapse it had arisen....

The reader is strongly encouraged to go through Carter's (who himself was one of the pioneers in establishing the physics of black holes) introductory chapters of his Les Houches lecture, to get the full flavor of the mood existed immediately after Penrose introduced the Cosmic Censorship hypothesis . As it turn out however, and to the surprise of every body, the reality turn out to be entirely different. The final black hole geometries are extremely simple. Only a few of the initial precollapsed attributes will be remembered in the final configuration and most of the initial irregularities accompanying the precollapsed star would either be radiated away or swallowed by the black hole. Perhaps one of the strongest accomplicements of relativists within the last thirty years or so, was the establishments of a few powerful theorems refereed as black hole Uniqueness Theorems (Heusler 1985). Those theorems are describing essentially all possible black hole equilibrium states resulting from the collapse of astrophysically realistic objects. We shall describe them here leaving aside rigorous mathematical definitions and complexities, so that their contend be accessible to astrophysicists and observers. From here on the term black hole will always refereed to as an asymptotically flat black hole solutions of the Einstein equations. Asymptotic flatness roughly implies that if the entire universe consists only of a single asymptotically flat black hole then the spacetime curvature (or equivalently gravitational field) decays in suitable way so as one is moving further and further away from the black hole, the decay rate is such that the spacetime looks more and more like Minkowski spacetime. Often in the sequel, we shall make use of the terms "black hole final equilibrium state" or the "final stationary state". Those two terms will be used intergeangeably and essentially refer to the state that the black hole will eventually settle in after all the transients effects of collapse, such as emission of gravitational waves, dumping of various asphericities etc have been diet away. On physical grounds, one expects that the final geometry of the black hole will be "time independent". This state is refereed as the final black hole equilibrium state or stationary state. It might be worth to stress here that the time scale needed for the black hole to settle in the stationary state appears to be very very short. Perturbations calculations (Doroskevich et al. 1966, Vishveshwara 1972. Price 1972) on the Schwarzschild black hole background shows that perturbations generally die away on short time scales, of the order of the Schwarzschild radious divided by the speed of light. It is generally believed that similar behavior will take place in the generic collapse, although at this moment this behavior is an implicit assumption.

Returning back to the possible final outcomes of gravitational collapse the black hole Uniqueness Theorems refereed above, take for granted validity of the Cosmic Censorship hypothesis conjecture. Validity of the Cosmic Censorship hypothesis permits relativists to look for the final black hole equilibrium states and decouple the early dynamical collapse phase from the final one. With a little use of mathematical terminology, relativists look for solutions of Einstein's equations, which are time independent, asymptotically flat space times, possessing a regular event horizon. (The term regular event horizon means a horizon free of any kind of singularities.)

It should be stressed however, that in their search for stationary black hole equilibrium states, no additional symmetries, such as spherical or axially symmetry are imposed in advance. The symmetries of final stationary state, predicted by the Uniqueness Theorems, are consequences of the dynamics of Einstein equations and not convenient choices of relativists.

Referring the interest reader to the references (Carter 1973, 1986, Hawking 1972, Heusler 1996, Israel 1967, 1968, Robinson 1975, Chrusciel 1996. Majur 1982, Bunting 1983) for further details, we first present two black hole Uniqueness Theorems:

Theorem 1 : The only vacuum, non rotating black hole solution of Einstein's equations is the Schwarzschild black hole .

Theorem 2: The only vacuum, rotating black hole solution of Einstein's equations is the Kerr black hole .

We may recall that in suitable coordinates the spacetime geometry in the exterior to the event horizon region of a Schwarzschild black hole takes the following form:

$$ds^2 = -fdt^2 + f^{-1}dr^2 + r^2(d\theta^2 + sin^2\theta d\phi^2) \tag{1}$$

where in geometrical units, $f = 1 - \frac{2m}{r}$ and $m > 0$ stands for the mass of the hole as measured from infinity. Its rotating Kerr counterpart in the exterior region takes the form:

$$ds^2 = -e^{2\nu}dt^2 + e^{2\psi}(d\phi - \omega dt)^2 + e^{2\lambda}dr^2 + e^{2\mu}d\theta^2 \tag{2}$$

where:

$$e^{2\nu} = \frac{\Delta A}{B} \quad e^{2\psi} = \frac{B sin^2\theta}{A} \quad \omega = \frac{2amr}{b} \quad e^{2\lambda} = \frac{A}{\Delta} \quad e^{2\mu} = A$$

with:

$$A = r^2 + a^2 cos^2\theta \quad B = (r^2 + a^2)^2 - \Delta a^2 sin^2\theta \quad \Delta = r^2 - 2mr + a^2$$

and $m > 0$ the mass of the hole and ma the angular momentum of the hole respectively subject to $a^2 \leq m^2$.

The first theorem derived by Israel in 1967 while the second one involved the combined efforts of many researchers [Hawking and Ellis 1973, Hawking 1972, Robinson 1975)]. In its final form presented by Robinson in 1975 while a modern stronger version that eliminated various weak points and assumptions presented in (Chrusciel 1996).

The above black hole Uniqueness Theorems for the Einstein vacuum equations likely are describing all possible black hole states of astrophysical interest. It is rather amazing and at the same time puzzling the resulting conclusions. Black holes are the simplest astrophysical objects in the universe. The only degree of freedom needed to describe a non rotating black hole is simply its mass while a rotating black hole posses only two degrees of freedom: its mass and the angular momentum. Of course the theorems refer only to vacuum black holes, but as we shall see further bellow from an astrophysicists point of view, this is rather adequate. But let us first examine the current status of non vacuum black hole solutions of Einstein's equations.

4. NON VACUUM BLACK HOLES

The other long range observable field in nature besides gravitation is the electromagnetic field. Even though large aggregation of matter appear electrically neutral and long range electric fields are rather absent from cosmos, it is not the case for the magnetic fields. Neutron stars and other astrophysical objects due to high conductivity, are carries of strong magnetic fields. Naturally then one may ask: Can black holes carry magnetic fields or and electric fields? If so what is their nature? The answer to those questions are by now well understood. Search for black hole solutions of the coupled Einstein-Maxwell system ie

$$G_{ab} = 8\pi[F_{ac}F_b^c - \frac{1}{4}g_{ab}F_{kl}F^{kl}] \tag{3}$$

$$\nabla_a F^{ab} = 0 \qquad \nabla_{[a}F_{ab]} = 0 \tag{4}$$

resulted in a generalization of the theorems (1,2) in the following form (Carter 1986, Chrusciel 1996, Israel 1968, Majur 1982, Bunting 1983):

Theorem 3 : The only electro-vacuum non rotating black hole solution of Einstein's equations is the Reissner-Nordsrom black hole .

Theorem 4: The only electro-vacuum, rotating black hole solution of Einstein's-Maxwell's equations is the Kerr-Newman family of black hole .

The explicit form of the geometry outside the corresponding event horizons for the Reissner-Nordstrom black hole takes the following form:

$$ds^2 = -fdt^2 + f^{-1}dr^2 + r^2(d\theta^2 + sin^2\theta d\phi^2) \tag{5}$$

where now $f = 1 - \frac{2m}{r} + \frac{e^2}{r^2}$ and e stands for the electric charge of the hole while and the corresponding Maxwell field is described by the the vector potential $A = -\frac{e}{r}dt$.

The corresponding metric for the Kerr-Newman black hole is more complicated and has the following form:

$$ds^2 = -(\frac{\Delta - a^2 sin^2\theta}{\Sigma})dt^2 - \frac{2asin^2\theta(r^2 + a^2 - \Delta)}{\Sigma}dtd\phi$$
$$+ \frac{(r^2 + a^2)^2 - \Delta a^2 sin^2\theta}{\sigma}sin^2\theta d\phi^2 + \frac{\Sigma}{\Delta}dr^2 + \sigma d\theta^2 \tag{6}$$

where $\Sigma = r^2 + a^2 sin^2\theta \quad \Delta = r^2 + a^2 + e^2 - 2mr$ and the electromagnetic field is described by the following vector potential:

$$A = -(\frac{er}{\Sigma})[dt - asin^2\theta d\phi] \tag{7}$$

The term electro-vacuum used in the above Theorems is a a standard terminology of relativists. It stands for solutions of the coupled Einstein-Maxwell system whenever the Maxwell field is source free. Theorems (3,4) imply that only the Coulomb monopole component of the initial electromagnetic field carried by the precollapsing star will be present in the final state. All other electromagnetic attributes are dumped to either radiated away or advected within the hole. Thus in contrast to other astrophysical objects and in the absence of magnetic poles, black holes cannot carry their own magnetic field (Israel 1968, Ginzburg et al. 1965, Wald 1972). But even in the unlike case that that the progenitor star was endowed with a net electric charge, it appears unlikely that the black hole solutions predicted but the above theorems would be astrophysically important. Preferential accretion of charged matter of the opposite sign of the black hole charge, would rapidly neutralize the hole and in fact detail calculations verify this expectation (Wald 1972).

Although theorems (1-4) describe all possible solutions of the Einstein vacuum or Einstein-Maxwells equations, and likely those solutions predicted by theorems (1,2) are the only ones of astrophysical significance, sceptical relativists and astrophycisists would raise additional question. It is well known that beside the long range forces discussed so far there exist also other short range one, responsible for nuclear binding and radioactive decays ie the strong and weak force. Would then those forces and associated generalized charges, attribute any imprints to the final black hole geometry? What is the fate of the total baryon number of the progenitor star? What is so special about the electric charge, mass and angular momentum that are registered by the final geometry?

Now days those interrelated questions are well understood. Only quantities that can be expressed as surface integral at infinity would survive the catastrophic collapse. As long as they are non vanishing at one time they would be non vanishing for ever. That explains the reason why the electric charge, mass and angular momentum survive the gravitational collapse. All three of them are expressible as Gaussian integrals involving fields in the asymptotic region. On the contrary, even though the baryon current is conserved, the total baryon number is not expressible as a two-surface integral at infinity, thus all informations about baryons are essentially lost. Everything that can be radiated in the process of gravitational collapse would be eventually radiated. Wheeler expressed this view in the graphical statement that black holes carry no hairs and often this statement is quoted in the literature as the "No hair conjecture". Wheeler's conjecture at first appeared to be correct as it was supported by model calculations involving linear fields (Bekenstein 1972, Chase 1970, Zannias 1995, 1998, Xanthopoulos 1991 Heusler 1992ab, Sudarsky 1995, 1998). Typically one starts with a star endowed with some form of a generalized charge acting as a source of some linear field assuming for the sake of simplicity to be a massless scalar field Φ. After the collapse has been completed and in accordance with the reasoning underlying the Uniqueness Theorems discussed above, one is looking for stationary black hole states of the following coupled system:

$$G_{ab} = 8\pi[\nabla_a\Phi\nabla_b\Phi - \frac{1}{2}g_{ab}\nabla_c\Phi\nabla^c\Phi] \tag{8}$$

$$\nabla_c\nabla^c\Phi = 0 \tag{9}$$

Surprisingly and in accordance with the No Hair Conjecture, this system does not contain any knew black hole solutions beyond those predicted by theorems (1,2) (Bekenstein 1972, Zannias 1995). All traces of the initial field Φ possessed by the precollapsing star had dissapearred from the final black hole stationary state. This conclusion appears rather generic as long as considerations are restricted to linear fields. It should be stressed however that recently whenever non linearities in the field configurations have been introduced either in the form of self interactions or in the form of coupled fields, black hole solutions of the non vacuum Einsteins equations have been found carrying informations about the fields of the precollapsed star. It appears that some form of non linearities are interlocked in such a manner so that manage to survive the inward pull of gravity and eventually peacefully coexist with the event horizon. For illustration purposes let us describe explicitly one such class of black holes.

It is reffered in the literature as the dilatonic class of black holes (Gibbons 1982, 1988, Garfinkle te al. 1991) and for the present purpose it is more convenient to describe the action functional S from which it originates rather explicitly recording the coupled Einstein's equations. It has following form:

$$S(g, \Phi, F) = \int [R + 2(\nabla \Phi_a \nabla \Phi^a) + e^{-\alpha \Phi} F_{ab} F^{ab}](-g)^{\frac{1}{2}} d^4 x \qquad (10)$$

ie it contains besides gravity a real scalar field Φ coupled to Maxwell field F. When the full set of Einsteins eqs are written out one discovers at least one black hole solution. Its spacetime geometry outside of the horizon, is described by:

$$ds^2 = -f(r)dt^2 + f(r)^{-1}dr^2 + g(r)d\Omega^2 \qquad (11)$$

where

$$f(r) = (1 - \frac{r_+}{r})(1 - \frac{r_-}{r})^{\frac{1-\alpha^2}{1+\alpha^2}} \qquad g(r) = r^2(1 - \frac{r_-}{r})^{\frac{2\alpha^2}{1+\alpha^2}} \qquad (12)$$

where without loss of generality the parameter $a \geq 0$, measures the strength of the dilaton coupling to the $U(1)$ field, while r_+, r_- are positive constants subject: $r_+ \geq r_-$. The corresponding Maxwell field is purely magnetic field (monopole in nature) and is obtained from the Maxwell field:

$$F = Q \sin\theta d\theta d\phi \qquad (13)$$

while the corresponding dilatonic field Φ is given by:

$$e^{2\Phi} = -(1 - \frac{r_-}{r})^{\frac{2a^2}{1+a^2}} \qquad (14)$$

The geometry of (11) describes a genuine black hole provided $r_+ > 0$, $a \geq 0$ and $r_+ > r_-$.

The above dilatonic black hole, non vacuo solution of Einstein's eqs, constitutes a typical example of the many classes of non vacuo black holes discovered within the last five years or so (Bison 1990, Kunzle et al. 1990, Sudarky et al. 1992). For the dilatonic case one can understand the underlying reason for the survival of the scalar degree of freedom. It is "clinched in" to the monopole Maxwell field the survival of which is guaranteed by the electric charge. However, there has been found other black hole solutions where such interpretation is not possible. (see for instance (Bison 1990, Kunzle et al. 1990, Sudarky et al. 1992)). We shall not expand any further on this subject, which is a current topic of investigation, but instead we shall the following question: Should one concerned about those new type of black holes? Could they for instance be astrophysically important? We shall demonstrate later on that as far as accretion is concerned likely for the moment and near future one should not worry. Besides the fact many of the new solutions are unstable, as far as accretion is concerned, the accreting plasma interacts directly with the background gravitational field and not with the additional fields that the black hole may carry. On the contrary to pass a verdict on those black holes additional work in different directions is required. How those additional fields influence the stellar equilibrium configurations? What are the maximum masses of neutron star models whenever additional fields are present? We believe that those questions should be on the top of the priority list of current research on the subject. Once those questions are understood, it would be then an easier task to pass a verdict on the astrophysical importance of those new black hole solutions. In short additional work is needed.

5. DISTORTED BLACK HOLES

As is clear from the discussion so far, the astrophysically important black hole Uniqueness Theorems discussed above, preassume absence of any additional material sources besides the collapsing star itself. However the Uniqueness Theorems still remain intact, provided the self gravity of the additional material sources is negligible in comparison to the gravitational field of the collapsing star. In the case where the external material sources are of comparable self gravity to the central black hole the situation is not clear. Surely one does not expect that a binary system consisting of two gravitationally bounded black holes to be described by the Kerr or the Schwarzschild solution. The spacetime geometry of such a system is highly dynamical, one black hole distorts the other and vice versa. To this point relativists are unable to describe analytically the geometry of such a system. One configuration however that a detailed description is possible and potentially of astrophysical importance concerns the following setting. Let us imagine that a ring or a torus of matter has been brought quasistatically on the equatorial plane of a Schwarzschild black hole . The quasistatic lowering of matter into the gravitational well of the hole results into quasistatic adjustments of the hole. The final product is an axially symmetric configuration consisting of a black hole and a torus of a possibly rotating matter on the equatorial plane of the hole. The combined system is not any longer described by the black hole solutions predicted by theorems (1,2). The spacetime geometry is more complex and in principle this class of black holes named for the obvious reasons distorted black holes (Israel 1973, Geroch et al. 1982) are described in principle by an infinite set of parameters. The additional parameters are describing the state of the matter outside of the hole. For the special case where the added matter is non rotating but axially symmetric the combined hole-external matter system has been understood and all black hole states have been classified (Israel 1973, Geroch et al. 1982). Let us briefly describe their classification. Because of the static axially symmetric character of the combined system and because near the horizon the spacetime is free of any material, the spacetime geometry near and on the event horizon can be described by appealing to Weyl's formalism (Synge 1971). The spacetime metric takes the following form:

$$ds^2 = -e^{2\lambda}dt^2 + e^{2(\nu-\lambda)}(dr^2 + dz^2) + r^2 e^{-2\lambda}d\phi^2 \qquad (15)$$

where here r, z, ϕ are cylindrical coordinates and the metric functions depend only upon (r, z). The usefulness and beauty of Weyl's formalism lies on the fact that once the Einstein vacuum equations are written out explicitly, for the line (15) one finds that λ is a harmonic function. It satisfies

$$\nabla^2 \lambda = 0 \qquad (16)$$

with the Laplacian operator ∇^2 formed out of the three Euclidean metric $ds^2 = dr^2 + dz^2 + r^2 d\phi^2$. The other metric function ν is essentially determined via quadratures in terms of the derivatives of the λ function. Although Einsteins field equations are highly non linear, Weyl's description of static axially symmetric gravitational fields allows a partial superposition of different solutions. It is this (partial) linear superposition that allows for the complete classification of all distorted axially symmetric black holes. The geometry of a Schwarzschild black hole expressed in Weyl's form is generated by choosing the harmonic function λ_s to be a solution of Laplace equation that is formally identical to the Newtonian potential of a of uniform density rod, lying along the $z-axis$ with end points at coordinates $(m, -m)$ respectively ie:

$$\lambda_s = \frac{1}{2} ln[\frac{R_+ + R_- - 2m}{R_+ + R_- + 2m}] \qquad (17)$$

where R_+, R_- are the Euclidean distances of the field point (r, z) from the end-points of the rod. If now the Schwarzschild black hole is to be distorted by an axially symmetric distribution of matter in the manner described above, then for the combined system one can choose another harmonic function V representing the added matter. Because of the linearity of the Laplace operator, the sum $\lambda_s + V$ is still harmonic and thus describes the combined system. Regularity of the geometry on the location of the horizon imposes the condition $V(0, m) = V(0, -m)$. In principle then, once a choice of V is made, then the other metric function ν is specified by simple integrations and thus the spacetime geometry near and on the event horizon is completely determined (for details see for instance (Israel 1973, Geroch et al. 1982)).

But could those distorted black holes be potentially useful in astrophysics? There have been lots of discussions in the present meeting dealing with various models of accretion disks around black holes. Up to-date all those models assume that the self gravity of the disk is negligible in comparison to the black hole mass. Although for the moment there is no reason for considering self gravitating disk, it is worth keeping in mind that should a need arise, the class of distorted black hole could be useful.

Let us end this brief survey of the currently available black holes solutions of Einstein's equations, by describing briefly a class of exotic black hole states. In all of the so far discussion, it has been implicitly assumed that the event horizon possess the "topology of a sphere". Relativists however, have also constructed black hole solutions with the property that the horizon is not any longer spherical but instead is topologically a torus ie it has the shape of a doughnut (Geroch et al 1982, Xanthopoulos 1983). Although mathematically those black hole solutions are interesting, since they provide a theoretical laboratory where the dynamics of Einsteins theory is tested, most likely, astrophysically are irrelevant. In order to be formed, some form of unphysical matter (negative energy densities) must be present (Geroch et al 1982, Xanthopoulos 1983).

6. BONDI ACCRETION ONTO BLACK HOLES

It would be of interest here to examine briefly whether accretion flows on the classes of the non vacuum black holes discussed so far would be markedly different than flows on a Schwarzschild or a Kerr black hole. If they are, then one may expect distinct observational signature to be associated with them. We shall briefly indicate here that actually that is not case. To make our point, we compare the dynamical equations describing the standard Bondi accretion on the background of a magnetically charged dilatonic black hole to the same equations for the same type of accretion on a Schwarzschild black hole . As is well known, for the Schwarzschild black hole, as long as adiabaticity is maintained, and the flow is subsonic at infinity, there always exist a unique regular transonic flow reaching the event horizon (Chakrabarti 1990, 1996). To check whether the same property holds true for the dilatonic case, let us consider a perfect fluid flow modeling the accreting matter. It is described by a conserved stress tensor

$$T_{\alpha\beta} = (\rho + P)u_\alpha u_\beta + P g_{\alpha\beta} \tag{18}$$

and a conserved baryon current $J_\alpha = n u_\alpha$ ie:

$$\nabla_\alpha T^{\alpha\beta} = 0 \qquad \nabla_\alpha(n u^\alpha) = 0 \tag{19}$$

In the absence of external supply of energy, and irrespectively of the equation of state, the first law of thermodynamics combined with the above conservation equations implies that the fluid evolves without its constituents exchanging any heat ie the flow is adiabatic. Imposing a polytropic equation of state $P = Kn^\Gamma$ and introducing the adiabatic speed of sound a as one of the flow variables, we found that the dynamical eqs. describing the flow are given by:

$$(\log v^2)' = (1 - v^2) \frac{2 D_1 - (\log f(r))' D}{D} \tag{20}$$

$$(\log a^2)' = -(\Gamma - 1 - a^2) \frac{(\log g(r))' D + D_1}{D} \tag{21}$$

where now D, D_1 are given by:

$$D = \frac{f(r)(v^2 - a^2)}{1 - v^2}, \qquad D_1 = f(r) (\log g(r))' \frac{a^2}{1 - v^2} - \frac{1}{2} \frac{df}{dr}, \tag{22}$$

and we have introduced the ordinary three velocity $v = \frac{dr}{dt}$ as measured by a local orthonormal observer tied to the coordinate system (11), instead of the radial component $u = u^r$ of the four velocity of the flow. The two are related via

$$u^2 = \frac{v^2 f(r)}{(1 - v^2)} \tag{23}$$

The flow equations (20,21) are valid for both, the dilatonic background and a Schwarzschild black hole background. One simply has to substitute the corresponding explicit form of (f, g) for the two backgrounds respectively (see eq. 11). Standard analysis shows that for both cases the critical point is of a saddle type, and as long as restrictions are restricted to non extremal dilatonic holes, regularity of the flow requires $v^2 = 1$ during horizon crossing. There is only a shift on the location of the critical point that eventually results into variations of the accretion rate of order unity. For both cases for a subsonic at infinity flow there is a unique regular over the event horizon transonic flow.

Similar considerations valid if instead of Bondi accretion one considers the standard geometrically thin optically thick accretion model on a dilatonic background. Many of the salient features of the disk are rather incentive to the changes of the background geometry (provided of course the latter posses an event horizon). Particularly the spectrum at infinity cannot really differentiate if the background is dilatonic or a Schwarzschild black hole. Even though we have mentioned explicitly the dilatonic black hole, the derivation of the central eqs (20, 21) are actually valid for every non rotating black hole solution of the Einstein's equations, and thus the above conclusions are of general validity.

Whenever, hopefully in the near future, the existence and clean identification of black holes would be completed, the question as to whether they belong to Schwarzschild class, Kerr, dilatonic or to any other type, could be raised. For the moment the Kerr family is sufficient for astrophysical modeling of accretion.

7. Acknowledgement

The author would like to thank all the participants of the meeting for the lively discussions and the efforts made to created a nice friendly atmosphere during the entire meeting. But most of all the author is grateful to prof Sandip Chakrabarti for the kind hospitality at the S.N. Bose National Center. The success of the meeting was the results of his efforts.

Finally I would like to thank Mr. Joaquin Estevez Delgado for his valuable help in preparing the manuscript. Part of this work was supported by a grant from Coordination Scientifica of UMSNH as well as by Catedra-Patrimonial fellowship administered by Conacyt-Mexico.

References

Bekenstein J.D. , (1972) *Phys.Rev. D*, **5 2941**
Bison P.(1990) *Phys. Rev. Lett.* **64,2844**
Bunting G. (1983) *Proof of the Uniqueness Conjecture for Black Holes,* **PHD thesis Unpublished.**
Carter B. (1973) *Black Holes , Les Houches* (**eds. C. DeWitt and B.DeWitt**)
Carter B. (1986) *Gravitation in Astrophysics, Gargese* (**eds. B.Carter and J.B. Hartle**)
Chakrabarti S.K.(1990) *Theory of Transonic Flows.* **World Scien.**
Chakrabarti S.K.(1996) *Phys. Rep.* **266, 229**
Chase J.E., (1970) *Commun. Math. Phys.* ,**19, 276**
Christodoulou D. (1984) *Commun. Math. Phys.* **93, 171**
Chrusciel P.T. ,(1996)*Journées Relativistes*, **gr-qc 9610010**
Doroskevitch A.G. , Ya. Zel'dovich and I. Novikov (1966) *Sov. Phys. J.E.T.P.* **22, 122**
Eardley D.M. and L.Smarr (1979) *Phys. Rev. D* **19, 2239**
Garfinkle D., G.Horowitz, A.Strominger (1991) *Phys.Rew.D* , **43,3140**
Geroch R. and J. B. Hartle (1982) *J. Math. Phys.* **23, 68**
Gibbons G.W.(1982) *Nucl.Phys.* , **B207,337**

Gibbons G.W.and K.Maeda (1988) *Nucl.Phys.* , **B298,741**
Ginzburg L. and L. M. Ozernoi (1965) *Zh Eksp Teor. Fiz*, **47, 1030**
Harrison B.K. , K.S. Thorne, M. Wakano, J.Wheeler (1965) *Gravitation theory and Gravitation Collapse* (**Univer, of Chicago Press**)
Hawking S.W. (1972) *Commun. Math. Phys.*, **25, 152**
Hawking S.W. and G.F.R. Ellis (1973) *The large scale structure of space time.* (**Cambridge Univ. Press**)
Heusler M. (1992) *J.Math. Phys.* , **33, 3497**
Heusler M. (1992) *Class. Quant. Grav.* , **12, 779**
Heusler M. (1996) *Black Hole Uniqueness Theorems.* (**Camb. Univ. Press**)
Israel W.(1967) *Phys. Rev*, **164, 1776**
Israel W. (1968) *Commun Math. Phys* , **8, 245**
Israel W.(1973) *Nuovo Cimento* , **6, 267**
Kunzle H.P. and A.K.M.Masood-ul-Alam, (1990) *J. Math. Phys.* **31, 928**
Lake K. and T.Zannias (1990) *Phys. Rev.* **D 41, 3866**
Lake K. and T.Zannias (1991) *Phys. Rev.* **D 43, 1978**
Mazur P. O. (1982) *J. Phys. A*, **15, 3173**
Ori A. and T. Piran (1987) *Phys. Rev. Lett.* **59, 2137**
Penrose R. (1965) *Phys. Rev. Lett.*, **14, 455**
Penrose R. (1969) *Riv. Nuovo Cimento.*, **1, 252**
Price R. (1972) *Phys. Rev*, **5, 2419**
Robinson D.C. (1975) *Phys. Rev. Lett.*, **34, 908**
Sudarsky D. (1995) *Class.Quant. Grav.*, **12, 579**
Sudarsky D. and T. Zannias (1998) *Phys.Rev. D*, **in Press**
Sudarsky D. and R.M.Wald, (1992) *Phys. Rev.* **D46 , 1453**
Synge J.L. (1971) *RELATIVITY The General Theory.* (**North Holland**)
Vishveshwara C.V. (1970) *Pys. Rev.*, **D1, 2870**
Wald R. M., (1974) *Phys. Rev. D*, **10, 1680**
Xanthopoulos B.C.(1983) *Proc. R. Soc. Lond. A* **388, 117**
Xanthopoulos B.C.and T.Zannias (1991) *J.Math. Phys.* , **32, 1875**
Yodzis P. , H.J. Seifert and M. zum Hagen (1973) *Commun. Math. Phys.* **37,29**
Zannias T.(1995) *J. Math. Phys.* , **9, 2643**
Zannias T.(1998) *J. Math. Phys.***in Press**

T. Zannias

M. Miyoshi

WATER MEGA MASERS IN NGC4258

M. MIYOSHI

National Astronomical Observatory Japan
2-12, Hoshigaoka, Mizusawa, Iwate 023-0861, Japan

1. Introduction

Maser phenomena in the universe are very general actions often occurring in molecular gases under the condition of population inversion. The emission mechanisms of astronomical masers, including mega masers, do not have any direct relations to the relativity or black hole physics. Because of the high brightness and spot-like features of astronomical masers, they are very excellent probes of the positions and movements of masing objects. The water mega masers in NGC4258 (Claussen et al. 1984), through the aid of high spatial resolutions of VLBI (very long baseline interferometer), gave us evidence of super massive black hole at the center of the galaxy surrounded by the rotating gas disk with high velocity.

2. Astronomical Masers

In 1965, curious strong line emissions were detected from radio telescope. They were later recognized as the first detection of the astronomical masers and as the 18 cm radio radiation from OH, hydroxyl radical (Weaver et al. 1965, Gundermann 1965). Maser (Microwave Amplification by Stimulated Emission of Radiation) is a phenomenon which is fundamentally the same as that of laser. Only difference is the wavelength of the radiation. Instead of light, powerful radio wave occurs in maser action. After the discovery of OH masers, astronomical maser phenomena were also detected in other molecules like water vapor (H_2O), silicon monoxide (SiO), and methanol (CH_3OH) in the universe. These astronomical masers are found in the molecular gases in circumstellar envelopes of evolved stars like Mira variables or around young stellar objects in star forming regions such as Orion KL or W49N.

3. Mega Masers, Powerful Maser Emissions from External Galaxies

Mega maser is the term named to very powerful extra galactic maser often found from active galactic nucleus. The first detection of mega maser was done by Baan et al. (1982), who detected broad OH maser emission in IC4553. The isotropic luminosity was in order of $10^3 L_\odot$. One can understand the powerfulness when it is compared with the most luminous water maser emission from W49N in our Galaxy which is only $1 L_\odot$. Dos Santos and Lepine (1979) found the first water mega maser from NGC4945, followed by detection in NGC4258, NGC1068 (Claussen et al. 1984), and NGC3079 (Henkel et al. 1984, Haschick and Baan 1985). Assuming isotropic emissions, the luminosities of water mega masers are found to be from hundreds to thousands of solar luminosities. The total luminosities H_2O emission from NGC 1068, NGC3079, NGC4258 are $350 L_\odot$, $520 L_\odot$, and $120 L_\odot$ respectively. The most powerful water mega maser emission reaching $L = 6100 \pm 900 L_\odot$ is detected in TXFS2226-184 (Koekemoer et al. 1996).

4. Nobeyama 45-m Monitoring Observations of Water Mega Masers

Nakai et al. (1993, 1995) performed monitoring observations of several water mega masers including NGC4258 using Nobeyama 45m telescope. The primary purpose was to check whether water mega masers have periodic time variations or not. The discovery leading to obtain the black hole evidence happened from an accidental setup of frequencies in radio spectrometers. Nobeyama 45m radio telescope has very wide band radio spectrometers. Nakai et al. (1993, 1995) used the AOS (Acoustic Optical Spectrometer)-High system which covers 320 MHz band width simultaneously with about 40 kHz frequency resolutions. In case of 22 GHz, H_2O maser observations, the frequency band width and resolution correspond to 4500 km/sec coverage and 0.5 km/sec resolution in velocities. For the monitoring observations, a 40 MHz band width, one subset of the AOS-High is sufficient to cover the known mega maser emissions. Nakai et al. (1993) happened to use all subsets of the AOS-High and set to cover the velocity range from +2000 km/sec to -2000 km/sec around the systemic velocities of observed galaxies. Nakai's monitoring observations including NGC4258, began in the autumn of 1991, but other works prevented him to check the data in detail. Six months had passed when he noticed the strange emissions in NGC4258, which are offset at about 70 MHz from the known water mega maser emission. If the lines are really water masers, the difference in velocity were about ±1000 km/sec. The figure 1 shows the spectra in NGC4258. On both sides of the central features at the systemic velocity of the galaxy (v_{sys}=460 km/sec), there are the stronger red shifted features with intensities of about one-half or one-third of the main features at v_{LSR}=1200-1440km/sec (LSR stands for local standard of rest) and the weaker blue shifted features with intensities of about one-tenth at v_{LSR}=-300 to -460 km/sec. The relative velocities of the new features to the systemic velocity of the galaxy NGC4258 are v_{rel}=740 to 980 km/sec and v_{rel} =-760 to -920 km/sec, respectively. No significant emission stronger than 51 mJy (3s) could be recognized outside the velocity range in Fig.1. The line width of individual components of the high velocity features is Δv= 1- 6 km/sec. The isotropic luminosities of individual high-velocity components are calculated to be L=0.1-6 $L_\odot$, which is comparable to the total luminosity of the most powerful maser source in our Galaxy, W49N, and the total luminosities of red shifted and blue shifted features are $23L_\odot$ and $1L_\odot$, respectively. The total luminosity of the main features in our observing epoch is $101L_\odot$. Where do these emissions come from? The beam size of the 45-m radio telescope is about 74 arcseconds at 22 GHz observations. They might be from the disk of NGC4258, not from the nucleus where known systemic water masers occur. Nakai changed slightly the direction of the telescope, the strange emissions disappeared together with known systemic masers from the spectrometer. It meant the position difference from the known systemic velocity water maser is within 12 arc-seconds.

5. KNIFE Observation of the new Emission in NGC4258

Hearing the news of the discovery in NGC4258, M. Morimoto strongly recommended us to perform VLBI observations. With the help of Dr. H. Takaba and Dr. T. Iwata in CRL (Communication Research Laboratory) at Kashima, we could get the chance to perform VLBI observations (KNIFE) in June 1992, only 1 month after the notice of the spectra. The Kashima Nobeyama Interferometer (KNIFE) has the 200 km east-west baseline, its minimum fringe spacing is 14 milli-arcseconds (mas) at 22 GHz . Though the spatial resolution is comparatively low, the large apertures of the 45-m telescope at Nobeyama and the Kashima 34-m telescope promised us high sensitivity to detect the weak emissions in NGC4258. NAOCO (New Advanced One unit Correlator) was used for initial data processing with help of Dr. S. Kameno and Dr. K. Matsumoto, quick data reduction was done with the help of Dr. K. Fujisawa (Miyoshi et al. 1994). Figure 2 shows the map constructed today from the first VLBI observation data of the emissions in NGC4258 (Miyoshi 1997). Compared with the result from VLBA (Figure 4), it is clear that the KNIFE observation caught substantially the positional relation between the red shifted high velocity features and the systemic maser features. This observation however was half a failure because we missed all of red shifted emissions except v_{sys}=1440km/sec component due to the wrong frequency setting. Anyway, we could find out that the high velocity emissions come from the nucleus as well as from systemic features. If we had captured all the emissions, the Keplerian motion would have been found from KNIFE in 1992. In Nakai et al. (1993), the upper limit of 50 mas was reported as the separation between the high velocity and the systemic features, which was obtained from relative fringe rate analysis. From the KNIFE observations, we understand two characteristic of the high velocity spectra. First, the emissions have the brightness temperature of order of $10^7 K$ at least, which demonstrated the emissions are not thermal line but maser emissions. Second, the emissions are within 50 mas (2 pc) from the known water mega masers. Nakai et al. (1993) put forward some interpretations on the spectra in NGC4258. One of possibilities suggested is that the emission is a new molecular emission line such as HDO, $HO^{37}Cl$. Another is that the emissions are caused by the effect of Raman scattering (Deguchi et al. 1993). It is reasonable however that the emissions are water maser lines frequency-shifted due to Doppler effect. First explanation of such Doppler shifts is that the emission comes from high velocity molecular outflow originated from a central black hole. Not only the water mega masers, the NGC4258 exhibits some activities. NGC4258 is a Seyfert 2 galaxy possessing a highly obscured central X-ray source with a 2-10 keV luminosity of $4\times 10^{40} erg/sec$ (Makishima et al. 1994). Hα images showed anomalous arms (Cortes and Cruvellier 1961) in NGC4258. Radio Continuum image at 1.4GHz shows that the anomalous arms are bright in non-thermal synchrotron emission (Van der Kruit, et al. 1972). Second explanation is that the emissions are from rotating molecular gas disk around massive black hole. If the separation between the high velocity features and systemic features, 2 pc (50mas) measured with KNIFE, is the radius of the disk, the central mass is estimated to be about $10^8 M_\odot$.

6. Several Results obtained before VLBA Observations

We collaborated with the Smithsonian group headed by J.M. Moran to investigate further detail of the high velocity water maser in NGC4258 using VLBA (VLBI Array). But we had to wait the completion of the VLBA till April 1994. I note here the three results about NGC4258 prior to our VLBA observation.

a) From single dish monitoring of the water mega masers in NGC4258 the systematic shifts of velocities about 9 km/sec/year in systemic maser features were found while the high velocity features both in red- and blue shifted showed no sign of such shifts in velocity (Fig. 3). The shifts reported are 7.5 km/sec/year in Haschik et al. (1994), 9.5±1.2km/sec/year

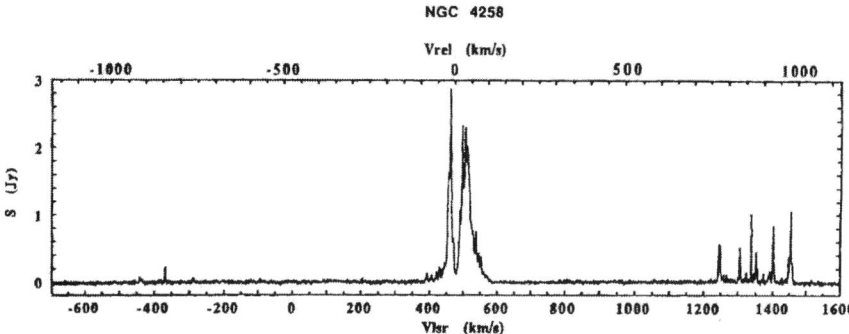

Fig. 1 : Detected H_2O maser emission in NGC 4258 in Nobeyama 45-m telescope. The maser emission comprises the systemic features (VLSR = 380 - 580 km/sec) near to VSYS = 476 km/sec and the blue- (VLSR = -515 - -285 km/sec) and red- (VLSR = 1240 - 1460 km/sec) shifted high-velocity features (Nakai et al. 1995)

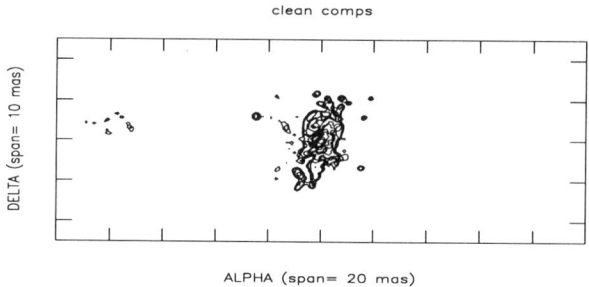

Fig. 2 : The oldest map of high velocity water maser features in NGC 4258. The center is systemic maser features and the left spots is the red shifted maser at VLSR = 1440 km/sec. The synthesized beam of KNIFE is about 10 mas, this figure is composed from Clean components with 0.5 mas restoring beams.

in Greenhill et al. (1995) and 9.6±1.0km/sec/year in Nakai et al. (1995).
b) Greenhill et al. (1995) performed the reanalysis of old VLBI observation data taken in 1984. The data included only the systemic water maser features in NGC4258. They show that the systemic features stay on a line when velocity is plotted against date.
c) Watson and Wallin (1994) found that edge-on rotating disk around mas-

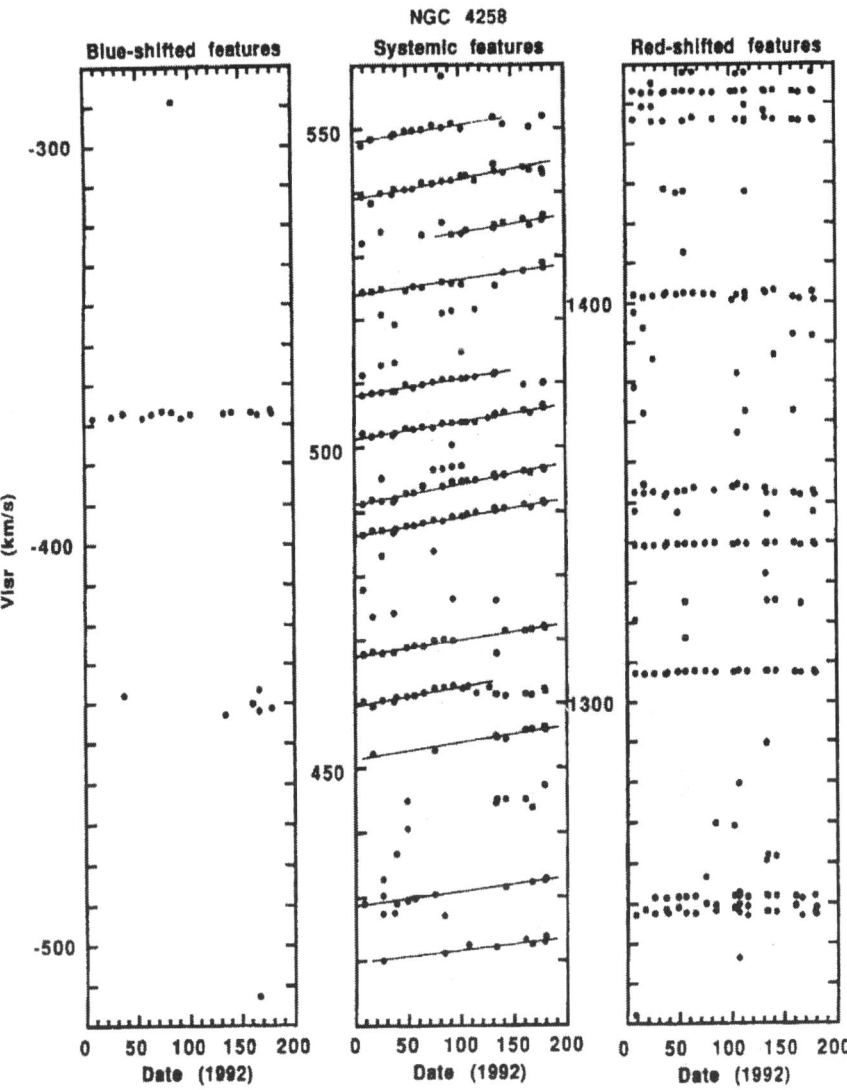

Fig. 3: Variations of the velocity of the prominent features in NGC4258. The (instrumental) error is less than 0.5 km/sec. The velocities of the systemic features are drifting, while the velocities of red- and blue shifted features are constant (Nakai et al 1995).

sive black hole explains very well these two facts mentioned above as well as the geometry of both the systemic and high velocity masers.

7. The VLBA Observation of the NGC4258 in April 1994

The VLBA consists of 10 radio telescopes with 25m diameter that are located all over the United States. The longest baseline between the St. Croix to Mauna Kea is 8000 km, which yields the minimum fringe spacing of 0.4 mas at 22 GHz. In case of our NGC4258 observation, the synthesized beam is 0.6×0.3 at a position angle of 7°. Participation of the phased VLA (Very Large Array) as one of VLBI stations gave us high sensitivity reaching $10mJy/ch(5\sigma)$. FX type correlator at Socorro produced the data of 4096 velocity channels with 0.2 km/sec resolution. We met some troubles mainly due to the initial failures of the VLBA system. Because of the limited IF bandwidth of VLA, we could not get whole spectra in NGC4258 all at once. Then we had two frequency setups, one was for red shifted features, and the other was for blue shifted features. This was not a serious problem. The trouble was that the setup for red shifted features at VLA did not work well. Then the sensitivities were different between the red shifted features (5σ-level, 15 mJy/ch) and the blue shifted features (5σ-level, 10mJy/ch). If the troubles were at blue set up, we could not get the image of weak blue shifted features.

Some of the correlation data sets were weird since the FX correlator was not perfectly completed at that time. J. R. Herrnstein checked all raw data sets and selected good ones. The AIPS (Astronomical Imaging Package) was also under improvement for VLBA data, and had some bugs in it. P. Diamond corrected the AIPS program every time when we sent him e-mails of finding new bugs in AIPS. In spite of such difficulties, the data reduction was finished within 6 months after the VLBA observations. This is exceptionally quick imaging as line VLBI observation. The distribution of water mega masers in NGC4258 has following physical characteristics. Both of the red and blue shifted high-velocity features are offset from the systemic features in a nearly planar structure, and the line of sight velocity decreases with distance from the center according as $r^{-0.5}$, exactly as expected from Keplerian motion, with the features occurring in a line perpendicular to the line of sight. These maser features arise in a disk that has inner and outer radii of 4 and 8 mas, respectively. The high velocity features define the systemic velocity (476km/sec), the binding mass ($3.6 \times 10^7 M_\odot$), the position angle of the disk (86 degree) and indicate that the disk is nearly edge-on. The inclination angle (83°) are estimated from the displacement of the systemic features, which are closer to us than the high-velocity features, show a systemic trend in velocity with position of 3.89 μarcsec per (km/sec) along the major axis (inset in Fig. 4). This very tight linear correlation and the binding mass derived above indicate that all these features lie at a nearly constant radius of 4.1 mas (0.13 pc at estimated distance of 6.4 Mpc), and

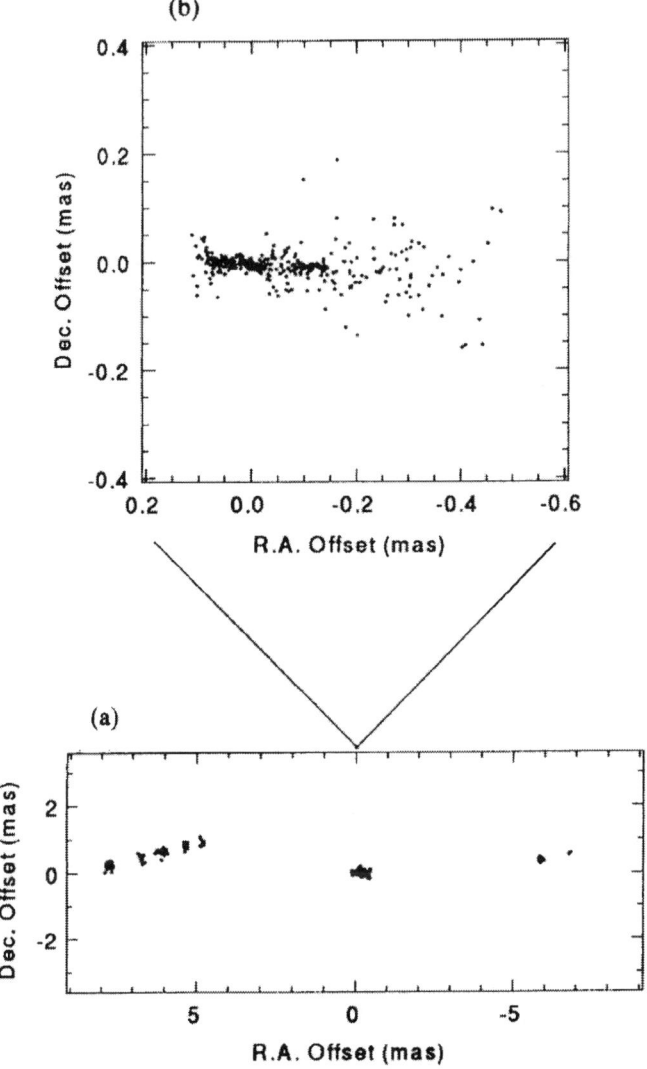

Fig. 4 : (a)Map of the distribution of the water masers in the NGC4258. All features greater than 5σ (20-40 mJy) are plotted. The position errors are (5 (10 (asec for the systemic features, and $\sim 20 \times 40 \mu$sec for high velocity features. (b) Expanded view of the systemic velocity features (Miyoshi et al. 1995).

the slope is caused by the change in projection of the velocity vector along the line of sight. The disk is very thin as the axial dimension is < 0.1 mas so that height / radius < 0.02 (see Fig. 4a,b), as expected for a disk in hydrostatic equilibrium. Because the binding mass must be confined to a sphere with radius smaller than 4.1 mas, the resulting mass density is more than $4 \times 10^9 M_\odot/pc^3$. It is unlikely that such a high density cluster could be stable for a long time. Even globular clusters have stellar densities of $<$ 105 $M_\odot/pc^3$. If the mass were composed of stars of 1 $M_\odot$, the mean stellar separation would be only 100 AU, the collision time would be less than 10^8yr; such collisions would disrupt the cluster system instantly. Therefore the central mass is probably a black hole. Other black-hole candidates have lower limits on their mass density (Maoz 1998). For example, the central mass and central density of M87 are $1.6 \times 10^9 M_\odot$, $9 \times 10^6 M_\odot/pc^3$ respectively (Marconi et al.1997). Those of NGC3115 are $10^9 M_\odot$, $3 \times 10^7 M_\odot/pc^3$ respectively (Kormendy et al. 1996). The central mass of the Galactic Center is $2.45 \times 10^6 M_\odot$, and the central density reaches $6.5 \times 10^9 M_\odot/pc^3$, which is higher than that of NGC4258, but the derivation of the mass and the density is performed with some assumptions (Eckart and Genzel, 1997). The mass of the central objects is much greater than the mass of the disk. The disk does not cause a measurable deviation in the Keplerian dependence of the high-velocity features, indicating that its mass is $< 4 \times 10^6 M_\odot$. If we assume a disk thickness of 0.1 mas (0.003pc) and a density of the disk of 10^{10} hydrogen molecules per cm^3, which is the maximum likely value to avoid thermal quenching of the maser (removing the inversion in population levels and stopping the maser), the mass is $2 \times 10^5 M_\odot$. Any radial motion (infall or outflow) is probably small compared to the rotational motion. The systematic velocity of the disk is very close to the systemic velocity of the galaxy $v_{SYS} = 472 \pm 4$km/sec, (Cecil et al. 1992), and to the systemic velocity estimated from the high-velocity features alone, which means that any radial motion in the systemic group is < 10km/sec. The observed line-of-sight velocities of $\sim$ 1000 km/sec of the high-velocity features are closely described by Keplerian motion, and are unlikely to be due to outflows. The restricted distribution of maser emission may be related to physical conditions in the disk. The velocity gradient is zero along the line of sight to the center of the disk and along the radial line (mid-line) through the disk perpendicular to the line of sight. Hence the high-velocity features appear where the path for amplification is naturally very long (10^{16} cm for a velocity change of 1 km/sec, the typical maser linewidth). The systemic features are clustered at the inner radius of the disk within a 8° angle, as viewed from the center of the disk. This suggests that they may amplify a background continuum source (as suggested for the masers in Arp220 and NGC3079) of angular size $\sim$ 0.5 mas, which means that their beam angle would be 7°. The striking absence of features in the range from v_{LSR} = -400 to 410 km/sec and from v_{LSR} = 530 to 1100 km/sec is probably caused by the lack of a background source to amplify in our direction and the unfavorably short gain length. The emission in this part of the disk is beamed radially in other directions. The clustering of the systemic velocity masers on the inner edge of the disk is interesting because the high-velocity features are distributed from radii of 4 to 8 mas. This clustering might be due to a beaming effect, as masers in the systemic group at the outer edge will be beamed into only 4° (because they are twice as far from the central background source as those on the inner radius) and their radiation might miss the Earth. Given a beam angle of 8°, the probability that we would see the maser emission from this galaxy (if it had a random inclination) is ∼ 6 %. This is approximately the detection rate of mega maser in active galaxies. The nature of the pumping mechanism is unclear; however Neufeld et al. (1994) suggest that the X-ray emission from the nucleus could lead to a layer of excited water molecules in a shielded region of the molecular envelope, as well as a population inversion of the 1.35-cm transition. The warp of the disk also plays an important part in generating and characterizing of the high velocity maser features (Neufeld and Maloney, 1995; Herrnstein 1997) The alignment of the spin axis of the molecular disk is offset from the spin axis of the galaxy by 119° (counter-rotating). Such a misalignment is not uncommon between galaxies and their cores. But the spin axis of the disk is parallel, at least in projection, to the H_α and radio synchrotron jets emanating from the nucleus on the scale of 500 pc. The molecular disk and helical strands have the same sense of rotation. There are systematic deviations from the simple disk model presented here, such as the departure from a linear dependence of the systematic features below 460 km/sec (Fig.5 inset) and the deviations in the declinations of the high-velocity features from planar geometry (Fig.4). The former effect is probably due to a slight fluctuation in the radii of the masers in that part of the disk. The latter effect is probably due to a warp or other distortion. It can be modeled as a series of concentric rings with constant inclination but with a progression in the line of nodes of 6°. Note that the parameters in Table 1 of Miyoshi et al. (1995) are significantly affected. The high-velocity features have velocity drift of < 1 km/sec (Greenhill et al., 1995; Haschik et al. 1994; Nakai et al. 1995) as expected if they lie on the mid-line of the disk. They should also have small proper motions. The systemic features, however, should show proper motions of 35μ arcsec/year. In fact these motions were measured in subsequent VLBA observations, and improved the definition of the disk to estimate the distance more accurately.

8. Result from following Observations of NGC4258 using VLBA

We have performed several VLBA observations of NGC4258 after the April 1994 observation. I note here four results obtained from these observations.

a) Detection of proper motions of systemic maser features: Herrnstein

(1997) detected the proper motions of 30 μarcsecond/year from the 14 systemic maser spots, which was predicted from circular rotation velocity at the radius 0.13pc where systemic masers reside. The red- and blue shifted high velocity features do not show detectable proper motions.

b) Geometric distance to the NGC4258: We calculated the distance to the NGC4258 from the apparent radius of the maser disk (θ), rotational velocity of the disk (V_{ROT}), and the shift of velocity in systemic maser features treated as the centripetal acceleration (α). From $\alpha = V_{ROT}^2/R$, where V_{ROT} is the rotational velocity at the inner radius R, since α and V_{ROT} are observable, we can deduce the real radius R, then comparing the apparent size of the radius, $R = \theta \times D$, we found out the distance to the NGC4258 ,$D = 6.4 \pm 0.9$Mpc (Miyoshi et al. 1995). The proper motion and acceleration measurements from the VLBA observations with the refinement of the disk model yielded better estimate of the distance to the NGC4258, 7.3 ± 0.3Mpc. These distance estimates are based on geometric method, and they are independent of the usual hierarchy of distance ladders.

c) Detection of continuum emission: Due to the use of phase referencing technique, continuum emission about 3mJy near the water masers in NGC4258 was detected and mapped (Fig. 6). The strongest emission is located about 0.5 mas (0.015pc) at the north of dynamical center of the disk. Another continuum emission was also found at 1 mas (0.03pc) south of the dynamical center. These continuum peaks are jets and definitely not the cores. In NGC4258, from the analysis of masing disk (Herrnstein et al 1997), we know well where the center of mass is.

d) Luminosity of the core: Herrnstein et al. (1998) report that the 3σ upper limit of the luminosity of the core is 220mJy at 22 GHZ. If the accretion flow into the core of NGC 4258 is advection dominated, this implies that the inner advection-dominated flow cannot extend significantly beyond 10^2 Schwarzschild radii.

9. Future Prospect of the Water Mega Maser in NGC4258 with VERA

Until now, I have discussed about the water mega maser in the galaxy NGC4258 and the central massive black hole. As for the activities concerning the central black hole, the details will be understood much more from future observations. Though there is no relation to the studies of black holes, here I would like to mention another possibility towards dynamics of clusters of galaxies using the water mega masers in NGC4258. The central massive black hole in NGC4258 is not visible, but from the analysis of the rotating disk around it, we can now understand where the position of massive black hole is, in all likelihood also the position of the center

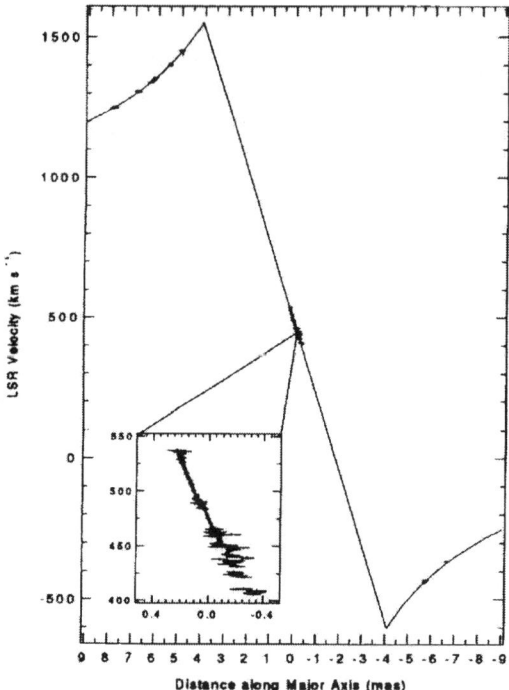

Fig. 5 : Line of sight velocity of maser features versus distance along the major axis (position angle, 86°. Inset, data near the systemic velocity of the galaxy. The position errors only visible on the scale of the inset. The line is derived from the model in Miyoshi et al. (1995). The high-velocity emission regions define Keplerian orbits and constrain the estimate of the enclosed mass. The linear dependence of the systemic emission is a consequence of the change in projection of the rotation velocity (Miyoshi et al. 1995).

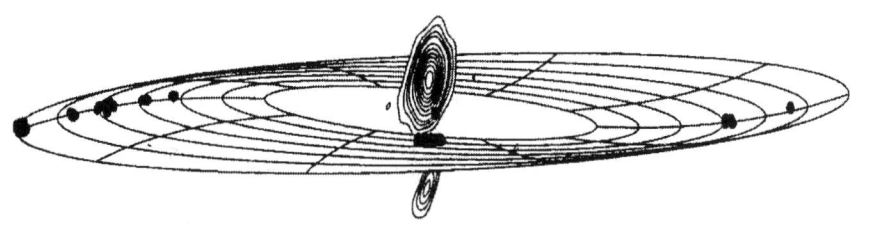

Fig. 6 : The continuum emission in NGC4258 (contours) and the disk model traced from maser analysis (dots in the disk). The dot at center is the dynamical center of the disk which is offset from continuum emission (Herrnstein et al. 1997).

of mass of the galaxy NGC4258, within accuracy of a few tens of micro-arcseconds relative to the position of the water mega masers (Herrnstein 1996). This is very useful in measuring the proper motion of the galaxy. The measurement of proper motions of galaxies, as well as distances of galaxies, will give us important information about formation and evolution of galaxies in the universe. NGC4258 is the best candidate so far and we will be able to measure its extra galactic proper motion in near future. Assuming the transversal velocity of NGC4258 is 500 km/sec, it corresponds to 15 micro-arcsecond/year in proper motion, which is half of the proper motion of systemic maser spots relative to high velocities. In ten years, the displacement of the position amounts to be 0.15 mas , which is sufficiently large to be detected using differential VLBI technique by comparing the fringe phases between those of nearby fixed reference sources like QSOs. In Japan, after the success of VSOP (VLBI Space Observatory Programme) conducted by ISAS, a new VLBI project, named VERA (VLBI exploration of radio astrometry) has been promoted by NAOJ (National Astronomical Observatory Japan). The VERA is the first VLBI network dedicated to differential VLBI observations. Using the pair antenna method for removing the fluctuations in the fringe phase due to the atmospheric turbulence, the VERA measure the positions, proper motions, annual parallaxes of thousands of maser sources in the Galaxy with 10 micro-arcsecond accuracy relative to distant quasars. The first detection of extra galactic proper motion will be accomplished with the VERA. The VERA will give a conclusion to another open question in NGC4258. A pair of X ray sources was found (Pietsch et al. 1994) around the NGC4258. Both of X ray blobs are 9 minutes apart from the center of NGC4258, and the line connecting them passes through the nucleus. In case the X ray blob pair are the ejection matters from the nucleus, the proper motions are about 7 mas/year (Ozernoy, 1996), which is the same amount as that of proper motion of Sgr A^* recently measured (Reid et al. 1997). If the X ray blobs emit strong radio emission and good reference source exists near them, we can measure the motion even today and distinguish whether the blobs are really ejected from the core in NG4258 or not.

10. References

Baan, W.A., Wood, P.A.D. and Haschick, A.D. (1982) *Ap. J.* **260**, L49
Burbidge,E.M. and Burbidge, G. (1997) *Ap. J.* **477**, L13
Cecil, G., Wilson, A. S. and Tully, R. B.,(1992) *Ap. J.*, **390**, 365
Claussen, M.J., Heiligman, G.M. and Lo, K.Y., (1984) *Nature* **310**, 208
Courtes, G. and Cruvellier, P. (1961) *Compt. Rend. Acad. Sci. Paris*, **253**, 218

(left to right) D. Ryu, P.J. Wiita, H.M. Lee, E. Salinas, T. Zannias & A. Burinski

L. Ho

SUPERMASSIVE BLACK HOLES IN GALACTIC NUCLEI

Observational Evidence and Some Astrophysical Consequences

LUIS C. HO
Harvard-Smithsonian Center for Astrophysics
60 Garden St., Cambridge, MA 02138, USA

Abstract.
I review the status of observational determinations of central masses in nearby galactic nuclei. Results from a variety of techniques are summarized, including ground-based and space-based optical spectroscopy, radio VLBI measurements of luminous water vapor masers, and variability monitoring studies of active galactic nuclei. I will also discuss recent X-ray observations that indicate relativistic motions arising from the accretion disks of active nuclei. The existing evidence suggests that supermassive black holes are an integral component of galactic structure, at least in elliptical and bulge-dominated galaxies. The black hole mass appears to be correlated with the mass of the spheroidal component of the host galaxy. This finding may have important implications for many astrophysical issues.

1. Motivation

The discovery of quasars in the early 1960's quickly spurred the idea that these amazingly powerful sources derive their energy from accretion of matter onto a compact, extremely massive object, most likely a supermassive black hole (SMBH; Zel'dovich & Novikov 1964; Salpeter 1964; Lynden-Bell 1969) with $M \approx 10^6 - 10^9$ $M_\odot$. Since then this model has provided a highly useful framework for the study of quasars, or more generally, of the active galactic nucleus (AGN) phenomenon (Rees 1984; Blandford & Rees 1992). Yet, despite its success, there is little empirical basis for believing that this model is correct. As pointed out by Kormendy & Richstone (1995, hereafter KR), our confidence that SMBHs must power AGNs largely rests on the implausibility of alternative explanations. To be sure, a number of characteristics of AGNs indicate that the central engine must be tiny and that relativistic motions are present. These include rapid X-ray variability,

VLBI radio cores, and superluminal motion. However, solid evidence for the existence of SMBHs in the centers of galaxies has, until quite recently, been lacking.

As demonstrated by Soltan (1982), simple considerations of the quasar number counts and standard assumptions about the efficiency of energy generation by accretion allow one to estimate the mean mass density of SMBHs in the universe. The updated analysis of Chokshi & Turner (1992) finds $\rho_\bullet \approx 2 \times 10^5 \epsilon_{0.1}^{-1}$ $M_\odot$ Mpc^{-3} for a radiative efficiency of $\epsilon = 0.1\epsilon_{0.1}$. Comparison of $\rho_\bullet$ with the B-band galaxy luminosity density of $1.4\times 10^8 h$ $L_\odot$ Mpc^{-3} (Lin et al. 1996), where the Hubble constant $H_0 = 100h$ km s^{-1} Mpc^{-1}, implies an average black hole mass per unit stellar luminosity of $\sim 1.4\times 10^{-3}\epsilon_{0.1}^{-1}h^{-1}$ $M_\odot/L_\odot$. A typical bright galaxy with $L_B^* \approx 10^{10}h^{-2}$ $L_\odot$ potentially harbors a SMBH with a mass $\gtrsim 10^7\epsilon_{0.1}^{-1}h^{-3}$ $M_\odot$. These very general arguments lead one to conclude that "dead" quasars ought to be lurking in the centers of many nearby luminous galaxies.

The hunt for SMBHs has been frustrated by two principal limitations. The more obvious of these can be easily appreciated by nothing that the "sphere of influence" of the hole extends to $r_h \simeq GM_\bullet/\sigma^2$ (Peebles 1972; Bahcall & Wolf 1976), where G is the gravitational constant and σ is the velocity dispersion of the stars in the bulge, or, for a distance of D, $\sim 1''(M_\bullet/2 \times 10^8$ $M_\odot)(\sigma/200$ km s$^{-1})^{-2}(D/5$ Mpc). Typical ground-based observations are therefore severely hampered by atmospheric seeing, and only the heftiest dark masses in the closest galaxies can be detected. The situation in the last few years has improved dramatically with the advent of the *Hubble Space Telescope (HST)* and radio VLBI techniques. The more subtle complication involves the actual modeling of the stellar kinematics data, and in this area much progress has also been made recently as well.

Here I will highlight some of the observational efforts during the past two decades in searching for SMBHs, concentrating on the recent advances. Since this contribution is the only one that discusses nuclear BHs aside from that in the Milky Way (Ghez, these proceedings) and in NGC 4258 (Miyoshi, these proceedings), I will attempt to be as comprehensive as possible, although no claim to completeness is made, as this is a vast subject and progress is being made at a dizzying pace. To fill in the gaps, I refer the reader to several other recent review papers, each of which has a slightly different emphasis (KR; Rees 1998; Richstone 1998; Ford et al. 1998; van der Marel 1998).

2. Early Clues from Photometry

The prospect of finding massive BHs in globular clusters motivated much early effort to investigate the distribution of stars resulting from the adiabatic growth of a BH in a preexisting stellar system. The central density deviates strongly from that of an isothermal core and instead follows a cuspy profile $\rho(r) \propto r^{-3/2}$ (Young 1980) or steeper if two-body relaxation (Peebles 1972; Bahcall & Wolf 1976) or different initial density profiles (Quinlan, Hernquist, & Sigurdsson 1995) are taken into account. The discovery that the centers of some giant elliptical galaxies obey this prediction generated much enthusiasm for the existence of SMBHs. In the well-known case of M87 (Young et al. 1978), Lauer et al. (1992) have since shown that the central cusp persists to the limit of the resolution of the HST ($0''.1$).

However, as emphasized by Kormendy (1993; see also KR), photometric signatures alone do not uniquely predict the presence of a SMBH. The cores of most galaxies are now known to be nonisothermal. And moreover, contrary to naïve expectations, galaxy cores with high central surface brightnesses and small core radii, far from being the ones most likely to host SMBHs, are in fact *least* expected to do so. This apparently contradictory statement can be most easily understood by considering the so-called fundamental-plane relations for the spheroidal component of galaxies (Faber et al. 1987; Bender, Burstein & Faber 1992). More luminous, more massive galaxies tend to have more massive central BHs (§ 7), but they also have larger, more diffuse cores. Indeed, high-resolution photometric studies of early-type galaxies (Nieto et al. 1991; Crane et al. 1993; Jaffe et al. 1994; Lauer et al. 1995) find that the central surface brightness profiles either continue to rise toward the center as $I(r) \propto r^{-\gamma}$, with $\gamma \approx 0.5$–1.0 (the "power-law" galaxies) or they flatten at some characteristic radius to a shallower slope of $\gamma \approx 0.0$–0.3 (the "core" galaxies). The power-law galaxies are invariably lower luminosity, lower mass systems compared to those with distinct cores.

In summary, photometric signatures alone cannot be used as reliable indicators for the presence of SMBHs. Instead, we must turn to the more arduous task of obtaining kinematic measurements.

3. Methods Based on Stellar Kinematics

Contrary to the ambiguity of light profiles, the Keplerian rise in the velocity dispersion toward the center, $\sigma(r) \propto r^{-1/2}$, is a robust prediction for a wide variety of dynamical models containing a central massive dark object (MDO; Quinlan et al. 1995). Sargent et al. (1978) noticed that the innermost velocities of M87 were consistent with such a prediction, and, assuming an isotropic velocity distribution, they inferred that the center of this galaxy contained a dark mass of $\sim 5 \times 10^9$ $M_\odot$, presumably in the form of a SMBH. The central rise in $\sigma(r)$, unfortunately, can be insidiously mimiced by an anisotropic velocity distribution, and therefore an MDO is *not* required by the data for this object (Duncan & Wheeler 1980; Binney & Mamon 1982; Richstone & Tremaine 1985; Dressler & Richstone 1990; van der Marel 1994a). This degeneracy presents a serious difficulty for many mass determinations based on stellar kinematic data. An extensive and lucid discussion of this vast subject was presented by KR, and many of the details will not be repeated here. Nonetheless, an abbreviated synopsis is needed to motivate the topic.

Following the notation of KR, the radial variation in mass can be expressed by the first velocity moment of the collisionless Boltzman equation,

$$M(r) = \frac{V^2 r}{G} + \frac{\sigma_r^2 r}{G}\left[-\frac{\mathrm{d}\ln\nu}{\mathrm{d}\ln r} - \frac{\mathrm{d}\ln\sigma_r^2}{\mathrm{d}\ln r} - (1 - \frac{\sigma_\theta^2}{\sigma_r^2}) - (1 - \frac{\sigma_\phi^2}{\sigma_r^2})\right]$$

where V is the rotational velocity, σ_r is the radial and σ_θ and σ_ϕ the azimuthal components of the velocity dispersion, and ν is the density of the tracer population. In practice, several simplifying assumptions are adopted: (1) the mass distribution is spherically symmetric; (2) the mean rotation is circular; and (3) ν is proportional to the luminosity density, or, equivalently, that M/L does not vary with radius.

A brief scrutiny of the above equation indicates that the effects of velocity anisotropy can have a large and complicated effect on the derivation of $M(r)$ because the terms inside the bracket significantly affect the $\sigma_r^2 r/G$ term. If $\sigma_r > \sigma_\theta$ and $\sigma_r > \sigma_\phi$, each of the last two terms will be negative and can be as large as –1. The central brightness distributions of the spheroidal component of most galaxies typically have $-(\mathrm{d}\ln\nu/\mathrm{d}\ln r) \approx$ +1.1 for luminous, nonrotating systems and $\gtrsim$ +2 for low to intermediate-luminosity systems (e.g., Faber et al. 1997). Since $-(\mathrm{d}\ln\sigma_r^2/\mathrm{d}\ln r) \leq$ +1, it is apparent that, under suitable conditions, all four terms can largely cancel one another. As emphasized by KR, all else being equal, smaller, lower luminosity galaxies such as M32 potentially yield more secure mass determinations than massive galaxies like M87 because less luminous systems tend to have (1) steeper central light profiles, (2) a greater degree of rotational support, and (3) less anisotropy.

The principles behind the stellar kinematics analysis are conceptually straightforward but in practice technically challenging. Given the set of observed quantities $I(r)$, $V(r)$, and $\sigma(r)$, the goal is to derive a range intrinsic values for these quantities after accounting for projection and the blurring effects of seeing. Much of the machinery for these tasks has been developed and extensively discussed by Kormendy (1988a, b) and Dressler & Richstone (1988). The sensitivity of the results to the effects of anisotropy are examined through maximum-entropy dynamical models (Richstone & Tremaine 1984, 1988) to see whether conclusions regarding the presence of MDOs can be obviated by a suitable exploration of parameter space.

Perhaps the most serious limitation of these maximum-entropy models is that they do not properly take flattening into account.

The last several years have seen a resurged interest in improving the techniques of analyzing stellar kinematics data. In the context of SMBH searches, Gerhard (1993), van der Marel et al. (1994a, b), Dehnen (1995), among others, have stressed the importance of utilizing the full information contained in the velocity profile or line-of-sight velocity distribution (LOSVD) of the absorption lines, which are normally treated only as Gaussians. A system with significant rotation, for instance, can leave a measurable skewness on the LOSVD, while various degrees of anisotropy would imprint symmetric deviations from a Gaussian line shape. Neglecting these subtleties can lead to systematic errors in the measurement of $V(r)$, but in the cases best studied so far these effects do not seem to have been severe (KR). Furthermore, the line profile should develop weak, high-velocity wings if a SMBH is present (van der Marel 1994b), although the currently available data do not yet have the requisite quality to exploit this tool.

Yet another advance has focused on the development of dynamical models with two-integral phase-space distribution functions, $f(E, L_z)$, E being the total energy and L_z the angular momentum in the symmetry axis (van der Marel et al. 1994b; Qian et al. 1995; Dehnen 1995). Such models are properly flattened, and they generate predictions for the LOSVDs; on the other hand, it is not clear whether imposing a special dynamical structure is too restrictive. This limitation will be eliminated by fully general, axisymmetric three-integral models (van der Marel et al. 1998; Cretton et al. 1998; Gebhardt et al. 1998).

There are currently 10 galaxies with published MDO measurements determined from stellar kinematical data (Table 1). Of these, only three (M81: Bower et al. 1996; NGC 3379: Gebhardt et al. 1998; NGC 4342: Cretton & van den Bosch 1998) come solely from *HST* data; the remaining ones, although many by now confirmed with *HST*, were initially discovered from high-quality ground-based observations (see KR for a detailed account of each object). Kormendy and collaborators, in particular, making use of the excellent seeing conditions and instrumentation on the CFHT, continue to make progress in this area. Two new MDOs have been reported recently based on CFHT data: the low-luminosity elliptical galaxy NGC 4486B has $M_{\rm MDO} = 6\times10^8$ $M_\odot$ (Kormendy et al. 1997b), and NGC 3377, another close cousin, has $M_{\rm MDO} = 2.3\times10^8$ $M_\odot$ (Kormendy et al. 1998). This demonstrates the important fact that even in the *HST* era ground-based observations continue to play an important role in SMBH searches.

The new observations with *HST*, thus far all acquired using the Faint Object Spectrograph (FOS), provide an important contribution by increasing the angular resolution by about a factor of 5 compared to the best

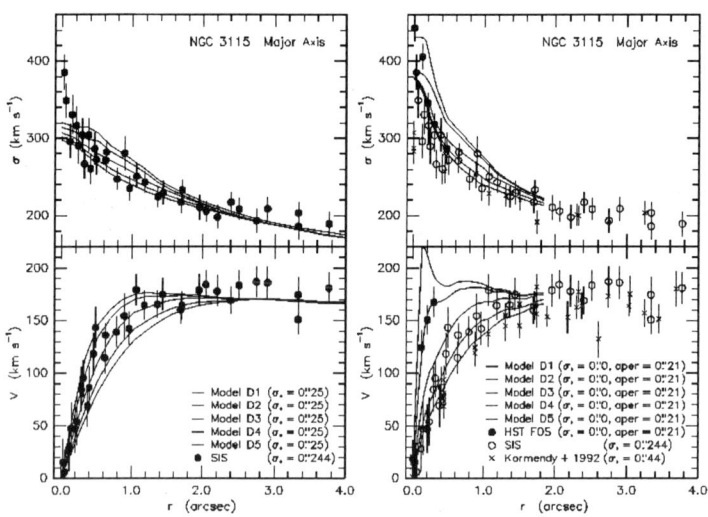

Figure 1. Stellar kinematic data for NGC 3115 compared with various dynamical models (see Kormendy et al. 1996). The *left* panel shows the best ground-based data, and the *right* panel the same data with new *HST* data superposed for comparison. Both V and σ rise much more steeply with radius in the new *HST* data.

ground-based data available. In all cases studied (NGC 3115: Kormendy et al. 1996; NGC 4594: Kormendy et al. 1997a; M32: van der Marel et al. 1997; M31: Ford et al. 1998), the velocity dispersions continue to rise toward smaller r and the maximum rotational velocity has generally increased (Fig. 1). In the case of NGC 3115, the FOS spectra are of sufficient quality to reveal wings in the LOSVD that extend up to $\sim$1200 km s^{-1} (Kormendy et al. 1996). The *HST* data thus considerably bolster the case for a MDO in these objects. The improvement in angular resolution additionally strengthens our confidence that the MDOs might indeed be SMBHs. A reduction of the size scale by a factor of 5 increases the central density by more than two orders of magnitude. Although in general this is still not enough to rule out alternative explanations for the dark mass (§ 6), it is clearly a step in the right direction.

I conclude this section with a few remarks on the dark mass in the Galactic Center (see Ghez, in these proceedings for more details), which, in my view, is now the most compelling case of a SMBH in any galactic nucleus. From analysis of an extensive set of near-IR radial velocities of individual stars, coupled with additional measurements from the literature, Genzel et al. (1996; see also Krabbe et al. 1995) found a highly statistically significant rise in the radial velocity dispersion between 5 and 0.1 pc from the dynamical center. Assuming an isotropic velocity distribution,

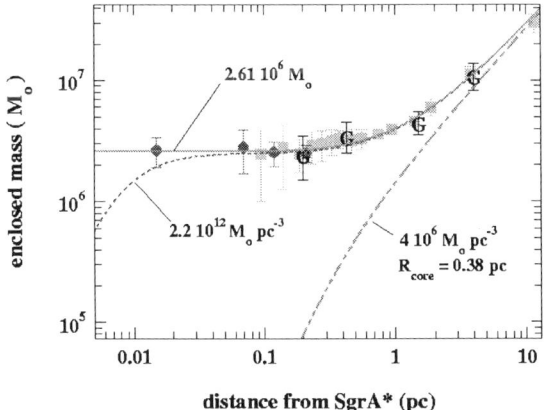

Figure 2. Enclosed mass versus radius for the Galactic Center derived from stellar radial velocities and proper motions (from Genzel et al. 1997). The points labeled with "G" come from gas kinematics. The *thick dashed* curve is a model for the stellar cluster with a total mass density of 4×10^6 $M_\odot$ pc^{-3}; the *solid* curve denotes the sum of this cluster and a point mass of 2.61×10^6 $M_\odot$; and *thin dashed* curve is the sum of the stellar cluster and an additional dark cluster with a central density of 2.2×10^{12} $M_\odot$ pc^{-3}

the observations require a dark mass of $\sim 3\times 10^6$ $M_\odot$ within $r = 0.1$ pc and $M/L_K \geq 100$; the dark mass must have a density in excess of 10^9 $M_\odot$ pc^{-3}, which argues strongly for it being a SMBH. These conclusions, and a suspicion nearly three decades old (Lynden-Bell & Rees 1971), have finally been vindicated by recent measurements of stellar proper motions within the central 1 pc region using high-resolution K-band astrometric maps (Eckart & Genzel 1996, 1997; Genzel et al. 1997; Ghez et al. 1998). The main results are the following: (1) the stellar radial velocities agree with the proper motions, which implies that on average the velocities are close to isotropic; (2) the combined velocities imply a dark mass (Fig. 2) within 0.006 pc of 2.61×10^6 $M_\odot$ (Genzel et al. 1997 quote a statistical error of ± 0.15 and a combined statistical and systematic error of ± 0.35); (3) the density, therefore, has an astonishingly high value of $>2\times 10^{12}$ $M_\odot$ pc^{-3}, which leaves almost no room to escape the conclusion that the dark mass must be in the form of a SMBH (§ 6). The presence of a large mass is also supported by the detection of several stars, within 0.01 pc from the central radio source Sgr A*, moving at speeds in excess of 1000 km s^{-1}. From the velocities of the fast-moving stars and the near stationarity of Sgr A*, Genzel et al. further use equipartition arguments to constrain the mass of the radio core itself ($\geq 10^5$ $M_\odot$), which, when combined with the exceedingly small upper limit for its size ($r < 4\times 10^{-6}$ pc), would imply a density of $>3\times10^{20}$ $M_\odot$ pc^{-3}.

4. Methods Based on Gas Kinematics

Unlike the situation for stars, gas kinematics are much easier to interpret if the gas participates in Keplerian rotation in a disklike configuration. But there are two caveats to remember. First, gas can be easily perturbed by nongravitational forces (shocks, radiation pressure, winds, magnetic fields, etc.). Indeed, in the case of the Galactic Center, it was precisely this reason that its central mass, which had been estimated for some time using gas velocities (Lacy et al. 1980), could not be accepted with full confidence prior to the measurement of the stellar kinematics. Second, there is no *a priori* reason that the gas should be in dynamical equilibrium, and therefore one must verify empirically that the velocity field indeed is Keplerian. The optically-emitting ionized gas in the central regions of some spirals show significant noncircular motions (e.g., Fillmore, Boroson, & Dressler 1986). NGC 4594 is a striking example. Kormendy et al. (1997a) showed that the emission-line rotation curve near the center falls substantially below the circular velocities of the stars, and hence the gas kinematics cannot be used to determine the central mass.

4.1. OPTICAL EMISSION LINES

The sharpened resolution of the refurbished *HST* has revealed many examples of nuclear disks of dust and ionized gas (Fig. 3). The nuclear disks typically have diameters ∼100–300 pc, with the minor axis often aligned along the direction of the radio jet, if present. Some examples include the elliptical galaxies NGC 4261 (Jaffe et al. 1993), M87 (Ford et al. 1994), NGC 5322 (Carollo et al. 1997), and NGC 315 (Ho et al. 1998), and the early-type spiral M81 (Devereux, Ford, & Jacoby 1997). I will highlight here only three cases; Table 1 gives a complete list of objects and references.

The first object for which the nuclear gas disk was used to determine the central mass was M87. Harms et al. (1994) used the FOS to obtain spectra of several positions of the disk and measured a velocity difference of ± 1000 km s^{-1} at a radius of $0\farcs25$ (18 pc) on either side of the nucleus. Adopting an inclination angle of $42°$ determined photometrically by Ford et al. (1994), the velocities were consistent with Keplerian motions about a central mass of $(2.4\pm0.7)\times10^9$ $M_\odot$. Since the implied $M/L_V \approx 500$, Harms et al. concluded that the central mass is dark, most likely in the form of a SMBH. The case for a SMBH in M87 has been considerably strengthened through the recent reobservation with *HST* by Macchetto et al. (1997), who used the long-slit mode of the Faint Object Camera to obtain higher quality spectra extending to $r = 0\farcs05$ (3.5 pc). The velocities

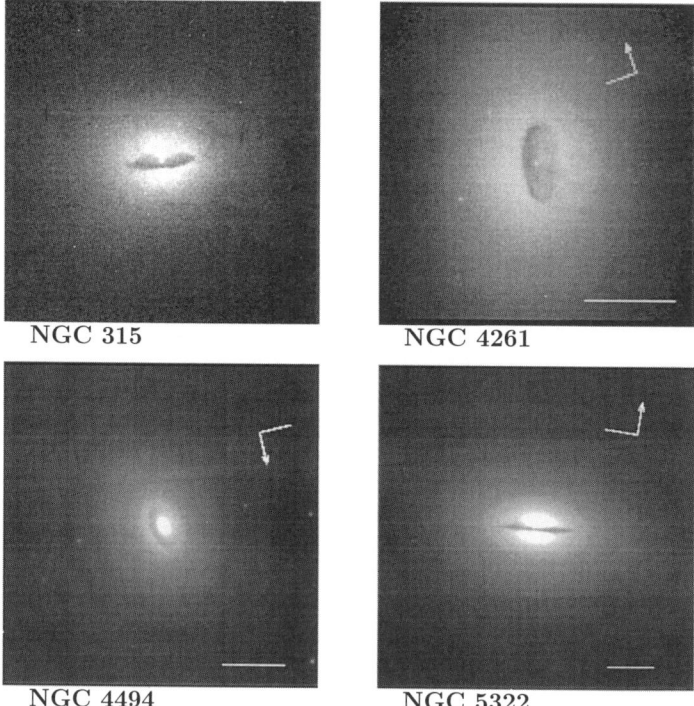

Figure 3. Nuclear disks from *HST* optical images. Each image is 35″ on a side.

in the inner few tenths of an arcsecond are well fitted by a model of a thin disk in Keplerian rotation (Fig. 4), although the inclination angle is not well constrained (47°-65°). The rotation curve at larger radii falls below the Keplerian curve, possibly indicating a warp in the disk (Macchetto et al. 1997) or substantial perturbations due to spiral shocks (Chakrabarti 1995). For $i = 52°$, $M_{\mathrm{MDO}} = (3.2\pm0.9)\times10^9$ $M_\odot$, and $M/L_V \gtrsim 110$. If, instead, a Plummer potential is assumed, the distributed dark mass can have a core radius no larger than ~ 5 pc. So, in either case, a density $\sim 10^7$ $M_\odot$ pc^{-3} is implied.

The mildly active nucleus of NGC 4261 contains a rotating disk of dust and ionized gas as well (Ferrarese, Ford, & Jaffe 1996); like M87, the disk is slightly warped and shows traces of weak spiral structure. Although the FOS data for this object are rather noisy, they indicate that the gas largely undergoes circular motions. The mass interior to $r = 15$ pc is $M_{\mathrm{MDO}} = (4.9\pm1.0)\times10^8$ $M_\odot$, and M/L_V has an exceptionally high value of 2×10^3.

The installation of the imaging spectrograph STIS in 1997 at long last gives *HST* an efficient means to obtain spatially resolved spectra of the central regions of galaxies. Much progress in the field is anticipated in the

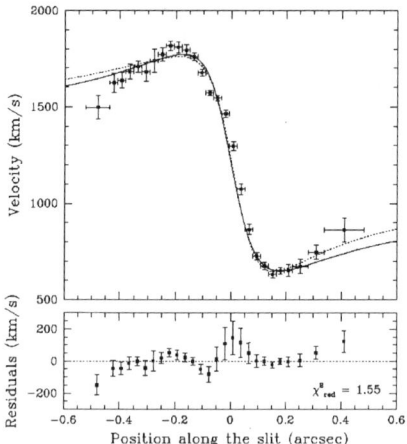

Figure 4. Optical emission-line rotation curve for the nuclear disk in M87. The two curves in the upper panel correspond to Keplerian thin disk models, and the bottom panel shows the residuals for one of the models (see Macchetto et al. 1997).

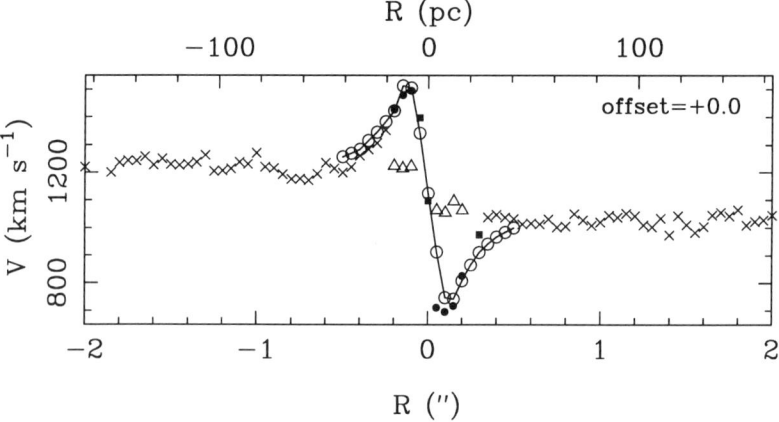

Figure 5. Optical emission-line rotation curve for the nuclear disk in M84 obtained with STIS on *HST* (Bower et al. 1998). The kinematics can be well fitted with a thin Keplerian disk model, which is plotted as open circles connected by the solid line.

near future. A taste of what might be expected can be seen in the early-release observations of M84 by Bower et al. (1998; Fig. 5). M84 is almost a twin of M87 in terms of luminosity, and its central dark mass (1.5×10^9 $M_\odot$), too, is similar.

Lastly, I mention an interesting, unconventional case. The radio galaxy Arp 102B belongs to a minority of AGNs that display so-called double-peaked broad emission lines. Several ideas have been proposed for the peculiar line profiles in this class of objects, but the favored explanation is that the lines originate from a relativistic accretion disk (Eracleous et al. 1997). During the course of a long-term optical monitoring of Arp 102B, the intensity ratio of the two peaks of the Hα line displayed sinusoidal variations with a period of 2.2 years for several years (Newman et al. 1997). The periodic signal was interpreted as arising from a "hot spot" in the accretion disk. By modeling the line profile from the epochs when the hot spot was quiescent, one can estimate the radius and inclination angle of the spot's orbit, and, combined with its period, the enclosed mass. The mass within $r = 0.005$ pc turns out to be 2.2×10^8 M$_\odot$, consistent with a moderately luminous ($M_B \approx -20$ mag) elliptical (see § 7).

4.2. RADIO SPECTROSCOPY OF WATER MASERS

Luminous 22-GHz emission from extragalactic water masers are preferentially detected in galaxies with active nuclei, where physical conditions, possibly realized in a circumnuclear disk (Claussen & Lo 1986), evidently favor this form of maser emission. With the detection in NGC 4258 of high-velocity features offset from the systemic velocity by $\sim \pm 900$ km s^{-1} (Nakai, Inoue, & Miyoshi 1993), Watson & Wallin (1994) already surmised that the maser spectrum of this Seyfert galaxy can be interpreted as arising from a thin Keplerian disk rapidly rotating around a mass of $\sim 10^7$ M$_\odot$. But the solid proof of this picture came from the high-resolution ($\Delta\theta = 0\farcs0006 \times 0\farcs0003$; $\Delta v = 0.2$ km s^{-1}) VLBA observations of Miyoshi et al. (1995) who demonstrated that the maser spots trace a thin (<0.003 pc), nearly edge-on annulus with an inner radius of 0.13 pc and an outer radius of 0.26 pc. The systemic features lie on the near side of the disk along the line-of-sight to the center (Fig. 6); the high-velocity features delineate the edges of the disk on either side and follow a Keplerian rotation curve to very high accuracy ($\lesssim 1\%$). The implied binding mass within 0.13 pc is 3.6×10^7 M$_\odot$, which corresponds to a density of $>4 \times 10^9$ M$_\odot$ pc^{-3}. In fact, one can place a tighter constraint on the density. The maximum deviation of the velocities from a Keplerian rotation curve limits the extent of the central mass to $r \lesssim 0.012$ pc (Maoz 1995), from which follows that the density must be $>5 \times 10^{12}$ M$_\odot$ pc^{-3}.

Two, possibly three, additional AGNs have H$_2$O megamasers suitable for tracing the central potential. The spectrum of the maser source in the Seyfert nucleus of NGC 1068 also exhibits satellite features (± 300 km s^{-1}) offset from the systemic velocity (Greenhill et al. 1996). The redshifted and blueshifted emission again lie on a roughly linear, 2-parsec feature passing through the systemic emission (Greenhill 1998). The rotation curve in this instance is sub-Keplerian, possibly because the disk has nonnegligible mass,

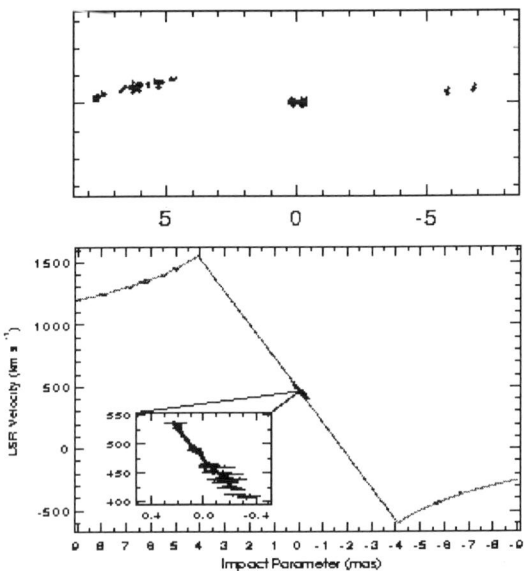

Figure 6. Water maser emission in NGC 4258 (Miyoshi et al. 1995). *Top:* spatial distribution of the maser features; *bottom:* rotation curve. Adapted from Greenhill (1997).

and the derived mass (1.7×10^7 $M_\odot$ within a radius of 0.65 pc) is less certain.

The maser in the nucleus of NGC 4945 shows a position and velocity distribution reminiscent of NGC 4258 as well: high-velocity features symmetrically straddle the systemic emission. Greenhill, Moran, & Herrnstein (1997) interpret the data, which in this case is considerably less accurate because of its location in the southern sky, in terms of an edge-on disk model and derive a central mass of 1.4×10^6 $M_\odot$ within $r = 0.3$ pc. This result is quite surprising because, as an Scd spiral, NGC 4945 is expected to be essentially bulgeless. If the dark mass in its center is truly in the form of a SMBH, then SMBHs evidently can form without a bulge.

The H_2O megamaser source in NGC 3079 is potentially useful for mass determination. Here, however, the complex spatial distribution of the emission regions and the large intrinsic widths of the lines complicate the analysis, and the interpretation of the data may not be unique. Trotter et al. (1998) tentatively assign a central mass of 1×10^6 $M_\odot$ to this galaxy.

4.3. DETERMINING CENTRAL MASSES OF ACTIVE GALACTIC NUCLEI

I mention one other method for determining masses in the central regions of galaxies, specifically in AGNs. Although AGNs largely provide the motivation for searching for SMBHs, ironically it is precisely in these objects that conventional techniques used to measure masses fail. The bright continuum emission of the active nucleus nearly always completely overpowers the stellar absorption lines near the center, and in many cases the narrow emission lines are significantly affected by nongravitational forces.

An approach taken in the past attempts to utilize the broad [$(1\text{–few})\times 10^3$ km s^{-1}] emission lines that are thought to arise from the so-called broad-line region (BLR), a tiny, dense region much less than a parsec from the central source. Assuming that the line widths trace gravity, the mass follows from $\eta v^2 r_{\rm BLR}/G$, where $\eta \approx 1\text{–}3$ depending on the kinematic model adopted. The BLR radius has traditionally been estimated from photoionization arguments (e.g., Dibai 1981; Wandel & Yahil 1985; Wandel & Mushotzky 1986; Padovani, Burg, & Edelson 1990), but recent variability studies indicate that the BLR is much more compact than previously thought (Netzer & Peterson 1997).

The continuum output from AGNs typically varies on timescales ranging from days to months in the UV and optical bands. Because the emission lines are predominantly photoionized by the central continuum, they vary in response to the changes in the continuum, but with a time delay (lag) that corresponds to the light-travel distance between the continuum source and the line-emitting gas. "Reverberation mapping" (Blandford & McKee 1982), therefore, in principle allows one to estimate the luminosity-weighted radius of the BLR, although in practice the complex geometry and ionization structure of the BLR complicate the interpretation of the "sizes" derived by this method (see Netzer & Peterson 1997 for a recent review).

If the widths of the broad emission lines reflect bound gravitational motions, as seems to be the case in most well-studied objects (Netzer & Peterson 1997; but see Krolik 1997), then, adopting a reasonable kinematic model (e.g., randomly moving clouds), the virial mass can be estimated from $v^2 r_{\rm BLR}/G$. If, instead, the clouds are infalling, as has been claimed in some cases, the mass will be smaller by a factor of 2. One of the uncertainties in the application of this simple formalism lies in the choice of v. What is appropriate? One reasonable choice might be $v = (\sqrt{3}/2)\text{FWHM}$, the full width at half-maximum of a representative broad line. Yet another ambiguity is which line to use, since not all broad emission lines have the same widths. Ultraviolet or high-ionization lines, for instance, generally have broader profiles than optical or low-ionization lines. For the purposes of this exercise, I simply chose the line for which the most data exist (Hβ) in order to obtain as large a sample as possible.

TABLE 1
MASSES DETERMINED FROM STELLAR AND GAS KINEMATICS

Galaxy (1)	Hubble Type (2)	D (Mpc) (3)	B_T^0 (mag) (4)	B/T (5)	Ref	M_B(bul) (mag) (6)	$M_{\rm MDO}$ ($M_\odot$) (7)	Radius (pc) (8)	Density ($M_\odot {\rm pc}^{-3}$) (9)	$(M/L_V)_\odot$ (10)	Method (11)	Ref
Milky Way	Sbc	0.008	—	—	—	−17.51	2.6×10^6	0.006	2.2×10^{12}	>25–500	S	5
M31	Sb	0.72	3.36	0.24	1	−19.38	7.5×10^7	0.30	6.6×10^8	>100	S	6
M32	E2	0.72	8.72	1.0	—	−15.57	3.4×10^6	0.30	3.0×10^7	>20	S	7
M33	Scd	0.795	5.75	—	2	−10.21	$<5 \times 10^4$	<0.39	—	—	S	2
M81	Sab	3.6	7.39	0.25	1	−18.88	4×10^6	2.6	5.4×10^4	13	G	8
M84	E1	16.8	10.01	1.0	—	−21.12	1.5×10^9	8.1	6.7×10^5	350	G	9
M87	E0	16.8	9.46	1.0	—	−21.67	3×10^9	3.5	1.7×10^7	>110	G	10
NGC 205	dE5	0.72	8.79	1.0	—	−15.49	$<9 \times 10^4$	0.35	—	—	S	11
NGC 1068	Sb	14.4	9.47	0.23	4	−19.73	1.7×10^7	0.65	1.5×10^7	—	G	12
NGC 3079	SBc	20.4	10.41	0.10	4	−18.59	1.3×10^6	0.64	1.2×10^6	—	G	13
NGC 3115	S0-	6.7	9.74	0.94	1	−19.39	2×10^9	1.7	9.7×10^7	50	S	14
NGC 3377	E5	8.1	11.07	1.0	—	−18.47	2.3×10^8	11.4	3.7×10^4	>10	S	15
NGC 3379	E1	8.1	10.18	1.0	—	−19.36	6×10^7	3.9	2.4×10^5	—	S	16
NGC 4258	SABbc	6.8	8.53	0.16	4	−18.64	3.6×10^7	0.012^a	4.9×10^{12}	—	G	17
NGC 4261	E2	35.1	11.36	1.0	—	−21.37	4.9×10^8	16.9	2.4×10^4	2000	G	18
NGC 4342	S0-	16.8	13.37	1.0	—	−17.76	3.0×10^8	20.3	8.6×10^3	—	S	19
NGC 4395	Sm	2.6	10.57	—	3	−7.27	$<8 \times 10^4$	0.35	—	—	S	3
NGC 4486B	E0	16.8	14.26	1.0	—	−16.87	6×10^8	12.2	7.9×10^4	20	S	20
NGC 4594	Sa	9.2	8.38	0.93	1	−21.36	1×10^9	4.4	2.8×10^6	>50	S	21
NGC 4945	SBcd	5.2	7.43	0.05	4	−17.89	1.4×10^6	0.3	1.2×10^7	—	G	22
NGC 5793	Sb:	46.5	13.19	0.23	4	−18.55	1.5×10^8	4.0	5.6×10^5	—	G	27
NGC 6251	E0	92.0	13.22	1.0	—	−21.59	7.5×10^8	44.4	2.0×10^3	—	G	23
NGC 7052	E4	55.5	12.73	1.0	—	−20.99	3.3×10^8	36.1	1.7×10^3	—	G	24
Circinus	Sb	4.0	8.32	—	—	-14.51^b	$<4 \times 10^6$	11.0	—	—	S	25
Arp 102B	E0	96.6	14.94	1.0	—	−19.98	2.2×10^8	0.0048	4.7×10^{14}	—	V	26

REFERENCES.— (1) Kormendy & Richstone 1995; (2) Kormendy & McClure 1993; (3) Filippenko & Ho 1998; (4) Simien & de Vaucouleurs 1986; (5) Genzel et al. 1997; (6) Ford et al. 1998; (7) van der Marel et al. 1997; (8) Bower et al. 1996; (9) Bower et al. 1998; (10) Macchetto et al. 1997; (11) Jones et al. 1996; (12) Greenhill 1998; (13) Trotter et al. 1998; (14) Kormendy et al. 1996; (15) Kormendy et al. 1998; (16) Gebhart et al. 1998; (17) Miyoshi et al. 1995; (18) Ferrarese, Ford, & Jaffe 1996; (19) Cretton & van den Bosch 1998; (20) Kormendy et al. 1997b; (21) Kormendy et al. 1997a; (22) Greenhill, Moran, & Herrnstein 1997; (23) Ferrarese, Ford, & Jaffe 1998; (24) van der Marel & van den Bosch 1998; (25) Maiolino et al. 1998; (26) Newman et al. 1997; (27) Hagiwara et al. 1997.

NOTE.— Cols.: (1) Galaxy name; (2) Hubble type from de Vaucouleurs et al. 1991; (3) distance from Tully 1988 or derived by assuming $H_0 = 75$ km s^{-1} Mpc^{-1}; (4) total apparent blue magnitude corrected for Galactic and internal extinction, from de Vaucouleurs et al. 1991; (5) ratio of bulge to total luminosity based on the photometric studies given in the adjacent column; (6) absolute blue magnitude of the bulge component; (7) mass of the MDO, taken from the study given in the last column; (8) radius containing $M_{\rm MDO}$; (9) density derived from radius and $M_{\rm MDO}$; (10) mass-to-light ratio (in the V band) of the MDO; (11) method used: "G" = gas kinematics; "S" = stellar kinematics; "V" = variability.

^{a}Upper limit on size based on arguments by Maoz 1995.

^{b}Bulge luminosity estimated assuming $M/L_B = 6$, where $M = 5\sigma_0^2 R_e/G$ is the mass of the bulge, σ_0 is the central velocity dispersion, and R_e is the effective radius. From Maiolino et al. (1998), $\sigma_0 \approx 80$ km s^{-1} and $R_e \approx 80$ pc.

TABLE 2
MASSES DETERMINED FROM REVERBERATION MAPPING[a]

Galaxy (1)	Hubble Type (2)	D (Mpc) (3)	B_T^0 (mag) (4)	B/T (5)	M_B(bul) (mag) (6)	r_{BLR} (ltd) (7)	FWHM(Hβ) (km s^{-1}) (8)	M_{MDO} ($M_\odot$) (9)
3C 120	S0:	132	13.63	0.24	−20.42	44	2300	3.4×10^7
Ark 120	S0/a	121	13.64	0.86	−21.61	39	5450	1.7×10^8
Fairall 9	S?	188	13.50	0.08	−20.13	23	4200	5.9×10^7
Mrk 79	SBb	89	13.32	0.20	−20.12	18	6200	1.0×10^8
Mrk 110	Pair?	141	16.00	—	—	20	2500	1.8×10^7
Mrk 279	S0	118	14.43	0.06	−17.87	10	5360	4.2×10^7
Mrk 335	S0/a	103	13.85	0.64	−20.73	17	1800	8.0×10^6
Mrk 509	comp	138	13.00	0.12	−20.39	80	2800	9.0×10^7
Mrk 590	Sa:	105	13.66	0.47	−20.63	21	2300	1.6×10^7
Mrk 817	S?	126	14.50	0.50	−20.88	16	4100	3.8×10^7
NGC 3227	SABa	20.6	11.18	0.52	−19.68	17	3900	3.8×10^7
NGC 3516	SB0:	38.9	12.14	0.61	−20.27	7	4760	2.3×10^7
NGC 3783	SBa	38.5	12.04	0.33	−19.68	8	2980	1.0×10^7
NGC 4151	SABab	20.3	10.71	0.36	−19.72	5	4670	1.6×10^7
NGC 4593	SBb	39.5	11.43	0.48	−20.76	4	3720	8.1×10^6
NGC 5548	S0/a	67.0	12.81	0.47	−20.50	19	5610	8.8×10^7
NGC 7469	SABa	65.2	12.64	0.40	−20.44	5	3388	9.1×10^6

NOTE.— Cols.: (1) Galaxy name; (2) Hubble type from de Vaucouleurs et al. 1991; (3) distance from Tully 1988 or derived by assuming $H_0 = 75$ km s^{-1} Mpc^{-1}; (4) total apparent blue magnitude corrected for Galactic and internal extinction, from de Vaucouleurs et al. 1991; (5) ratio of bulge to total luminosity (host galaxy plus AGN); Mrk 110 is too disturbed to yield a reliable photometric decomposition; (6) absolute blue magnitude of the bulge component; (7) lag (in light days) between the continuum and the Hβ light curves; (8) FWHM of the broad Hβ emission line; (9) virial mass derived from r_{BLR} and FWHM(Hβ).

[a] All Seyfert 1 nuclei studied to date with reverberation mapping of the Hβ line. Optical monitoring data exist for 3C 390.3, but it was not included because the profiles of its Balmer lines are very complicated. References for the data entries have been omitted for brevity.

Table 2 lists the derived masses for the 17 Seyfert 1 galaxies that have been monitored extensively in the optical; eight of the objects appear in the compilation of Peterson et al. (1998). Since the masses of MDOs derived from gas and stellar kinematics show a loose correlation with the bulge or spheroidal luminosity of the host galaxies (§ 7; Fig. 8a), we can ask whether those derived from reverberation mapping follow such a correlation. I have estimated the B-band luminosities of the bulges of the Seyferts based on published surface photometry of the host galaxies (taking care to exclude the contribution of the AGN itself, which often can be significant). Figure 8b indicates that, at a fixed bulge luminosity, the masses from reverberation mapping are *systematically lower* than the masses obtained using conventional techniques, on average by about a factor of 5. It is encouraging that this admittedly crude method of mass estimation is not *too* far off the mark. Notably, the scatter of M_{MDO} at a fixed luminosity is quite comparable in the two samples, and the constant offset suggests that one of the underlying assumptions in the mass estimate is incorrect. Since the line width affects the mass quadratically, it is conceivable that some measure of the line profile other than the FWHM is more appropriate.

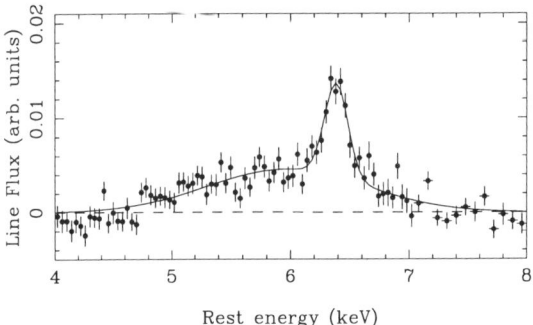

Figure 7. The Fe Kα line in the composite spectrum of Seyfert 1 nuclei (Nandra et al. 1997). The solid line is a fit to the line profile using two Gaussians, a narrow component centered at 6.4 keV and a much broader, redshifted component.

5. Indirect, but Tantalizing Evidence

Lastly, one additional piece of evidence, although it does not give a direct measure of the central mass, cannot go unmentioned — namely the recent detection in AGNs of the broad iron Kα line at 6.4 keV. This line has been known for some time to be a common feature in the hard X-ray spectra of AGNs, and it is thought to arise from fluorescence of the X-ray continuum off of cold material, presumably associated with the accretion disk around the SMBH (e.g., Pounds et al. 1990). The spectral resolution of the existing data, however, was insufficient to test the predicted line profile (Fabian et al. 1989). The *ASCA* satellite provided the much-awaited tell-tale signs in the deep exposure of the Seyfert 1 galaxy MCG–6-30-15 (Tanaka et al. 1995): the Fe Kα line exhibits Doppler motions that approach relativistic speeds ($\sim$100,000 km s^{-1} or $0.3c$!) as well as an asymmetric red wing consistent with gravitational redshift. The best-fitting disk has an inner radius of 6 Schwarzschild radii. The relativistic Fe Kα line, now seen in a large number of sources (Nandra et al. 1997; Fig. 7), provides arguably the most compelling evidence to date for the existence of SMBHs. Other mechanisms for generating the line profile are possible, but implausible (Fabian et al. 1995). Detailed modeling of the line asymmetry has even the potential to measure the spin of the hole, but this is still very much a goal of the future given the current data quality and uncertainties in the modeling itself (e.g., Reynolds & Begelman 1997; Rybicki & Bromley 1998).

6. Are the Massive Dark Objects *Really* Black Holes?

Thus far we have rigorously shown only that many galaxies contain central MDOs, not that the dark masses must be in the form of SMBHs.

Direct proof of the existence of SMBHs would require the detection of relativistic motions emanating from the vicinity of the Schwarzschild radius, $R_S = 2GM_\bullet/c^2 \approx 10^{-5}(M_\bullet/10^8 \, M_\odot)$ pc. Even for our neighbor M31, R_S subtends 3×10^{-6} arcseconds, and the Galactic Center only a factor of 2 larger. We are clearly still far from being able to achieve the requisite angular resolution and in the meantime must rely on indirect arguments.

One approach seeks to identify some observational feature that might be taken as a fingerprint of the event horizon or of physical processes uniquely associated with the environment of a BH. One such "signature" might be the broad Fe Kα line discussed in § 5; another is the high-energy power-law tail observed in some AGNs and Galactic BH candidates (Titarchuk & Zannias 1998). And yet a third possibility is the advection of matter into the event horizon (Menou, Quataert, & Narayan 1998).

A different strategy appeals to the dynamical stability of the probable alternative sources of the dark mass (Goodman & Lee 1989; Richstone, Bower, & Dressler 1990; van der Marel et al. 1997; Maoz 1998). The absence of strong radial gradients in the stellar population, as measured by variations in color or spectral indices, implies that the large increase in M/L toward the center cannot be attributed to a cluster of ordinary stars. On the other hand, the underluminous mass could, in principle, be a cluster of stellar remnants (white dwarfs, neutron stars, and stellar-size BHs) or perhaps even substellar objects (planets and brown dwarfs). To rule out these possibilities, however exotic they might seem, one must show that the clusters cannot have survived over the age of the galaxy, and hence finding them would be highly improbable.

As most recently discussed by Maoz (1998), the two main processes that determine the lifetime of a star cluster are evaporation, whereby stars escape the cluster as a result of multiple weak gravitational scatterings, and physical collisions among the stars themselves. Exactly which dominates depends on the composition and size of the cluster, and its maximum possible lifetime can be computed for any given mass and density. Maoz (1998) shows that in two galaxies, namely the Milky Way and NGC 4258, the density of the dark mass is so high ($\gtrsim 10^{12} \, M_\odot \, \mathrm{pc}^{-3}$) that it cannot possibly be in the form of a stable cluster of stellar or substellar remnants: their maximum ages [$\sim(1$–few$)\times 10^8$ yr] are much less than the ages of the galaxies. The only remaining constituents allowed appear to be subsolar-mass BHs and elementary particles. This constitutes very strong evidence that the MDOs — at least in two cases — are most likely SMBHs. In the following discussion, I will adopt the simplifying viewpoint that all MDOs are SMBHs, bearing in mind that at the current resolution limit we cannot yet disprove the dark-cluster hypothesis for the majority of the objects.

7. The Black-Hole Mass/Bulge Mass Relation

Does $M_\bullet$ depend at all on other properties of the host galaxies? A much-discussed possibility is that $M_\bullet$ scales with the mass of the spheroidal component of the host (Kormendy 1993; KR; Faber et al. 1997; Magorrian et al. 1998; Richstone 1998; Ford et al. 1998; van der Marel 1998). The significance of the scatter in the correlation, or whether any correlation exists at all, is not yet certain. It is somewhat disconcerting that different authors plotting the same objects do not always arrive at the same conclusion. The discrepancies can often be traced to different assumptions about distances, source of bulge-to-disk decomposition, and even apparent magnitudes adopted for the host galaxies (e.g., extinction is not always corrected). The set of host galaxy parameters I adopt is compiled in Table 1.

Figure 8a illustrates that there indeed appears to be a trend of $M_\bullet$ increasing with bulge mass (luminosity). It is encouraging to note that the central masses derived from gas and stellar kinematics do not show any obvious systematic offsets relative to one another. No obvious differentiation by Hubble type is evident either. As has been noted by others, the scatter of $M_\bullet$ at a given luminosity is considerable, at least a factor of 10, perhaps up to 100. The scatter may have been exacerbated slightly by four possibly anomalous points. NGC 4486B is a companion to M87, and it appears to have been tidally truncated; its original luminosity was probably higher. On the other hand, the bulge luminosity of NGC 4945 could very well have been overestimated. Its bulge-to-disk ratio was found using the relation of Simien & de Vaucouleurs (1986), which may be inappropriate for a galaxy of such late Hubble type (Scd). Finally, the masses of M81 and NGC 3079 are quite uncertain and probably have been underestimated.

The trend is much more significant when five upper limits are included. NGC 205, a dwarf elliptical companion of M31, contains a blue, compact nucleus with characteristics resembling an intermediate-age globular cluster. Its core radius, determined from *HST* photometry, combined with a ground-based measurement of its velocity dispersion yields an upper limit of 9×10^4 $M_\odot$ for any dark mass (Jones et al. 1996). The bulgeless, late-type (Scd) spiral M33 also has a stringent upper limit on its central mass. Its nuclear cluster is extremely tiny (core radius $\lesssim 0.39$ pc), and its central velocity dispersion is 21 km s^{-1}; Kormendy & McClure (1993) put an upper limit of $M_\bullet \leq 5 \times 10^4$ $M_\odot$. NGC 4395 in some ways resembles M33, but it is even more extreme. The nucleus is optically classified as a type 1.8 Seyfert (broad Hα and Hβ present), emits a largely nonstellar featureless continuum that extends into the UV (Filippenko, Ho, & Sargent 1993), and displays variable soft X-ray emission and a compact flat-spectrum radio core (Moran et al. 1998). These properties alone would be unremarkable

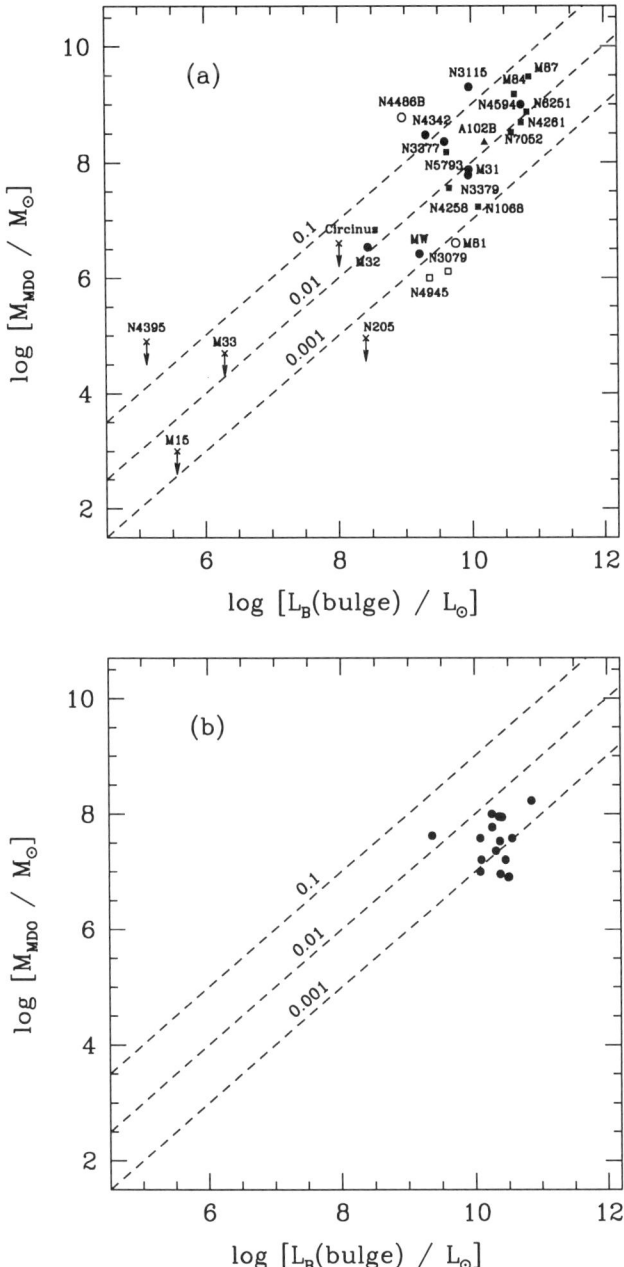

Figure 8. (a) Log $M_\bullet$ versus log L_B(bulge) for the objects listed in Table 1. The typical uncertainty of $M_\bullet$ is probably about a factor of 2. Open symbols denote points that may have an exceptionally large uncertainty in either of the two variables (see text). Masses derived from stellar kinematics are plotted as *circles*, those from gas kinematics as *squares*, the unconventional case of Arp 102B as a *triangle*, and five upper limits as *crosses*. Lines of constant mass to luminosity ratio are also shown. (b) Same as in (a), but for the Seyfert galaxies listed in Table 2 (except Mrk 110).

were it not for the fact that the nucleus has an absolute blue magnitude of only –9.8 and lives in a Magellanic spiral 2.6 Mpc away! Filippenko & Ho (1998) detected the Ca II infrared triplet lines in absorption from echelle spectra taken with the Keck telescope, from which they were able to estimate the strength of the stellar component contributing to the nuclear light ($M_B = -7.3$ mag) and the central stellar velocity dispersion ($\sigma \approx 30$ km s^{-1}). Combining the velocity dispersion with a cluster size ($r \lesssim 0.7$ pc) obtained from *HST* images, Filippenko & Ho limit the central mass to $\lesssim 8\times10^4$ M$_\odot$. The Circinus galaxy is thought to house a Seyfert nucleus, and if it contains a SMBH, its mass within $r \approx 10$ pc has been constrained to be $\lesssim 4\times10^6$ M$_\odot$ (Maiolino et al. 1998). The last upper limit shown in the figure pertains to the globular cluster M15; following KR, I adopt an upper limit of $M_\bullet = 1\times10^3$ M$_\odot$.

However, before reading too much into this diagram, we should ask whether the apparent correlation might arise from selection effects. The absence of points on the upper left-hand corner is probably real; there is nothing preventing us from detecting a massive BH in a small galaxy. Yet, we should be cautious, because very few low-mass galaxies have been studied so far, most of the effort having been focused on luminous, early-type systems. On the other hand, the empty region on the lower right-hand corner could be an artifact. Small masses are difficult to detect at large distances, and most luminous galaxies are far away. So the apparent correlation *could* be an upper envelope. Future observations are needed to settle this issue.

The median value of $M_\bullet/L_B$(bul) for the 20 detected objects is 0.012, which translates into a mass ratio of 0.002 for $M/L_B \approx 6$ typical for old stellar populations (van der Marel 1991). That is, on average about 0.2% of the bulge mass is locked up in the form of a SMBH. Magorrian et al. (1998) constructed axisymmetric $f(E, L_z)$ models for a sample of 32 early-type (mostly E and S0) galaxies having both *HST* photometry and ground-based stellar kinematics data, and they concluded that the data are consistent with nearly all of the galaxies having SMBHs. The 29 detected objects have a median $M_\bullet/M_{\rm bul} \approx 0.005$, higher than found here. However, as Magorrian et al. realize, the assumption of a two-integral distribution function may have caused them to overestimate $M_\bullet$ (cf. van der Marel 1998). Interestingly, quasars possibly also obey a similar $M_\bullet$-$M_{\rm bul}$ relation. McLeod (1998) finds that, for the most luminous quasars, there exists a minimum host luminosity that increases with nuclear power. Assuming that the quasar luminosities correspond to energy generation at the Eddington rate, $M_\bullet/M_{\rm bul}$ is again $\sim$0.002 (McLeod 1998).

With regard to the dead quasar prediction discussed in § 1, recall that we expect to find on average a 10^7 M$_\odot$ BH for every $L_B \approx 10^{10}$ L$_\odot$ galaxy,

SUPERMASSIVE BLACK HOLES IN GALACTIC NUCLEI

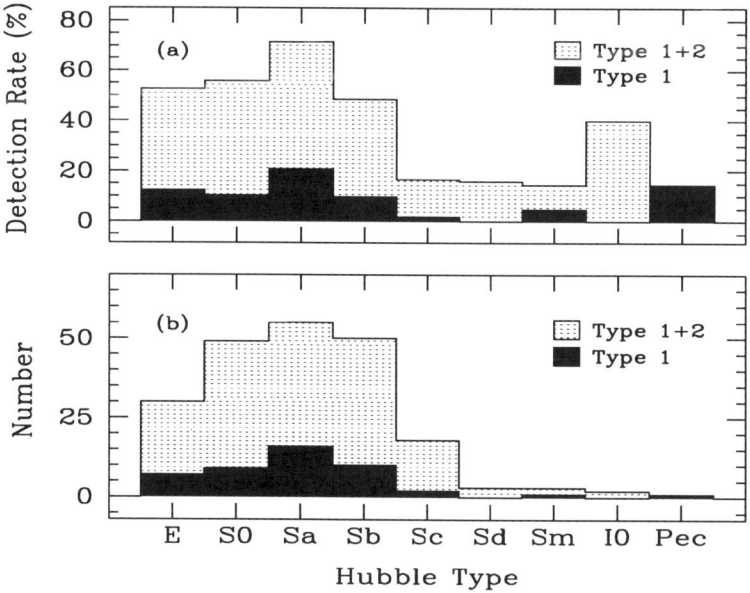

Figure 9. (a) Detection rate and (b) number distribution of AGNs as a function of Hubble type in the spectroscopic survey of Ho et al. (1995, 1997). "Type 1" AGNs (those with broad Hα) are shown separately from the total population (types 1 and 2).

or $M_\bullet/L_B(\mathrm{bul}) \approx 3.3 \times 10^{-3}$ $M_\odot/L_\odot$ since bulges contribute typical 30% of the galaxy light in B (Schechter & Dressler 1987). Evidently, if $\epsilon = 0.1$, we have already found about three times that value. This implies that either ϵ is smaller than 0.1, or that quasars do not make up all of the AGN population.

8. Are Supermassive Black Holes Ubiquitous?

They certainly have not been found in every case that has been looked. Kormendy has undertaken a systematic survey of a modest sample of galaxies (E–Sb), and his detection rate has been about 20% (KR). But, of course, many factors conspire against the detection of MDOs, and this estimate should be regarded as firm lower limit. If one takes seriously the $M_\bullet$-M_{bul} relation described above, it is possible that *every* bulge contains a SMBH with an appropriately scaled size. This view is supported by the statistical analysis of Magorrian et al. (1998). In fact, the detection of an MDO in NGC 4945 (§ 4.2) and the presence of a *bona fide* AGN in NGC 4395 indicate that perhaps even some galaxies without bulges may have SMBHs.

Additional support for this picture comes from the growing evidence that nonstellar nuclear activity is very common in galaxies, much more so than conventionally believed based on the statistics of bright AGNs and quasars. A recent spectroscopic survey of a large, statistically complete sample of nearby galaxies finds that over 40% of all bright ($B_T \leq 12.5$ mag) galaxies have nuclei that can be classified as "active," and the percentage is even higher among early-type systems (E–Sbc), approaching 50%–75% (Ho, Filippenko, & Sargent 1997). Most of the nearby AGNs have much lower luminosities than traditionally studied active galaxies, and a greater heterogeneity in spectral types is found (low-ionization nuclei, or LINERs, are common, for example), but the evidence is overwhelming that many of these nuclei are truly accretion-powered sources (see Filippenko 1996; Ho et al. 1997). Moreover, intrinsically weak, compact radio cores are known to be present in a significant fraction of elliptical and S0 galaxies (Sadler, Jenkins, & Kotanyi 1989; Wrobel & Heeschen 1991), almost all of which spectroscopically qualify as AGNs (Ho 1998).

Within the conventional AGN paradigm, the observed widespread nuclear activity implies that SMBHs are a generic component of many, perhaps most, present-day bulge-dominated galaxies, consistent with the picture emerging from the kinematic studies. This is a remarkable statement. It implies that SMBHs should not be regarded as "freaks of nature" that exist in only a handful of galaxies; rather, they must be accepted and understood as a normal component of galactic structure, one that arises naturally in the course of galaxy formation and evolution.

9. Some Implications and Future Directions

The SMBH hunting game is rapidly becoming a rather mature subject. I think we have progressed from the era of "the thrill of discovery" to a point where we are on the verge of using SMBHs as astrophysical tools. In this spirit, let me remark on a few of the ramifications of the existing observations and point out some of the more urgent directions that should be pursued.

A. The $M_\bullet$-$M_{\rm bul}$ relation. The apparent correlation between the mass of the central BH and the mass of the bulge, if borne out by future scrutiny, has significant implications (see below). From an observational point of view, the highest immediate priority is to populate the $M_\bullet$-$M_{\rm bul}$ diagram with objects spanning a wide range in luminosity, with the eventual aim of deriving a mass function for SMBHs. The samples should be chosen with the following questions in mind. (1) Is the apparent trend a true correlation or does it instead trace an upper envelope? (2) If the relation is real, is it linear? (3) What is the magnitude of the intrinsic scatter? And (4) is there a minimum bulge luminosity (mass) below which SMBHs do not exist?

In the near future, the most efficient way to obtain mass measurements for relatively large numbers of galaxies is to exploit the capabilities of STIS on *HST*. Several large programs are in progress. Although VLBI spectroscopy of H_2O masers delivers much higher angular resolution, this technique is limited by the availability of suitably bright sources. Conditions which promote H_2O megamaser emission evidently are realized in only a tiny fraction of galaxies (Braatz, Wilson, & Henkel 1996).

B. The formation of SMBHs. The $M_\bullet$-$M_{\rm bul}$ relation offers some clues to the formation mechanism of SMBHs. How does a galaxy know how to extract a constant, or at least a limiting, fraction of its bulge mass into a SMBH? An attractive possibility is by the normal dynamical evolution of the galaxy core itself. The spheroidal component of nearby galaxies can attain very high central stellar densities — up to 10^5 $M_\odot$ pc^{-3} (Faber et al. 1997) — and some with distinct nuclei have even higher concentrations still (Lauer et al. 1995). Although most galaxy cores are unlikely to have experienced dynamical collapse (Kormendy 1988c), the *innermost* regions have much shorter relaxation times, especially when considering a realistic stellar mass spectrum because the segregation of the most massive stars toward the center greatly accelerates the dynamical evolution of the system. Lee (1995, and these proceedings) shows that, under conditions typical of galactic nuclei, core collapse and merging of stellar-size BHs can easily form a seed BH of moderate mass. Alternatively, the seed object may form via the catastrophic collapse of a relativistic cluster of compact remnants (Quinlan & Shapiro 1990). In either case, subsequent accretion of gas and stars will augment the central mass, and, over a Hubble time, may produce the distribution of masses observed. It is far too premature to tell whether SMBHs form through the secular evolution of galaxies, as suggested here, through processes associated with the initial formation of galaxies (e.g., Rees 1984, 1998; Silk & Rees 1998), or both, depending on the galaxy type (elliptical vs. spiral galaxies). But the stage is set for a serious discussion. More sophisticated modeling of the growth of SMBHs that take into account a wider range of initial conditions in galactic nuclei (e.g., relaxing the adiabatic assumption or adopting more realistic density profiles for the stellar distribution) may eventually yield testable predictions (see, e.g., Stiavelli 1998).

C. Influence of SMBHs on galactic structure. Norman, May, & Van Albada (1985) showed through N-body simulations that a massive singularity in the center of a triaxial galaxy destroys the box-like stellar orbits and hence can erase the nonaxisymmetry, at least on small scales. This has several important consequences. First, it implies that the presence of a SMBH can influence the *global* structure and dynamics of galaxies. Second, the secular evolution of the axisymmetry of the central potential points to a natural mechanism for galaxies to self-regulate the transfer of angular momentum of the gas from large to small scales. This negative-feedback process may limit the growth of the central BH *and* the accretion rate onto it, and hence may serve as a promising framework for understanding the physical evolution of AGNs. Merritt & Quinlan (1998) find that the timescale for effecting the transition from triaxiality to axisymmetry depends strongly on the fractional mass of the BH; the evolution occurs rapidly when $M_\bullet/M_{\rm bul}$ $\gtrsim$ 2.5%, remarkably close to the observational upper limit (Fig. 8a).

D. *The origin of central cores.* It is not understood why giant ellipticals have such shallow central light profiles. "Cores" do not develop naturally in popular scenarios of structure formation, and even if they form, they are difficult to maintain against the subsequent acquisition of the dense, central regions of satellite galaxies that get accreted (Faber et al. 1997). Moreover, the very presence of a SMBH, whether it grew adiabatically in a preexisting stellar system or the galaxy formed by violent relaxation around it (Stiavelli 1998), ought to imprint a more sharply cusped light profile (see § 2) than is observed. An intriguing possibility is that cores were created as a result of mergers, where one or more of the galaxies contains a BH. As the single (Nakano & Makino 1998) or binary (Makino 1997; Quinlan & Hernquist 1997) BH sinks toward the center of the remnant due to dynamical friction, it heats the stars, thereby producing a "fluffy" core. If this interpretation is correct, it would provide a simple, powerful tool to diagnose the formation history of galaxies.

E. *Why are the black holes so black?* It has been somewhat puzzling how the BHs can remain so dormant. No doubt the dwindled gas supply in the present epoch, especially in ellipticals, is largely responsible for the inactivity. Yet the accretion rate cannot be zero; even in the absence of inflow from the general interstellar medium, some gas is shed through normal mass loss from the innermost evolved stars, and occasionally such stars get tidally disrupted (see below). If SMBHs are indeed present, the radiative efficiency of the accretion flow must be orders of magnitude lower than that of "standard" optically thick, geometrically thin disks. Such a situation may be realized in accretion flows where advection becomes important when the accretion rate is highly sub-Eddington (Narayan & Yi 1995; Abramowicz et al. 1995; Nakamura et al. 1996; Chakrabarti 1996). Sgr A* at the Galactic Center has a bolometric luminosity of only $\sim 10^{37}$ ergs s^{-1}, or $L_{\rm bol}/L_{\rm Edd}$ $\approx 3\times 10^{-8}$ (Narayan, Yi, & Mahadevan 1995); in the case of the LINER nucleus of M81 (Ho, Filippenko, & Sargent 1996), $L_{\rm bol} \approx 10^{41}$ ergs s^{-1} and $L_{\rm bol}/L_{\rm Edd} \lesssim 10^{-4}$. The spectral energy distributions emitted by both objects differ dramatically from those of luminous AGNs and can be approximately matched by advective-disk models.

F. *Tidal disruption of stars.* The prevalence of SMBHs suggested by the existing evidence predicts a relatively high incidence of tidal disruptions of stars as they scatter into nearly radial orbits whose pericenters pass within the tidal radius of the BH (Rees 1998, and references therein). For a typical stellar density of 10^5 stars pc^{-3}, $M_\bullet = 10^6$–10^8 $M_\odot$, and $\sigma = 100$–300 km s^{-1}, a solar-type star will be disrupted once every 10^2–10^4 years. (BHs more massive than 10^8 $M_\odot$ will swallow the star whole.) Roughly half the debris becomes unbound and half gets captured into an accretion disk which undergoes a bright flare ($\sim 10^{10}$ $L_\odot$) lasting a few months to a year. The spectrum is expected to be mainly thermal and to peak in the extreme-UV and soft X-rays. The contribution to the near-UV and optical bands is uncertain; it depends on assumptions concerning the geometry of the accretion disk (thick or thin) and on whether an optically thick envelope can form. For plausible parameters, Ulmer (1998) estimates that a 10^7 $M_\odot$ BH will produce a flare with an absolute magnitude of about -20 in U and -18.5 in V. The realization that SMBHs may be even more common than previously thought provides fresh motivation to search for such stellar flares; some observational strategies are mentioned by Rees (1998). Here, I wish to stress that quantifying the rate of stellar disruptions can be used as a tool to study the demography of SMBHs out to relatively large distances and hence should be regarded as complementary to the kinematic searches.

Acknowledgements

I am grateful to S. K. Chakrabarti for the invitation to participate in this workshop and his help in arranging a pleasurable visit to India. I thank G. A. Bower, S. Collier, R. Genzel, L. J. Greenhill, J. Kormendy, R. Maiolino, D. Maoz, E. Maoz, and K. Nandra for contributing to, or for providing comments that have improved the presentation of, the material in this paper. This work was supported by a postdoctoral fellowship from the Harvard-Smithsonian Center for Astrophysics and by NASA grants from the Space Telescope Science Institute (operated by AURA, Inc., under NASA contract NAS 5-26555).

References

Abramowicz, M. A., Chen, X., Kato, S., Lasota, J.-P., & Regev, O. 1995, ApJ, 438, L37

Bahcall, J. N., & Wolf, R. A. 1976, ApJ, 209, 214

Bender, R., Burstein, D., & Faber, S. M. 1992, ApJ, 399, 462

Binney, J., & Mamon, G. A. 1982, MNRAS, 200, 361

Blandford, R. D., & McKee, C. F. 1982, ApJ, 255, 419

Blandford, R. D., & Rees, M. J. 1992, in Testing the AGN Paradigm, ed. S. Holt, S. Neff, & M. Urry (New York: AIP), 3

Bower, G. A., et al. 1998, ApJ, 492, L111

Bower, G. A., Wilson, A. S., Heckman, T. M., & Richstone, D. O. 1996, in The Physics of LINERs in View of Recent Observations, ed. M. Eracleous et al. (San Francisco: ASP), 163

Braatz, J. A., Wilson, A. S., & Henkel, C. 1996, ApJS, 106, 51

Carollo, C. M., Franx, M., Illingworth, G. D., & Forbes, D. A. 1997, ApJ, 481, 710

Chakrabarti, S. K. 1995, ApJ, 441, 576

Chakrabarti, S. K. 1996, ApJ, 464, 664

Chokshi, A., & Turner, E. L. 1992, MNRAS, 259, 421

Claussen, M. J., & Lo, K.-Y. 1986, ApJ, 308, 592

Crane, P., et al. 1993, AJ, 106, 1371

Cretton, N., de Zeeuw, P. T., van der Marel, R. P., & Rix, H.-W. 1998, ApJ, in press

Cretton, N., & van den Bosch, F. C. 1998, ApJ, in press

Dehnen, W. 1995, MNRAS, 274, 919

de Vaucouleurs, G., de Vaucouleurs, A., Corwin, H. G., Jr., Buta, R. J., Paturel, G., & Fouqué, R. 1991, Third Reference Catalogue of Bright Galaxies (New York: Springer) (RC3)

Devereux, N. A., Ford, H. C., & Jacoby, G. 1997, ApJ, 481, L71

Dibai, E. A. 1981, Soviet Astron., 24, 389

Dressler, A., & Richstone, D. O. 1988, ApJ, 324, 701

Dressler, A., & Richstone, D. O. 1990, ApJ, 348, 120

Duncan, M. J., & Wheeler, J. C. 1980, ApJ, 237, L27

Eckart, A., & Genzel, R. 1996, Nature, 383, 415

Eckart, A., & Genzel, R. 1997, MNRAS, 284, 576

Eracleous, M., Halpern, J. P., Gilbert, A. M., Newman, J. A., & Filippenko, A. V. 1997, ApJ, 490, 216

Faber, S. M., et al. 1987, in Nearly Normal Galaxies, ed. S. M. Faber (New York: Springer), 175

Faber, S. M., et al. 1997, AJ, 114, 1771

Fabian, A. C., Nandra, K., Reynolds, C. S., Brandt, W. N., Otani, C., Tanaka, Y., Inoue, H., & Iwasawa, K. 1995, MNRAS, 277, L11

Fabian, A. C., Rees, M. J., Stella, L., & White, N. E. 1989, MNRAS, 238, 729

Ferrarese, L., Ford, H. C., & Jaffe, W. 1996, ApJ, 470, 444

Ferrarese, L., Ford, H. C., & Jaffe, W. 1998, ApJ, in press

Filippenko, A. V. 1996, in The Physics of LINERs in View of Recent Ob-

servations, ed. M. Eracleous et al. (San Francisco: ASP), 17
Filippenko, A. V., & Ho, L. C. 1998, ApJ, submitted
Filippenko, A. V., Ho, L. C., & Sargent, W. L. W. 1993, ApJ, 410, L75
Fillmore, J. A., Boroson, T. A., & Dressler, A. 1986, ApJ, 302, 208
Ford, H. C., et al. 1994, ApJ, 435, L27
Ford, H. C., Tsvetanov, Z. I., Ferrarese, L., & Jaffe, W. 1998, in IAU Symp. 184, The Central Regions of the Galaxy and Galaxies, ed. Y. Sofue (Dordrecht: Kluwer), in press
Gebhardt, K., et al. 1998, AJ, in press
Genzel, R., Eckart, A., Ott, T., & Eisenhauer, F. 1997, MNRAS, 291, 219
Genzel, R., Thatte, N., Krabbe, A., Kroker, H., & Tacconi-Garman, L. E. 1996, ApJ, 472, 153
Gerhard, O. E. 1993, MNRAS, 265, 213
Ghez, A. M., et al. 1998, in IAU Symp. 184, The Central Regions of the Galaxy and Galaxies, ed. Y. Sofue (Dordrecht: Kluwer), in press
Goodman, J., & Lee, H. M. 1989, ApJ, 337, 84
Greenhill, L. J. 1997, in IAU Colloq. 159, Emission Lines in Active Galaxies: New Methods and Techniques, ed. B. M. Peterson, F.-Z. Cheng, & A. S. Wilson (San Francisco: ASP), 394
Greenhill, L. J. 1998, in IAU Colloq. 164, Radio Emission from Galactic and Extragalactic Compact Sources, ed. A. Zensus, G. Taylor, & J. Wrobel (San Francisco: ASP), in press
Greenhill, L. J., Gwinn, C. R., Antonucci, R., & Barvainis, R. 1996, ApJ, 472, L21
Greenhill, L. J., Moran, J. M., & Herrnstein, J. R. 1997, ApJ, 481, L23
Hagiwara, Y., Kotaro, K., Ryohei, K., & Nakai, N. 1997, PASJ, 49, 171
Harms, R. J., et al. 1994, ApJ, 435, L35
Ho, L. C. 1998, ApJ, submitted
Ho, L. C., et al. 1998, in preparation
Ho, L. C., Filippenko, A. V., & Sargent, W. L. W. 1995, ApJS, 98, 477
Ho, L. C., Filippenko, A. V., & Sargent, W. L. W. 1996, ApJ, 462, 183
Ho, L. C., Filippenko, A. V., & Sargent, W. L. W. 1997, ApJ, 487, 568
Jaffe, W., Ford, H. C., Ferrarese, L., van den Bosch, F., & O'Connell, R. W. 1993, Nature, 364, 213
Jaffe, W., et al. 1994, AJ, 108, 1567
Jones, D. H., et al. 1996, ApJ, 466, 742
Kormendy, J. 1988a, ApJ, 325, 128
Kormendy, J. 1988b, ApJ, 335, 40

Kormendy, J. 1988c, in Supermassive Black Holes, ed. M. Kafatos (Cambridge: Cambridge Univ. Press), 219

Kormendy, J. 1993, in The Nearest Active Galaxies, ed. J. Beckman, L. Colina, & H. Netzer (Madrid: CSIC Press), 197

Kormendy, J., et al. 1996, ApJ, 459, L57

Kormendy, J., et al. 1997a, ApJ, 473, L91

Kormendy, J., et al. 1997b, ApJ, 482, L139

Kormendy, J., Bender, R., Evans, A. S., & Richstone, D. 1998, AJ, 115, 1823

Kormendy, J., & McClure, R. D. 1993, AJ, 105, 1793

Kormendy, J., & Richstone, D. O. 1995, ARA&A, 33, 581 (KR)

Krabbe, A., et al. 1995, ApJ, 447, L95

Krolik, J. H. 1997, in IAU Colloq. 159, Emission Lines in Active Galaxies: New Methods and Techniques, ed. B. M. Peterson, F.-Z. Cheng, & A. S. Wilson (San Francisco: ASP), 459

Lacy, J. H., Townes, C. H., Geballe, T. R., & Hollenbach, D. J. 1980, ApJ, 241, 132

Lauer, T. R., et al. 1992, AJ, 103, 703

Lauer, T. R., et al. 1995, AJ, 110, 2622

Lee, H. M. 1995, MNRAS, 272, 605

Lin, H., et al. 1996, ApJ, 464, 60

Lynden-Bell, D. 1969, Nature, 223, 690

Lynden-Bell, D., & Rees, M. J. 1971, MNRAS, 152, 461

Macchetto, F., Marconi, A., Axon, D. J., Capetti, A., Sparks, W. B., & Crane, P. 1997, ApJ, 489, 579

Magorrian, J., et al. 1998, AJ, 115, 2285

Maiolino, R., Krabbe, A., Thatte, N., & Genzel, R. 1998, ApJ, 493, 650

Makino, F. 1997, ApJ, 478, 58

Maoz, E. 1995, ApJ, 447, L91

Maoz, E. 1998, ApJ, 494, L181

McLeod, K. K. 1998, in Quasar Hosts, ed. D. Clements & I. Perez-Fournon (Berlin: Springer-Verlag), in press

Menou, K., Quataert, E., & Narayan, R. 1998, in Proc. of the 8th Marcel Grossmann Meeting on General Relativity (Jerusalem), in press

Merritt, D., & Quinlan, G. 1998, ApJ, 498, 625

Miyoshi, M., Moran, J., Herrnstein, J., Greenhill, L., Nakai, N., Diamond, P., & Inoue, M. 1995, Nature, 373, 127

Moran, E. C., Filippenko, A. V., Ho, L. C., Belloni, T., Shields, J. C.,

Snowden, S. L., & Sramek, R. A. 1998, ApJ, in press
Nakai, N., Inoue, M., & Miyoshi, M. 1993, Nature, 361, 45
Nakamura, K. E., Matsumoto, R., Kusunose, M., & Kato, S. 1996, PASJ, 48, 761
Nakano, T., & Makino, J. 1998, ApJ, in press
Nandra, K., George, I. M., Mushotzky, R. F., Turner, T. J., & Yaqoob, T. 1997, ApJ, 477, 602
Narayan, R., & Yi, I. 1995, 452, 710
Narayan, R., Yi, I., & Mahadevan, R. 1995, Nature, 374, 623
Netzer, H., & Peterson, B. M. 1997, in Astronomical Time Series, ed. D. Maoz, A. Sternberg, & E. M. Leibowitz (Dordrecht: Kluwer), 85
Newman, J. A., Eracleous, M., Filippenko, A. V., & Halpern, J. P. 1997, ApJ, 485, 570
Nieto, J.-L., Bender, R., Arnaud, J., & Surma, P. 1991, A&A, 244, L2
Norman, C. A., May, A., & Van Albada, T. S. 1985, ApJ, 296, 20
Padovani, P., Burg, R., & Edelson, R. A. 1990, ApJ, 353, 438
Peebles, P. J. E. 1972, ApJ, 178, 371
Peterson, B. M., Wanders, I., Bertram, R., Hunley, J. F., Pogge, R. W., & Wagner, R. M. 1998, ApJ, in press
Pounds, K., Nandra, K., Stewart, G. C., George, I. M., & Fabian, A. C. 1990, Nature, 344, 132
Qian, E. E., de Zeeuw, P. T., van der Marel, R. P., & Hunter, C. 1995, MNRAS, 274, 602
Quinlan, G. D., & Hernquist, L. 1997, New Astron., 2(6), 533
Quinlan, G. D., Hernquist, L., & Sigurdsson, S. 1995, ApJ, 440, 554
Quinlan, G. D., & Shapiro, S. L. 1990, ApJ, 356, 483
Rees, M. J. 1984, ARA&A, 22, 471
Rees, M. J. 1998, in Proc. Chandrasekhar Memorial Conf., Black Holes and Relativity, ed. R. Wald (Chicago: Chicago Univ. Press), in press
Reynolds, C. S., & Begelman, M. C. 1997, ApJ, 488, 109
Richstone, D. O. 1998, in IAU Symp. 184, The Central Regions of the Galaxy and Galaxies, ed. Y. Sofue (Dordrecht: Kluwer), in press
Richstone, D. O., Bower, G., & Dressler, A. 1990, ApJ, 353, 118
Richstone, D. O., & Tremaine, S. 1984, ApJ, 286, 27
Richstone, D. O., & Tremaine, S. 1985, ApJ, 296, 370
Richstone, D. O., & Tremaine, S. 1988, ApJ, 327, 82
Rybicki, G. B., & Bromley, B. C. 1998, ApJ, in press
Sadler, E. M., Jenkins, C. R., & Kotanyi, C. G. 1989, MNRAS, 240, 591

Salpeter, E. E. 1964, ApJ, 140, 796

Sargent, W. L. W., Young, P. J., Boksenberg, A., Shortridge, K., Lynds, C. R., & Hartwick, F. D. A. 1978, ApJ, 221, 731

Schechter, P. L., & Dressler, A. 1987, AJ, 94, 563

Silk, J., & Rees, M. J 1998, A&A, 331, L1

Simien, F., & de Vaucouleurs, G. 1986, ApJ, 302, 564

Soltan, A. 1982, MNRAS, 200, 115

Stiavelli, M. 1998, ApJ, 495, L91

Tanaka, Y., et al. 1995, Nature, 375, 659

Titarchuk, L., & Zannias, T. 1998, ApJ, 493, 863

Trotter, A. S., Greenhill, L. J., Moran, J. M., Reid, M. J., Irwin, J. A., & Lo, K.-Y. 1998, ApJ, 495, 740

Tully, R. B. 1988, Nearby Galaxies Catalog (Cambridge: Cambridge Univ. Press)

Ulmer, A. 1998, ApJ, in press

van der Marel, R. P. 1991, MNRAS, 253, 710

van der Marel, R. P. 1994a, MNRAS, 270, 271

van der Marel, R. P. 1994b, ApJ, 432, L91

van der Marel, R. P. 1998, in IAU Symp. 186, Galaxy Interactions at Low and High Redshift, ed. D. B. Sanders & J. Barnes (Dordrecht: Kluwer), in press

van der Marel, R. P., Cretton, N., de Zeeuw, P. T., & Rix, H.-W. 1998, ApJ, 493, 613

van der Marel, R. P., de Zeeuw, P. T., Rix, H.-W., & Quinlan, G. D. 1997, Nature, 385, 610

van der Marel, R. P., Evans, N. W., Rix, H.-W., White, S. D. M., & de Zeeuw, P. T. 1994a, MNRAS, 271, 99

van der Marel, R. P., Rix, H.-W., Carter, D., Franx, M., White, S. D. M., & de Zeeuw, P. T. 1994b, MNRAS, 268, 521

van der Marel, R. P., & van den Bosch, F. C. 1998, AJ, in press

Wandel, A., & Mushotzky, R. F. 1986, ApJ, 306, L61

Wandel, A., & Yahil, A. 1985, ApJ, 295, L1

Watson, W. D., & Wallin, B. K. 1994, ApJ, 432, L35

Wrobel, J. M., & Heeschen, D. S. 1991, AJ, 101, 148

Young, P. J. 1980, ApJ, 242, 1232

Young, P. J., Westphal, J. A., Kristian, J., Wilson, C. P., & Landauer, F. P. 1978, ApJ, 221, 721

Zel'dovich, Ya. B., & Novikov, I. D. 1964, Sov. Phys. Dokl., 158, 811

TIDAL DISRUPTION OF A STAR BY A MASSIVE BLACK HOLE

HYUNG MOK LEE
Department of Earth Sciences, Pusan National University, Pusan 609-735, Korea; e-mail: hmlee@uju.es.pusan.ac.kr

Abstract. Tidal force of a black hole often destroys a star in galactic nuclei. If the entire stellar mass is accreted by a black hole with a significant efficiency, the average luminosity gives relatively small M/L which is not observed in nearby galaxies with convincing dynamical evidences for the central black holes. However, there are many uncertainties regarding the fate of stellar material after the tidal disruption. It is not clear whether the stellar debris will form an accretion disk. We discuss the hydrodynamic process that might happen after the tidal disruption.

1. Introduction

Evidences are mounting up for the existence of massive black holes in the central parts of the galaxies. Recent high resolution studies reveal that the mass of the central black hole is roughly proportional to the mass of the hot component (bulge for disk galaxies and the entire galaxy for the elliptical galaxies) of the galaxies (e.g., Magorrian et al. 1998). The stellar motions in the central parts of our Galaxy strongly suggest the presence of a central black hole of $M \sim 2.5 \times 10^6 \, M_\odot$ (Eckart & Genzel 1996).

A central black hole is usually surrounded by a dense stellar system. Stars can occasionally become so close to the black hole so that they are tidally torn apart. The rate of stellar disruption depends on various parameters such as the mass of the black hole and distribution of stars. For physical parameters of our Galaxy, the disruption rate is estimated to be about 10^{-4} yr^{-1} (e.g., Rees 1984, Cannizzo, Lee & Goodman 1990). If the stellar material is accreted to a black hole, the average luminosity is about $6 \times 10^{41} (\epsilon/0.1)$ erg/sec, where ϵ is the efficiency of the conversion of the mass to energy.

However, galaxies with convincing dynamical evidences for central black hole usually do not reveal strong activity in the central parts. It is possible that the 'dark mass' implied by spectroscopic observations might be in the form of dark stars (Goodman & Lee 1989) although high resolution observations in the near future would enable us to eliminate some candidates for the constituents of the dark cluster.

It is, however, too early to rule out the existence of a central massive black hole on the grounds of low level activity in galactic nucleus. We simply do not understand well enough the physical processes after the disruption of a star. The evolution of the stellar debris after the disruption is a challenging problem, because it involves extreme physical conditions and a large dynamic range in the physical parameters. The key question is whether the stellar debris will be efficiently accreted onto the black hole (Rees 1994). Since the initial orbit of debris is extremely eccentric, accretion time scale will be longer than typical interval between two successive tidal disruption events (Gurzadyan & Ozernoi 1980, 1981). On the other hand, if the orbits of the debris become circularized, the accretion time scale can be as short as a few tens of years (Cannizzo, Lee & Goodman 1990). More recently, Loeb & Ulmer (1997) argued that the debris would form a thick disk that radiates thermally at Eddington rate for tens of years. Since typical effective temperature is expected to be around 10^4 K, these events will be easily detectable in large scale optical surveys. Thus circularization is a crucial issue in predicting the observational consequences of tidal disruption.

The circularization mechanism is thought to be the shocks produced by two intersecting streams of different orbital periods. The stellar debris forms a long and thin stream because of large spread in orbital energy. The outgoing stream collides with the incoming stream of longer orbital periods, because of relativistic precession of orbits of the debris (Rees 1988, Cannizzo, Lee & Goodman 1990, Monaghan & Lee 1994, Kochanek 1994, Laguna et al. 1994). However, supersonic collisions between two gas streams have not been investigated in enough detail to tell us how the gas flows evolve after the collision. If the collision efficiently dissipates the orbital energy, a circular disk will form. The problem becomes more complicated since it is rather difficult to determine the relevant parameters at the time of the collision. The Lens-Thirring precession of the orbital plane also can make stream collisions weak if the central black hole is rotating.

In this paper, we review the physical processes that are related to the problem of stellar disruption and the subsequent evolution of the stellar debris. Readers are referred to Rees (1994) for the previous review on this subject.

2. Simple Estimates for the Stellar Disruption

2.1. THE RATE OF DISRUPTION

Suppose that a black hole of mass M_B was introduced in a stellar system of density n_* and isothermal velocity dispersion of σ. If the black hole existed for a long time (compared to the relaxation time), the distribution of the stars should be modified. The radius below which the black hole dominates the gravitational potential can be defined as

$$r_H \equiv \frac{GM_B}{\sigma^2}. \tag{1}$$

Let's assume that the equilibrium density distribution of stars around the black hole is a power law: i.e., $n_*(r) = n_0 (r_H/r)^s$. The power-law index s can be determined by demanding that the energy flow via relaxation is independent of r. The resulting s is 7/4.

The flow of energy at radius r is $N(r)E(r)/t_r$ where $N(r)$ is the number of stars within r, $E(r)$ is the binding energy of a star of mass m_* at r, which is simply $-GM_B m_*/r$, and t_r is the local relaxation time. Such an energy flow should be balanced with the energy loss rate at r_t due to tidal disruption of stars, where r_t is the 'tidal distance' below which stars are destroyed by the tidal force of the black hole (see eq. [5]). Also, the energy flow should not depend on the location in equilibrium state. Therefore the energy balance equation becomes

$$\frac{N(r)E(r)}{t_r} = \frac{dN}{dt} \frac{GM_B}{r_t}. \tag{2}$$

Noting that the relaxation time is $t_r \approx v^3/G^2 m_*^2 n$, $v^2 = GM_B/r$, and $s = 7/4$ we obtain,

$$\frac{dN}{dt} \propto M_B^{7/3} m_*^{5/3} R_*^1 n_0^2 \sigma^{-7}, \tag{3}$$

where r_* is the stellar radius. The above estimate should be compared with the result obtained by integrating the Fokker-Planck equation by Cohn & Kulsrud (1978):

$$\frac{dN}{dt} = 1.2 \times 10^{-3} \, \text{yr}^{-1} \left(\frac{10^7 \, M_\odot}{M_B}\right)^{2.33} \left(\frac{5 \times 10^4 \, \text{pc}^{-3}}{n_0}\right)^{1.6}$$

$$\times \left(\frac{\sigma}{100 \, \text{km/sec}}\right)^{5.76} \left(\frac{M_\odot}{m_*}\right)^{1.06} \left(\frac{R_\odot}{r_*}\right)^{0.4}. \tag{4}$$

Here we can see some differences between eqs. (3) and (4). Such a difference was caused by the fact that the stars approaching close to the black hole are those residing far away from the black hole with low angular momentum while the simple estimate assumes that the stars near r_t are drifting inward. In any case, we can see from eq. (1) that the disruption rate in typical galactic nuclei is about 10^{-3} to 10^{-4} per year.

2.2. TIDAL DISRUPTION

Roughly speaking, a star will be disrupted when the tidal force exceeds the self-gravity of the star. This happens when the distance from the black hole is smaller than the 'tidal distance', i.e.,

$$r < r_t = \left(\frac{2M_B}{m_*}\right)^{1/3} R_*$$
$$\sim 2 \times 10^{13} \left(\frac{M_B}{10^7\,\mathrm{M_\odot}}\right)^{1/3} \left(\frac{\mathrm{M_\odot}}{m_*}\right)^{1/3} \left(\frac{R_*}{\mathrm{R_\odot}}\right) \mathrm{cm}. \quad (5)$$

Since the star is made of gas, hydrodynamic effects must be taken into account in order to have a better estimate. Actual tidal distance should be a function of the detailed structure of the stars, but the above estimate is sufficiently good for most purposes. We note here that the tidal distance is smaller than the gravitational radius for solar type stars if $M_B \gtrsim 10^8\,\mathrm{M_\odot}$. Therefore, tidal disruption occurs for black holes of $\lesssim 10^8\,\mathrm{M_\odot}$.

Since the stars approaching to the black hole are usually unbound to the black hole, the condition for the disruption can be expressed that the pericenter of the stellar orbit is smaller than r_t. If R_p is close to r_t, the star is barely destroyed. On the other hand, if $R_p \ll r_t$ the star becomes severely compressed and may even experience explosive nuclear reaction (e.g., Luminet & Carter 1986). However, such a strong collision is much rarer than the marginal events since the differential cross section is proportional to R_p. In this review, we concentrate on marginal disruptions only. The strong collision may lead to the similar fate of the stellar material (Rees 1994), but more detailed studies are necessary.

2.3. GENERAL BEHAVIOR OF THE STELLAR DEBRIS

The stellar material will move along the near Keplerian orbit according to the binding energy. There is a wide spread in binding energy because the typical random velocity at the time of disruption is of order of escape velocity from the star. Suppose that the star was approaching to the black hole along a parabolic orbit with R_p. The velocity at the pericenter is then

$$v_p = \sqrt{GM_B/R_p} \sim 8.4 \times 10^4\,\mathrm{km/sec} \left(\frac{M_B}{10^7\,\mathrm{M_\odot}}\right)^{1/2} \left(\frac{2 \times 10^{13}\mathrm{cm}}{R_p}\right) \quad (6)$$

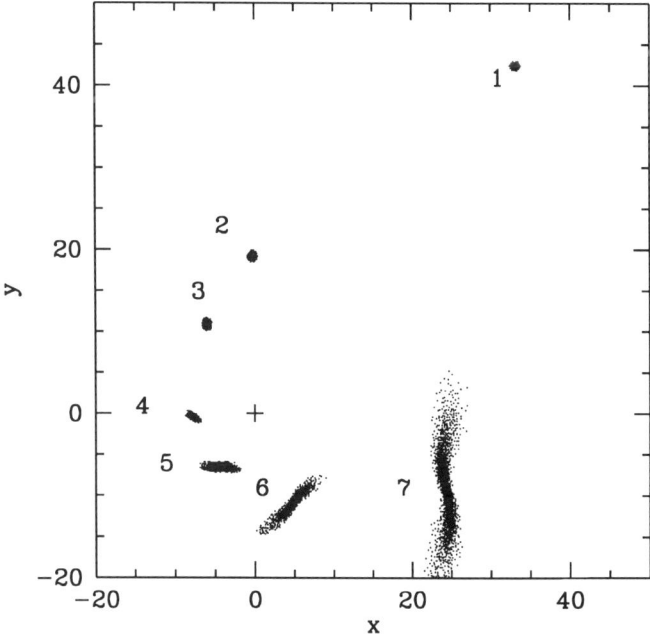

Fig. 1 : The process of disruption of a solar type star by a black hole of $M_B = 10^7 \, M_\odot$. The size of the star was amplified by 10 times. The unit of length is GM_B/c^2.

which is much greater than v_* ($\sim 600 \, \text{km/sec}$) where v_* is the escape velocity from the stellar surface. Therefore, the spread in energy $\Delta E \sim 2v_p v_*$ is much greater than the binding energy of the star itself. This means that about half of the material will be immediately unbound from the black hole and the remaining half will be very tightly bound to the hole.

The distribution of binding energy of stellar debris is roughly constant in unit energy interval. Energy distribution function can be translated into the orbital period distribution function: $f(P)dP = f(E)\frac{dE}{dP}dP = const \times P^{-5/3}$. This means that the rate of material returning to the pericenter is proportional to $t^{-5/3}$. Such an expectation was verified by numerical integration of SPH by Evans & Kochanek (1989) and Monaghan & Lee (1994).

Because of the large spread in binding energy, the material is quickly spread along a narrow line. Figure 1 shows the process of disruption of a solar type star by a black hole of $10^7 \, M_\odot$ with $R_p = 10^{13}$ cm. The relativistic precession angle θ_p for an elongated orbit is about $1.5\pi R_p/R_s$, where R_s is the gravitational radius of the black hole. For the case shown in Figure 1, $R_p \approx 10 R_s$ and thus $\theta_p \approx 27°$. Therefore, the incoming stream of long period particles will cross with the outgoing stream of the short period particles. The location of the collision depends on the mass of the black hole and the pericenter of the incoming stellar orbit. Figure 2 shows the velocity vector of the stream near the stream collision point which is about 80 times of the gravitational radius of the black hole for the case shown in Figure 1. The internal temperature of the stream is likely to be very low, and therefore the stream collision will be intrinsically highly supersonic. Such a hydrodynamic collision will play an important role in determining the fate of the disrupted material. If the collision efficiently transfers the orbital energy into thermal energy and then radiated away, the collision will be very sticky. The disrupted material will quickly settle into circular orbits.

If the hydrodynamic collision efficiently removes the orbital energy, the time scale for the circularization is essentially the orbital time scale of stellar debris. For the case discussed above, 90% of the material have orbital period less than 4.2 years.

Once the circularization takes place, the subsequent evolution can be characterized as the secular evolution driven by the viscosity. The time-dependent accretion disk was studied by Cannizzo, Lee & Goodman (1989). The later phase of the evolution is nearly self-similar and the accretion luminosity becomes a power-law of $t^{-19/16}$. Since the photons of the disk become softer with time, the luminosity in short wavelengths falls faster than than in long wavelengths.

3. Numerical Studies of Stream Collisions

It is almost impossible to follow the evolution of the entire debris using a single numerical method. Kochanek (1994) has used the approximate method to estimate the physical parameters of the streams. Since the stream collision is the key process modifying the dynamics of the stream, we concentrate on the hydrodynamics of stream collisions.

Lee, Kang & Ryu (1996) have studied supersonic collisions between three-dimensional streams using TVD (see Ryu in this volume). As a first step, they have considered the adiabatic case which can be done with a small number of parameters. Although the stream's orbital energy changes into the thermal energy just after the shock, adiabatic expansion of the shocked medium quickly follows. The thermal energy is transferred back to the expansion energy and the conversion rate to thermal energy is very

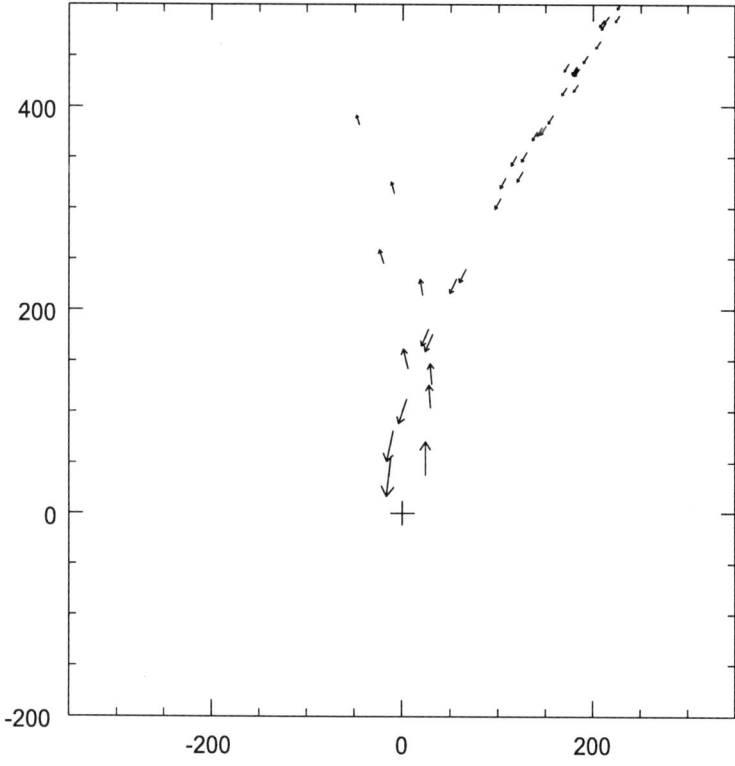

Fig. 2 : The motion of the stream of stellar debris near the collision point. Here the particles are assumed to be 'collisionless'.

small. However, the shock transforms the incoming and outgoing stream into an expanding bubble of low density. Since the collision destroys the stream geometry, there would be no more stream coming into the black hole until the tail of stream makes the collision. There will be a pause of the stream collision until the next stream reaches collision point after passing the pericenter. If we ever observe the stream collision as a burst of radiation, it will be likely to be an intermittent phenomena, although the visibility of the stream collision remains very uncertain.

Since the shocked gas has very high temperature, the radiation process is likely to be very important. Kim, Lee & Park (1998) have considered the stream collisions including the radiative effects. Unlike the adiabatic case, the introduction of radiative processes requires some dimensional constraints. Kim, Lee & Park have taken the collision parameters from Kochanek (1994): the stream debris was assumed to have come from the encounter between a 10^6 $M_\odot$ black hole and a solar type star, which initially has a parabolic orbit with a pericenter radius of 100 $R_\odot$. Velocities and mass fluxes of two streams were obtained from the values at $t \approx 1750\, t_D$ of Kochanek's Figure 6 ($v_1 = 0.0147\,c$, $v_2 = 0.0083\,c$, $\dot{M}_1 = 1\,M_\odot \mathrm{yr}^{-1}$, and $\dot{M}_2 = 0.5\,M_\odot \mathrm{yr}^{-1}$). The collision angle was about $135°$. However, Kochanek's results for the widths of two streams Δ_1 and Δ_2 (subscript 1 is for the stream making its first orbit, stream 1, and 2 is for the stream making its second orbit, stream 2) at a certain epoch very much depend on the way of treating the viscosity. The standard model has round shape cross sections (i.e., $\Delta_1 = H_1$ & $\Delta_2 = H_2$), and the mass flux and the radius of stream 1 are 2 and 5/3 times larger than stream 2, respectively. All the simulations are carried out using a TVD code.

The computational box of the simulations consisted of 101^3 cells. The physical length of each cell l_{cell} is 5.93×10^{10} cm. The initial pressure of the streams is set such that Mach number of the stream is 300. This gives an initial temperature of about 2×10^4 K for stream 1 and 7×10^3 K for stream 2. The initial density of the ambient medium is 10^{-4} times smaller than the average density of the streams, and the initial temperature of the ambient medium is set to be slightly smaller than that of the streams. The initial density profile of stream's cross section has a functional form of $\exp(-4y^2/\Delta^2 - 4z^2/H^2)$.

3.1. RADIATIVE PROCESSES

The average particle number density of the streams is of order 10^{17} cm^{-3}, and the characteristic radius of the streams is of order 10^{12} cm. For the typical column density of $\sim 10^{29}$ cm^{-2}, the electron scattering optical depth τ_{es} is much larger than unity, and so is the effective optical depths $\tau_* \equiv \sqrt{3\tau_{ff}(\tau_{ff} + \tau_{es})}$, where τ_{ff} is the Rosseland mean free-free absorption optical depth. Thus the post-collision region may be approximated to be in thermal equilibrium (TE) where matter and radiation is strongly coupled. The total energy density, u, is now the sum of the gas thermal energy density, u_g, and the radiation energy density, u_r:

$$u = u_g + u_r = \frac{nkT}{\gamma_g - 1} + aT^4, \qquad (7)$$

where n is the gas number density, k the Boltzmann constant, a the radiation constant, and γ_g the adiabatic index for gas (5/3 for monatomic ideal gas). The combined gas plus radiation has an effective adiabatic index γ_{eff} between 5/3 and 4/3, depending on the ratio u_g/u_r or equally by the ratio P_g/P_r, where P_g and P_r are the pressure of the gas and the radiation respectively.

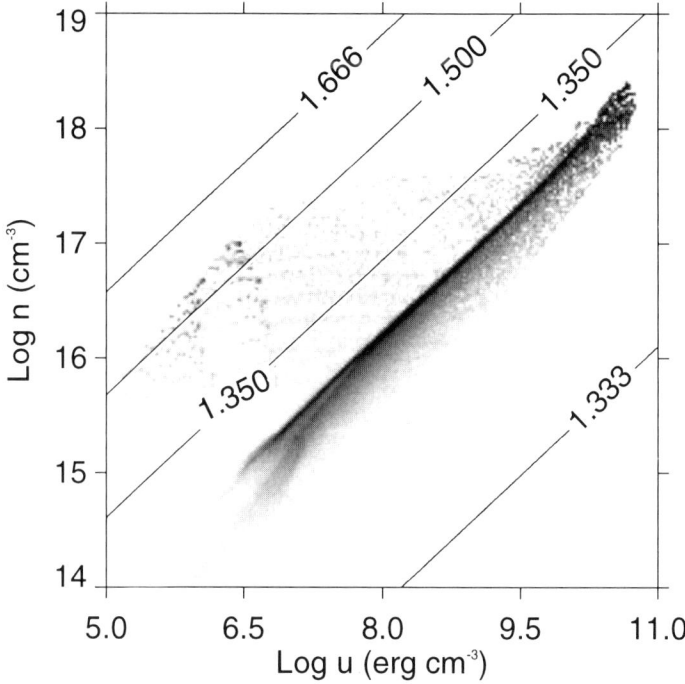

Fig. 3 : Density weighted $u - n$ distribution of post-shock gas. The values labeled on the counter are γ_{eff}. Most of the gas satisfies $\gamma_{eff} \sim 4/3$.

The simulation shows that the post-collision region has physical conditions $P_r \gg P_g$. Figure 3 shows the density weighted distribution of u and n for the standard model at $t = 4 \times 10^4$ sec. Diagonal lines represent the γ_{eff} values, which is defined by

$$\frac{P_{tot}}{\gamma_{eff} - 1} \equiv \frac{P_g}{\gamma_g - 1} + \frac{P_r}{4/3 - 1}. \tag{8}$$

It is clear that the most of the thermalized material falls onto the region where γ_{eff} is very close to $4/3$. So effective adiabatic index γ_{eff} was fixed to be $4/3$ in numerical simulations.

The temperature ($\equiv u/\frac{3}{2}nk$) of the post-collision region ranges from 10^5 to 10^7 K. So gas is completely ionized and is cooled only by bremsstrahlung. Between 10^4 K and 10^5 K, line cooling by various atoms can be significant compared to bremsstrahlung. However, line photons are immediately absorbed by nearby atoms and therefore do not contribute much to actual cooling.

Photons resulting from bremsstrahlung can not escape freely either in most cases. They have to diffuse out from the inner, generally hot, region to the outer, generally cool, region. As in the stellar interior, the amount of diffusion depends on the temperature gradient and the temperature gradient itself is maintained by the amount of energy transported from one part of the gas to the other. However, one important difference from the case of stellar interior is that part of the bulk kinetic energy can always be converted to the thermal energy which will ultimately be radiated away, whereas energy generation is usually confined in the nuclear burning core in stellar interior. Hence, the correct way to handle this kind of dynamic mixture of gas and radiation is to fully solve the three-dimensional radiation hydrodynamics. This will tell how much radiation is generated and transported from one part to the other and, therefore, determine the physical state of the gas and radiation in the next time step. However, this is almost an impossible task and we need to find a simpler way to treat this.

One reasonable and efficient way for this type of hydrodynamic calculation is to use a volume cooling rate which can approximate the radiation transport process. The following form for the cooling was adopted:

$$\epsilon = \eta \beta \frac{T^4}{R_*} \frac{1 - e^{-\tau_*}}{e^{-\tau_*} + (1 - e^{-\tau_*})\eta \tau_*/12\sigma}, \tag{9}$$

where

$$\eta \equiv \frac{\epsilon_{ff}}{\alpha_{ff}T^4} = 8.2 \times 10^{-3}\frac{\bar{g}_B}{\bar{g}_R};$$

$$\beta \equiv \sqrt{\frac{\alpha_{ff}}{3(\alpha_{ff} + \alpha_{es})}}. \tag{10}$$

Here ϵ is the cooling rate per unit volume, σ is the Stefan-Boltzmann constant, $\bar{g}_B$ is the frequency average of the velocity averaged free-free Gaunt factor, $\bar{g}_R$ is the Rosseland mean free-free Gaunt factor, and α_{ff} and α_{es} are the Rosseland mean free-free absorption and electron scattering coefficients, respectively. This form is valid both for effectively optically thin ($\tau_* \ll 1$) and for effectively optically thick ($\tau_* \gg 1$) cases (Liang & Wandel 1991; Wandel & Liang 1991). The amount of cooling is calculated explicitly by equation (9), and is subtracted from the thermal energy of each cell after every hydrodynamic step. The effective photon travel length R_* is determined by $\tau_*/\sqrt{3\alpha_{ff}(\alpha_{ff} + \alpha_{es})}$, where local values were used for α_{ff} and α_{es}.

For effectively optically thick ($\tau_* \gg 1$) cases, equation (9) reduces to the cooling rate by diffusion

$$\epsilon = \frac{u_r(T)}{\tau_{tot}R_{tot}/c} \quad \text{for } \tau_* \gg 1, \tag{11}$$

where τ_{tot} is the total optical depth $\tau_{ff} + \tau_{es}$ and $R_{tot} \equiv \tau_{tot}/\sqrt{\alpha_{ff} + \alpha_{es}}$. For effectively optically thin ($\tau_* \ll 1$) cases, bremsstrahlung emission directly determines the cooling rate:

$$\epsilon = \epsilon_{ff} \quad \text{for } \tau_* \ll 1, \tag{12}$$

where ϵ_{ff} is the bremsstrahlung emission rate per unit volume.

Energy of bremsstrahlung photons can be changed via Compton scattering since electron scattering optical depth is usually quite high. However, this happens only when the typical length scale for absorption is longer than that for Compton upscatter. Otherwise, photons will be absorbed and thermalized before being significantly upscattered. Since the mean free path due to electron scattering is $(n_e \sigma_{es})^{-1}$ and photons need to be scattered on the order of $4 m_e c^2 / kT$ times to gain significant energy boost, the typical length scale for upscattering is $\lambda_{\text{Compt}} \equiv (n_e \sigma_{es} 4kT/m_e c^2)^{-1}$. In the above, n_e is the electron number density, σ_{es} the Thompson cross section, and m_e the mass of electron. In the thermalized regions of all simulations, the absorption length scale $\lambda_{\text{abs}} = \alpha_{ff}^{-1}$ is much larger than λ_{Compt} and Compton scattering can be neglected.

4. Results from Numerical Calculations

4.1. MORPHOLOGY AND STRUCTURE

The thermalized gas emerges from the narrow, slab-like shock region, and expands out into a larger volume in two opposite directions due to the post-shock pressure. Thus the post-collision material forms two expanding streams several times thicker than the pre-shock streams. Figure 4 shows the density, temperature and velocity field near the collision point. The temperature of the shocked material is kept well below 10^6 K because cooling becomes more efficient at higher temperature. On the other hand, τ_* is greater than 10 in most regions because of the high Thompson opacity, and goes up as high as $\sim 10^6$ in the regions that are not thermalized yet. Cooling is most prominent at the shock slab through diffusion, but the ratio of the cooling to the pre-cooling thermal energy in that region during one simulation time step is only $\sim 1\%$, which shows that the radiative cooling time scale is considerably larger than the hydrodynamical time scale.

4.2. ENERGY CONVERSION

The key question of the hydrodynamic collision is the how much fraction of stream's kinetic energy is converted into thermal energy and how much fraction of such thermal energy escapes through radiative cooling. The typical duration of stream collision is 10^6 seconds. Since the numerical simulations usually are limited within a small region around the collision point, the duration of simulations with reasonable accuracy is only the crossing time over the computational domain, which is a few percents of the duration of the first collision phase. For this reason we need to extrapolate the results to make some estimates on the energy budget problem during one whole collision phase.

Since mass elements are continuously coming into and going out from the calculation volume, and since it is very difficult to keep track of mass elements in the simulations based on fixed data points (grids), calculating the amount of energy conversion of a mass element which experiences collision is not trivial. Let's define $E_{in,sum}$ as the cumulated (from $t = 0$) input energy ($E_{in,sum} = 0$ at $t = 0$), $E_{th,sum}$ as the summation of the thermal energy inside the calculation box at a certain epoch and cumulated thermal energy which has escaped from the box until that epoch, and $E_{r,sum}$ as the cumulative energy which has been radiated away before the same epoch.

The thermal energy of a mass element before collision is negligible compared to its kinetic energy. Thus the ratio $R_{th} \equiv (E_{th,sum} + E_{r,sum})/E_{in,sum}$ is the fraction of total thermalized energy converted from the total input kinetic energy. And the ratio $R_r \equiv E_{r,sum}/E_{in,sum}$ is the fraction of total radiated energy out of the total input energy. After a rapid increase in the beginning, R_{th} decreases to an asymptotic value slowly. This relatively low thermalization rate in the later part is due to a growth of accumulated mass elements in the collision area which enlarges the effective collision cross section and lessens the relative velocity between two streams. Although R_{th} reaches as high as 16 % in the beginning, it quickly decreases to an asymptotic value of about 9 %. On the other hand, the cooling fraction R_r of the same simulation has a gentle maximum at $t \approx 1.5 \times 10^4$ sec and is converged to ~ 0.3 %. The maximum of R_r is slightly behind that of R_{th}, but this lag does not represent the actual time for diffusion from the collision center to the surface because rather simplified cooling treatment tends to make the radiation occur sooner although it calculates about right amount of radiation. From these ratios, we expect that the overall amount of energy that escapes from the calculation region in forms of radiation during the whole collision phase is ~ 0.3 % of the total energy input or ~ 3 % of total thermalized energy.

4.3. REDISTRIBUTION OF ENERGY AND ANGULAR MOMENTUM

The orbits of the shocked gas is is determined by E_k and L_z values in the post-shock expanding phase. The stellar material within the stream has narrow distribution of energy and angular momentum. The shock redistributes the energy and angular momentum significantly. The shock pro-

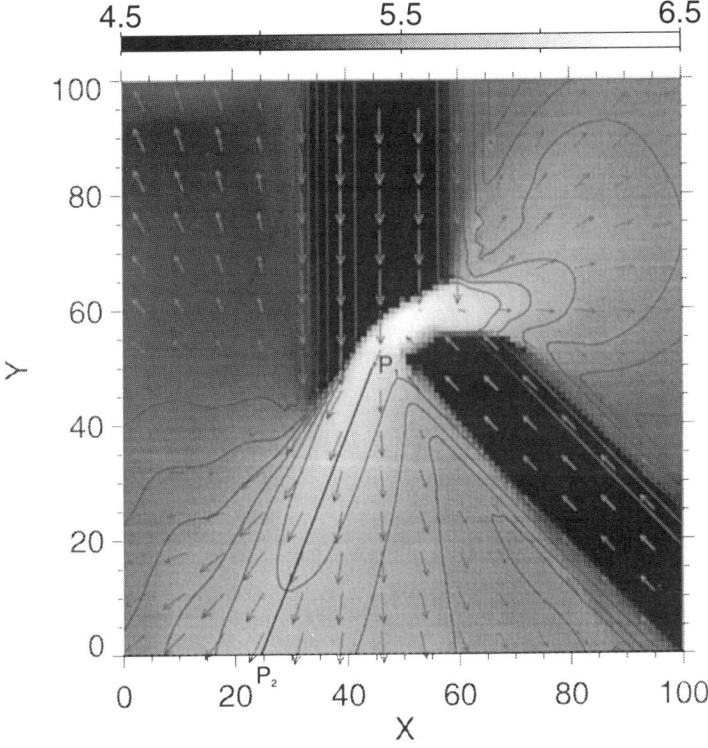

Fig. 4 : Density, temperature and velocity vector map near the stream collision. The contour is for density, grey-scale map is for the temperature and the arrows represent the velocity field. The unit of grid is 5.93×10^{10} cm. The logarithmic scale of temperature is shown as a bar on top.

duces essentially two blobs that are expanding in opposite directions. The blob expanding forward becomes less tightly bound compared to the blob expanding backward.

The hydrodynamic collision changes the angular momentum more efficiently. The angular momentum of the backward blob changes sign so that they move on retrograde orbit. Those on prograde orbits will eventually collide with those on retrograde orbits, although the density of the shocked gas is much smaller. The redistribution of angular momentum provides next step for the circularization of the stellar debris. The subsequent evolution of the shocked material would require more elaborate treatment of the hydrodynamics.

The shock heats up the gas which expands nearly adiabatically when the optical depth is large. The photons diffuse outward and the hot blob can be approximated as a black-body. In this phase, the effective temperature is about 10^5 K and the peak luminosity is around 10^{40} ergs/s. As the blob expands, the bremsstrahlung dominates the cooling. The luminosity rises to about 10^{42} ergs/s because photons can escape more freely during this phase.

Tidal disruption of a single star can potentially generate energy comparable to the supernova explosion but lasts for much longer time ($\sim$ a few tens of years). In order to produce significant luminosity, the stellar material has to be settle down to an accretion disk (or torus) in short time scale. The hydrodynamic collision of stream formed by stellar debris could provide an efficient way of circularizing the initially loosely bound material. The numerical studies of stream collisions give some hints on the final fate of stellar debris.

Thermalized gas emerges out from the narrow, slab-like shock region and expands out into a larger volume in two opposite directions forming two expanding streams that are several times thicker than the pre-shock streams. The angle of the shock slab, which determines the effectiveness of the collision, depends on the heights and widths of the incoming and outgoing streams.

Approximately 5 to 10% of the initial kinetic energy of the streams is converted into thermal energy during the collision depending on the stream parameters, and again 3 to 4% of the thermalized energy is almost immediately radiated away in form of radiation. The collision alters the angular momentum of the stream more than it changes the kinetic energy. These changes in energy and angular momentum can transform the orbits of the debris stream into more tightly bound ones in a relatively short time scale (less than a year or so).

The thermalized gas stream experiences cooling in two phases: the immediate cooling dominated by the diffusion from optically thick, adiabatically expanding gas, and the subsequent cooling by Bremsstrahlung and recombination from extended, optically thin gas. While the shock heated gas will emit like a blackbody radiation with effective temperature of 10^5 K and bolometric luminosity of 10^{40} erg/sec during the first phase, the luminosity during the second cooling phase will be around 10^{42} erg/sec. Such a flare up event will be intermittent with the period of about a month because the collision itself is likely to be intermittent.

Acknowledgements

This work was supported in part by the Cray Research and Development Grant in 1996. Numerical calculations were carried out by Seungsoo Kim. The numerical code of TVD was kindly provided by D. Ryu.

5. References

Cannizzo, J. K., Lee, H. M., & Goodman, J., 1990, ApJ, 351, 38
Cohn, H. & Kulsrud, R. M., 1978, ApJ, 226, 1087.
Eckart, A., & Genzel, R., 1996, Nature, 333, 415
Evans, C. R., & Kochanek, C. S., ApJL, 346, L13
Goodman, J. & Lee, H. M. 1989, ApJ, 337, 84
Gurzadyan, V. G. & Ozernoi, L. M., 1980, A&AP, 86, 315
Gurzadyan, V. G. & Ozernoi, L. M., 1981, A&Ap, 95, 39
Harten, A. 1983, J. Comp. Phys., 49, 357
Kim, S. S., Lee, H. M., & Park, M-G., 1998, in preparation.
Kochanek, C. S. 1994, ApJ, 422, 508
Kormendy, J. & Richstone, D. O. 1992, ApJ, 393, 559
Laguna, P., Miller, W. A., Zurek, W. H., & Davies, M. B., 1993, ApJ, 410, L83
Lee, H. M., Kang, H., & Ryu, D. 1995, ApJ, 464, 131
Liang, E. P. & Wandel, A. 1991, ApJ, 376, 746
Loeb, A., & Ulmer, A., 1997, ApJ, 489, 573.
Luminet, J. P., & Carter, B., 1986, ApJS, 61, 219.
Monaghan, J. J. & Lee, H. M., 1994, in *Nuclei of Normal Galaxies*, NATO ASI Series C:445, eds. R. Genzel & A. I. Harris, (Kluwer), p449
Magorrian, J., Tremaine, S., Richstone, D., Bender, R., Bower, G., Dressler, A., Faber, S. M., Gebhardt, K., Green, R., Grillmair, C., Kormendy, J., & Lauer, T., 1998, AJ, in press.
Rees, M. J., 1988, *Nature*, 333, 523
Rees, M. J., 1994, in *Nuclei of Normal Galaxies*, NATO ASI Series C:445, eds. R. Genzel & A. I. Harris, (Kluwer), p453
Wandel, A. & Liang, E. P. 1991, ApJ, 380, 84

(left to right) I. Chattopadhyay, T.K. Das, A. Ray & B. Mukhopadhyay

L. Takalo

MULTIFREQUENCY MONITORING OF BLAZARS

LEO O. TAKALO
Tuorla Observatory
FIN-21500 Piikkiö, Finland

Abstract: We discuss the results from recent multifrequency monitoring campaigns of selected blazars covering the whole electromagnetic spectrum from radio bands all the way up to TeV-energies. Special attention will be given to simultaneous observations at different frequencies. The importance of long term monitoring at different spectral regions will be discussed in connection with the current theoretical models for blazar radiation.

1. Introduction

Blazars are a subsample of AGNs, consisting of BL Lacertae objects and high polarization quasars (HPQs) (e.g. Urry 1996; Urry & Padovani 1995 and references therein). They are thought to be "normal" AGN, except for the viewing angle; in ordinary QSOs we are looking along the accretion disk, in blazars perpendicular to the disk (Urry & Padovani 1995). BL Lac objects can be divided into (at least) two groups according to their discovery technique; namely radio selected BL Lacs (RBL) and X-ray selected BL Lacs (XBL). There exists small differences in the variability behaviour of RBL's and XBL's (e.g. Heidt and Wagner 1998). Possible other differences (e.g. host galaxies and environments) are being presently studied in various institutes.

Blazars are found at the centers of elliptical galaxies (e.g. Wurtz et al. 1997), where the central engine is thought to be a supermassive black hole, with a jet coming from the center. In blazars we are looking almost directly along this relativistic jet. Since we are looking down the jet, understanding blazars will give excellent opportunity for studying the close environments of supermassive black holes. Most blazars are also superluminal radio sources. Characteristic to them is large variability in all observed frequencies from radio bands to TeV energies in timescales from tens of minutes to years (e.g. Wagner & Witzel, 1995; Takalo 1994). The bulk of the observed radiation is synchrotron radiation coming from the relativistic jet. The variable radio emission is known to be due to shocks in this jet (Marscher 1998). Part of the optical and IR emission could be also due to these shocks (e.g. Takalo et al. 1998a), but some of it is coming from a different emitting region (e.g Sillanpää et al. 1996a,b). The high energy emission is thought to be due to upscattered radio and/or optical photons at the base of the jet (e.g. Ghisellini et al. 1998).

2. Multiwavelength monitoring

Multiwavelength monitoring can be used to investigate the emission mechanisms and locations in these objects. The aims in these programs include the following items:

- To study the variability characteristics in different frequencies and their correlations.

- To study the object spectrum across the frequency space.

- To study the spectral evolution and its possible correlation with the object brightness.

These multiwavelength monitoring programs can be divided into two different categories; long term programs, lasting years, and short campaigns lasting from few days to few weeks. Good examples of the long term programs are the Michigan and Metsähovi radio monitoring programs (Aller et al. 1996; Teräsranta et al. 1998) and the Florida and Tuorla optical programs (e.g. Smith 1996; Takalo et al. 1998b). These long term programs have provided valuable information about the quiescent levels and the activity timescales in the observed blazars. Unfortunately it seems that, at least, in the optical bands the temporal coverage of these monitoring programs is not good enough. The intensive monitoring of 3C66A (see Takalo et al. 1996) and OJ 287 (see Sillanpää et al. 1996a,b) during the OJ-94 project (Takalo 1996) has shown that these objects are active all the time. One should notice that both objects were in outburst during this monitoring.

During the last few years the short term monitoring campaigns have usually been centered around some satellite observations of a selected object. Mostly these campaigns have been organized during EGRET or/and

X-ray satellite (XTE, ASCA, SAX) observations (e.g. Wehrle et al. 1998). A lot of such programs have been organized, utilizing the Target of Opportunity time at these satellites. Such campaigns are usually performed when a blazar is seen in outburst in optical and/or radio bands (e.g. Bloom et al. 1997). Campaigns have also been organized involving the TeV telescopes (Whipple and HEGRA) to monitor Mk 421 and Mk 501 (e.g. Catanese et al. 1997; Petry et al. 1998).

3. Some results

Since blazars are so unpredictably variable in all observed frequencies, the behaviour of an object cannot usually be predicted during a monitoring campaign. Really the only time that the behaviour of a blazar has been successfully predicted are the optical outbursts in OJ 287 during November 1994 and December 1995 (Sillanpää et al. 1996a,b; these proceedings). Note also that there was no radio outburst during the 1994 optical outburst, but there was one in 1995 (Valtaoja 1998), indicating that perhaps there are two separate mechanism responsible for the outbursts.

A blazar can also show very different behaviour during different campaigns. For example PKS 2155-304 showed simultaneous variability in X-ray, UV and optical bands during a campaign in 1991. But in 1994 there was a clear time delay between the variability in X-rays and UV- region, the X-rays leading the UV by a few days (e.g Urry 1997 and references therein).

Similarly correlations can sometimes be seen between optical and radio variability in some blazars, but in other times such correlation is not present in these objects (e.g. Takalo et al. 1998a; Tornikoski et al. 1994; Valtaoja 1998). This kind of "dual" behaviour indicates again, that there are at least two emitting regions in blazars responsible for the optical radiation. It has been clearly shown that the radio outbursts are produced by the shocks in the jet, so part of the optical emission is also coming from these shocks, but part from some other location.

Claims have also been presented, that the gamma-ray activity is related to optical behaviour (e.g. Bloom et al. 1997; Wagner et al. 1995) or to radio behaviour (Valtaoja & Teräsranta 1996). Unfortunately the EGRET data (Mukherjee these proceedings) has so poor temporal coverage, that nothing definite can be said about such correlations. The same applies also to the TeV observations. New multifrequency campaigns are in progress, which may improve these correlation studies.

In 3C 279 a possible correlation can be seen between gamma-ray and X-ray emission during an outburst in 1996. Simultaneous UV and optical data do not show any major outbursts (Wehrle et al. 1998), but show clear flux increase at this time. The radio emission displayed rapid high amplitude variability.

4. Discussion

Even though multifrequency monitoring campaigns have produced a lot of very useful and interesting results, there are still a lot more that can be done. Major problems in organizing these campaigns are:

- The need for simultaneous observations in ALL covered frequencies. In optical and radio bands weather conditions are the most serious problem. These can be overcome by involving more observatories at different locations. Lately the use of electronic communications has improved the situation dramatically.

- Temporal coverage. In order to be able to perform correlation analysis between the variability in two (or more) frequencies the time coverage in all these bands must be good enough. In optical (and radio) bands this can be achieved by increasing the number of telescopes. In X-rays and gamma-rays the problem is really the small number of detected photons.

- Spectral coverage. This problem can be at least partly solved by increasing the number of telescopes in the program, minimizing the problems caused by local weather conditions.

- Lack of (optical/IR) polarization data. Polarization data with good temporal coverage is still missing from the monitoring campaigns. This is mostly due to lack of suitable instrumentation in the monitoring telescopes. But, since most of the blazar emission is due to synchrotron radiation, polarization is essential for understanding blazars.

- Lack of observations with good temporal coverage of blazars in quiescent level in several frequencies. Only in radio, and in some cases in optical, bands do we have good data of some objects at low flux levels. Of course in higher energies the problem is the small number of detected photons at quiescent levels. Understanding the emission mechanisms at "normal" flux levels is very important, since it enables us then to separate the outbursts from the base-emission. So we will get a better handle of the outburst energetics.

During last few years a lot of progress has been made in planning and organizing new multifrequency monitoring campaigns. This is mainly due to increased collaboration between blazar observers around the world, the use of internet and email for communication. Also new automatic telescopes are becoming operational, reducing thus the manpower needed for observations. Right now any detection of unusual behaviour of a blazar in any frequency will be relayed to other institutes, with a request of further observations of this source. In this way we can study in greater detail all observed flares in blazars. So we can create a collection of flare shapes and spectra, that can then be used for detailed investigation on the emission mechanism and location(s). Of course also the long term monitoring outside the flaring events is very important. This will give us the base-emission level and an estimate of the outburst time scales. Several groups are conducting this kind of monitoring right now, some with automatic optical telescope. So within the next few years I expect that we will know much more about these fascinating objects and about the black holes in their centers.

References

Aller, M., et al. (1996), in *Blazar Continuum Variability*, ed. H.R. Miller, C. Noble and J. Webb (PASP conference series), ,
Bloom, S., et al. (1997),, *ApJ*, **490**, L145
Catanese, M., et al. (1997), *ApJ*, **487**, L143
Ghisellini, G., et al. (1998), *MNRAS*, submitted
Heidt, J., and Wagner, S. (1998), *A&A*, **329** , 853
Marscher, A., Gear, W.K., and Travis, J.P. (1991) in *Variability of Blazars*, ed. E. Valtaoja and M: Valtonen (Cambridge Univ. Press), p 85
Petry, D., et al. (1998), in preparation
Sillanpää, A., et al. (1996a), *A&A*, **305**, L17
Sillanpää, A., et al. (1996b), *A&A*, **315** , L13
Smith, A., 1996, in *Blazar Continuum Variability*, ed. H.R. Miller, C. Noble and J. Webb (PASP conference series), **110** , 3
Takalo, L.O., (1994) OJ 287; *Vistas in Astronomy*, **33**, 77
Takalo, L.O., 1996, in *Blazar Continuum Variability*, ed. H.R. Miller, C. Noble and J. Webb (PASP conference series), **110**, 70
Takalo, L.O., et al. (1996), *A&AS*, **120**, 313
Takalo, L.O., et al. (1998a), *A&AS*, in press
Takalo, L.O., et al. (1998b), in *OJ-94 Annual Meeting 1997*, ed. G. Tosti and L. Takalo (Perugia University Observatory publications), **Vol. 3**, 97
Teräsranta, H., et al. (1998), in preparation
Tornikoski, M., et al. (1994), *A&A*, **286**, 60
Urry, M., et al. (1997), *ApJ*, **486**, 799
Urry, M. (1996), in *Blazar Continuum Variability*, ed. H.R. Miller, C. Noble and J. Webb (PASP conference series), **110**, 391

Urry, M., and Padovani, P. (1995) *PASP*, **107**, 803
Valtaoja, E. (1998) in *OJ-94 Annual Meeting 1997*, ed. G. Tosti and L. Takalo (Perugia University Observatory publications), **Vol. 3**, 62
Valtaoja, E., & Teräsranta, H., (1996) *A&AS*, **120**, 491
Wagner, S., et al. (1995) *ApJ*, **454**, L97
Wagner, W., Witzel, A. (1995) *ARA&A*, **33**, 163
Wehrle, A., et al. (1998), *ApJ*, submitted
Wurtz, R., et al. (1997) *ApJ*, **480**, 547

THE OJ287 SUPERMASSIVE BINARY BLACK HOLE MODEL AND THE NEW UNIFIED SCHEME FOR THE AGNS

AIMO K. SILLANPÄÄ
Tuorla Observatory
FIN-21500 Piikkiö, Finland

1. Introduction

It has been suggested already a very long time ago that supermassive binary black holes (SMBBHs) may exist in the nuclei of the Active Galactic Nuclei (AGN) (e.g. Begelman, Blandford and Rees 1980). To prove this observationally is an extremely difficult task. There are basically three different methods to do this:

1. To prove that the regular helical structures in the VLBI and VLA jets are caused by the binary system (e.g. Gower et al. 1982).

2. With studies of the peculiar double-peaked Balmer lines in some radio galaxies and proving that these doubles come from the individual broad-line regions associated with black holes of a SMBBH system (e.g. Gaskell 1996).

3. To search for strict periodicities in the energy outputs of the AGNs (e.g. Sillanpää et al. 1988). A binary system is the only known system which can produce such a strict periodicity in any extragalactic object.

In this review I will present evidences for the clear periodicity in the historical optical light curve of the BL Lac object OJ287. This periodicity is interpreted by the effects of the SMBBH system in the nucleus of the host of OJ287. There are already several papers published giving slightly different models to explain the periodicity. Some of these models don't need strict periodicity but some are based on the very strict cycles.

Nowadays there are also some examples of similar periodicities in other blazars (ON231, Tosti et al. 1998; Mk421, Liu et al. 1997). The problem to find these periodicities is simply that they are so long (at least 10-100 years) and we have systematic data on quasars only from the last 35 years.

The overall importance of the possible SMBBH systems in the whole class of the AGNs is also briefly discussed in the second part of this review.

2. The case of the blazar OJ287

OJ287 is a prototype of the BL Lac objects. Its redshift is 0.306. It is a very strong emitter from gamma-rays to centimeter radio-bands showing also superluminal motion in its VLBI knots. Because of its suitable position on the sky quite close to the ecliptic, its historical light curve covers already more than 100 years. Based on this light curve Sillanpää et al. (1988) proposed a SMBBH model for the optical outbursts being repeated every 12 years. In this model the outbursts were explained by tidally triggered mass inflow in the accretion disk around the primary black hole that is disturbed by a less massive secondary black hole. This disturbation occur when the secondary makes its closest approach during its eccentric orbit around the primary black hole.

Based on the historical light curve Sillanpää et al. (1988) predicted that the next outburst will occur in the fall 1994 and that there will be also a secondary outburst about 1.2 year later. A large international monitoring project, OJ-94, was set up with the aim of monitoring the blazar OJ287 during the predicted times. Both of these outbursts were also detected almost exactly at the expected time (Sillanpää et al. 1996a,b). The whole historical light curve including also the last two outbursts is shown in Fig.1.. From this plot it is easy to see that the object has been extremely variable all the time, also between the major outbursts. The whole amplitude of the variations has been about 5.5 magnitudes which means in the linear scale that the ratio between the brightest and faintest level has been about 150!

Because of the huge variability it is not possible to use normal periodicity search methods (Stothers and Sillanpää 1997) or those methods are at least very uncertain. The only way to check if the suggested periodicity is strict (like it should be if it is caused by the supermassive binary system) is to plot the best observed periods into one plot. This is shown in Fig.2. where I have plotted last five periods in which we can say almost exactly the real outburst time. I have used the time difference between the last two outbursts, 11.86 year, because this can be determined with an accuracy of about two weeks. The vertical line in the middle of the graph shows the exact time when the outbursts should occur. It is very easy to see that the accuracy has been extremely good with the only exception being the

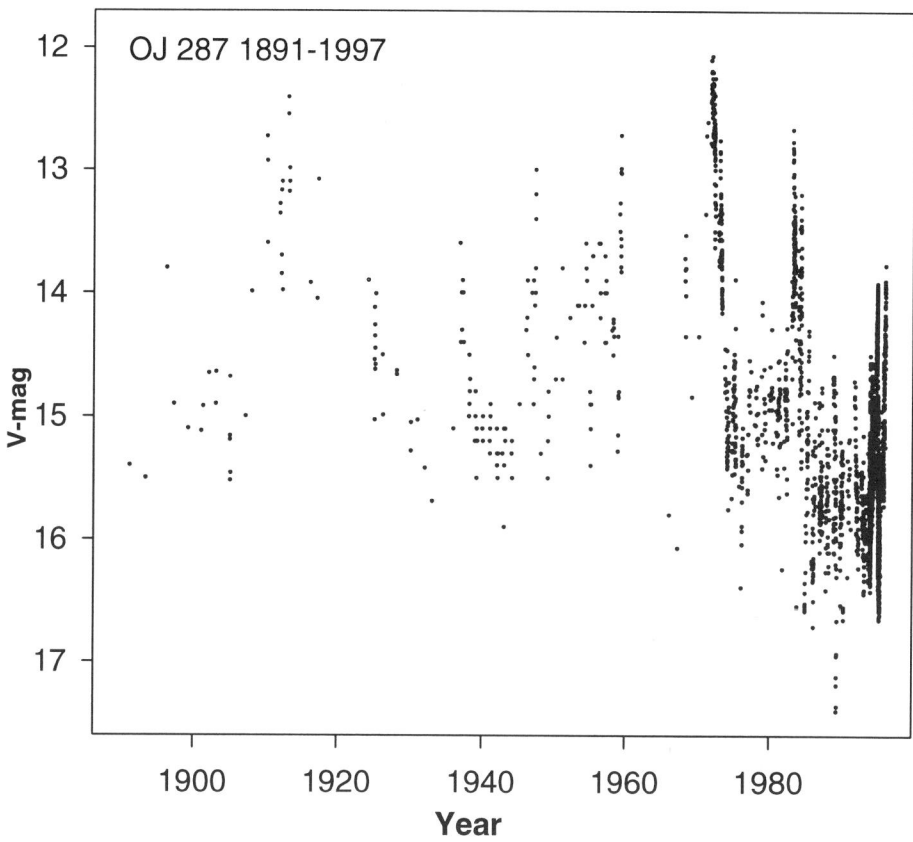

Figure 1. The historical V-band light curve of the blazar OJ287.

outburst in 1970's. However, also during this period OJ287 has been very bright and it is possible that we have only missed the real outburst because of the lack of the data points just at the correct time. This has maybe also caused the misunderstanding that the periods are not strict! From the other periods we can say that the outburst accuracy has been better than three weeks! This is a very strong proof for the SMBBH system behind the outbursts. From the earlier periods before 1940's we don't have enough data to see the exact outburst times.

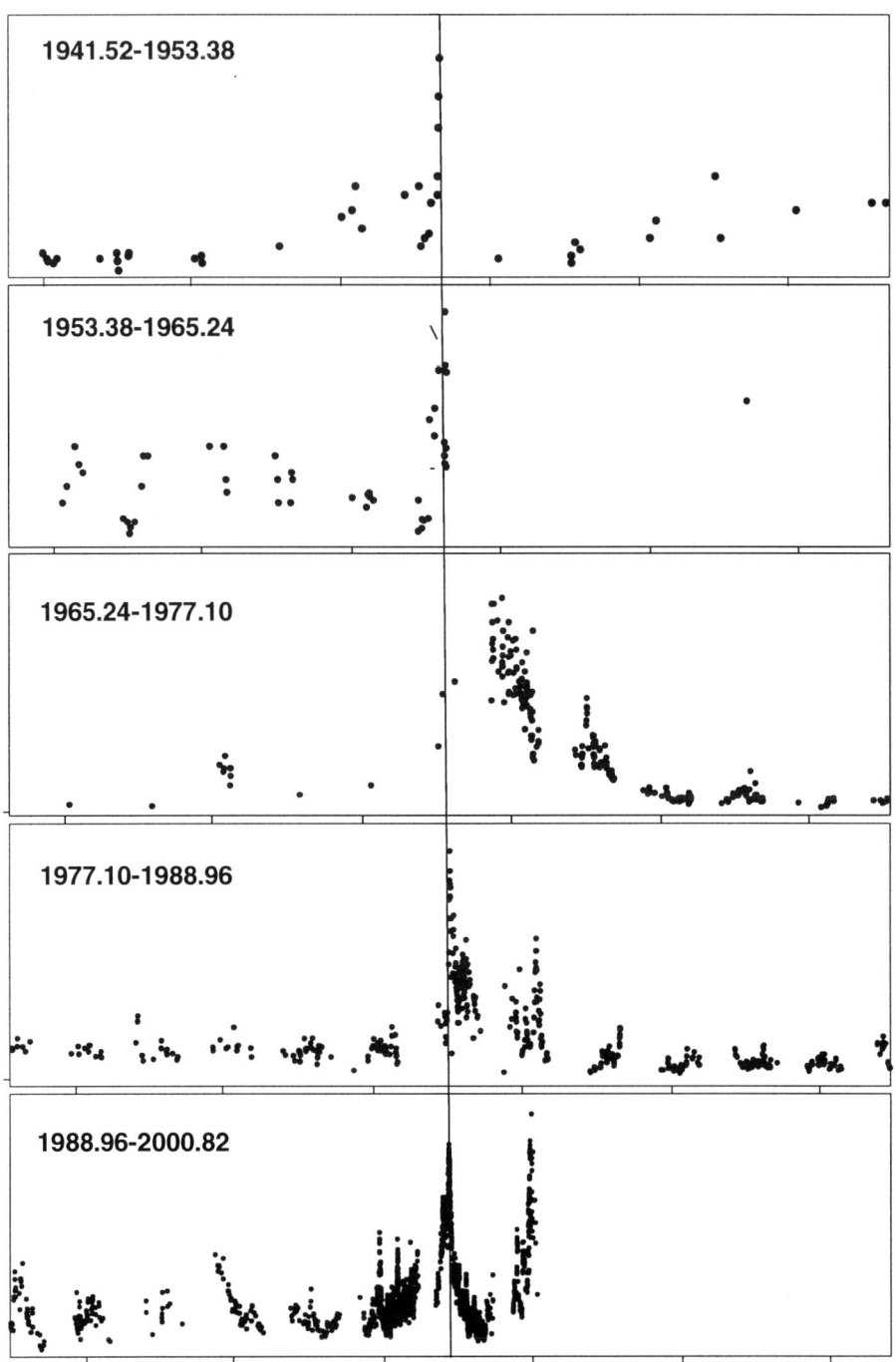

Figure 2. The last five 11.86 year optical outburst periods. The vertical line in the middle of the graph shows the proposed outburst times.

3. The new unified model for the AGNs

3.1. THE OLD MODEL AND THE BASIC DIFFERENCES BETWEEN RADIO-QUIET AND RADIO-LOUD AGNS.

The old Unified Scheme for the radio-loud Active Galactic Nuclei says that they are in principle similar objects but their relativistic jets are viewed from different directions (e.g. Antonucci 1993). If the viewing angle is smaller than about 45 degrees we see a quasar but with the larger angles we see a radio galaxy. In the case of the BL Lac objects the jets are oriented almost exactly towards us.

BUT only about 10 per cent of the AGNs are radio-loud so the old scheme does not tell us basically anything about the 90 per cent of the AGNs which are radio-quiet! However, the only real difference between the radio-loud and radio-quiet AGNs is the radio emission. All the other properties are similar at infrared, optical, ultraviolet and X-rays (e.g. Steidel and Sargent 1991).

There are still at least four basic differences between radio-loud and radio-quiet AGNs:

1. The radio-loud objects produce large-scale radio jets with a kinetic power of the jets being a significant fraction of the total luminosity but the radio-ejecta of the radio-quiet ones are insignificant.
2. The radio-loud objects are always associated with elliptical galaxies but the radio-quiet objects prefer almost always spiral hosts.
3. The absolute magnitudes of the hosts of radio galaxies and radio-loud quasars are almost the same but the hosts of the radio-quiet quasars are about one magnitude less luminous (e.g. Véron-Cetty and Woltjer 1990).
4. Radio-quiet AGNs tend to be in low galaxy-density environments whereas those hosting radio-loud objecst are found mostly in high-density environments (Smith and Heckman 1990).

3.2. THE NEW MODEL

In my new Unified Scheme for the AGNs the visible difference between radio-quiet and radio-loud objects is still the same as in the old one: in the quiet objects we don't have any radio jet! But now I can say also the reason for that: in radio-quiet objects we have only a single supermassive black hole in the nucleus of the host but in the radio-loud ones we have a SMBBH system. This system is the only mechanism which can collimate and accelerate the matter outflow from the primary black hole. I have at least three reasons to believe that this is the situation:

1. The radio-loud AGNs are always found in elliptical galaxies and these galaxies are assumed to be formed by merging of two spiral galaxies. The typical merging rate in rich cluster is about 1 every 10^9 year. On the other hand Ho (these proceedings) has shown that almost all galaxies (at least massive enough) have a SMBH in their nuclei. Where are the BHs of the two merging galaxies now: in the nucleus of the formed elliptical!

2. The main problem with the SMBBH systems has been the very short life time before the final coalescence but some very recent simulations by Makino (1997) have shown that they can live at least 10^9 year. This is almost exactly the same as normally assumed to be the life time of a radio-loud AGN!

3. There is also a clear analogy with Galactic mini-quasars where we ALWAYS have a binary system behind their relativistic radio jets emerging from the system (Mirabel and Rodriguez 1994). So why the situation should be different if we scale the system to the SMBBHs?

As a brief summary of the new model we can say that in radio-loud AGNs we have a SMBBH system in their nuclei. This system collimates the matter outflow from the primary via two possible scenarios:

1. The production of the binary system causes a very rapid rotation of the BH and this rotation collimates the jet (Wilson and Colbert 1995).

2. The secondary black hole creates a magnetic field which collimates and accelerates the matter outflow and produces a radio jet.

References

Antonucci, R. (1993), *ARA&A*, **31**, 473
Begelman, M., Blandford, R., Rees, M. (1980), *Nat*, **287**, 307
Gaskell, C. M. (1996), *ApJ*, **464**, L107
Gower, A., et al. (1982), *ApJ*, **262**, 478
Liu, F. K., Liu, B. F., Xie, G. Z. (1997), *A&AS*, **123**, 569
Makino, J. (1997), *ApJ*, **478**, 58
Mirabel, I. F., Rodriguez, L. F. (1994), *Nat*, **371**, 46
Sillanpää, A., et al. (1988), *ApJ*, **325**, 628
Sillanpää, A., et al. (1996a), *A&A*, **305**, L17
Sillanpää, A., et al. (1996b), *A&A*, **315**, L13
Smith, E. P., Heckman, T. M. (1990), *ApJ*, **348**, 38
Steidel, C. C., Sargent, W. L. W. (1991), *ApJ*, **382**, 433
Stothers, R. B., Sillanpää, A. (1997), *ApJ*, **475**, L13
Tosti, G., et al. (1998), *A&A, in press*
Véron-Cetty, M. P., Woltjer, L. (1990), *A&A*, **236**, 69
Wilson, A. S., Colbert, E. J. M. (1995), *ApJ*, **438**, 62

HIGH ENERGY GAMMA-RAY EMISSION FROM BLAZARS:

*EGRET OBSERVATIONS**

R. MUKHERJEE
Barnard College & Columbia University
Dept. of Physics & Astronomy
3009 Broadway, 506 Altschul
New York, NY 10027

Abstract. We will present a summary of the observations of blazars by the Energetic Gamma Ray Experiment Telescope (EGRET) on the Compton Gamma Ray Observatory (CGRO). EGRET has detected high energy γ-ray emission at energies greater than 100 MeV from more that 50 blazars. These sources show inferred isotropic luminosities as large as 3×10^{49} ergs s^{-1}. One of the most remarkable characteristics of the EGRET observations is that the γ-ray luminosity often dominates the bolometric power of the blazar. A few of the blazars are seen to exhibit variability on very short time-scales of one day or less. The combination of high luminosities and time variations seen in the γ-ray data indicate that γ-rays are an important component of the relativistic jet thought to characterize blazars. Currently most models for blazars involve a beaming scenario. In leptonic models, where electrons are the primary accelerated particles, γ-ray emission is believed to be due to inverse Compton scattering of low energy photons, although opinions differ as to the source of the soft photons. Hardronic models involve secondary production or photomeson production followed by pair cascades, and predict associated neutrino production.

1. Introduction

One of the most striking accomplishments of the Energetic Gamma Ray Experiment Telescope (EGRET) instrument on the Compton Gamma-Ray Observatory (CGRO) is the detection of high-energy γ-rays from active galaxies whose emission at most wavebands is dominated by non-thermal

---
*Invited review paper to appear in *Observational Evidence for Black Holes in the Universe*, ed. S. K. Chakrabarti (Dordrecht: Kluwer).

processes. These objects, called "blazars," are highly variable at most frequencies and are bright radio sources. Prior to the launch of CGRO, 3C 273, discovered by COS-B (Swanenburg et al. 1978), was the only known extragalactic source of γ-rays. Since then, EGRET has detected more than 50 blazars in high energy (> 100 MeV) γ-rays (Mukherjee et al. 1997; Thompson et al. 1995; 1996).

The blazars detected by EGRET all share the common characteristic that they are radio-loud, flat-spectrum radio sources, with radio spectral indices $\alpha_r \geq -0.6$ (von Montigny et al. 1995). Several of these blazars are known to demonstrate superluminal motion of components resolved with VLBI (3C 279, 3C 273, 3C 454.3, PKS 0528+134, for example). The blazar class of active galactic nuclei (AGN) includes BL Lac objects, highly polarized quasars (HPQ), or optically violent variable (OVV) quasars and are characterized by one or more of the properties of this source class, namely, a non-thermal continuum spectrum, a flat radio spectrum, strong variability and optical polarization. For many of the EGRET-detected blazars, the γ-ray energy flux is dominant over the flux in lower energy bands. The redshifts of these sources range from 0.03 to 2.28 and the average photon spectral index, assuming a simple power law fit to the spectrum, is ~ 2.2. Many of the blazars exhibit variability in their γ-ray flux on timescales of several days to months. In addition, blazars exhibit strong and rapid variability in both optical and radio wavelengths.

Of the 51 blazars reviewed here, 14 are BL Lac objects, and the rest are flat spectrum radio quasars (FSRQs). BL Lac objects generally have stronger polarization and weaker optical lines. In fact, some BL Lac objects have no redshift determination because they have no identified lines above their optical continuum. FSRQs are generally more distant and more luminous compared to the BL Lac objects.

This review summarizes the present knowledge on γ-ray observations of blazars by EGRET. A brief description of the EGRET instrument and data analysis techniques, and the list of blazars detected by EGRET is given in §2. Temporal variations and γ-ray luminosity of blazars are discussed in §§3 & 4. Section 5 describes the spectral energy distribution of blazars and summarizes the various models that have been proposed to explain the γ-ray emission in blazars.

2. EGRET observations and analysis

2.1. THE EGRET INSTRUMENT

EGRET is a γ-ray telescope that is sensitive in the energy range ~ 30 MeV to 30 GeV. It has the standard components of a high-energy γ-ray instrument: an anticoincidence dome to discriminate against charged particles, a spark chamber particle track detector with interspersed high-Z material to convert the γ-rays into electron-positron pairs, a triggering telescope to detect the presence of the pair with the correct direction of motion, and an energy measurement system, which in the case of EGRET is a NaI(Tl) crystal. EGRET has an effective area of 1500 cm^2 in the energy range 0.2 GeV to 1 GeV, decreasing to about one-half the on-axis value at 18° off-axis and to one-sixth at 30°. The instrument is described in details by Hughes et al. (1980) and Kanbach et al. (1988, 1989) and the preflight and postflight calibrations are given by Thompson et al. (1993) and Esposito et al. (1998), respectively.

Although EGRET records individual photons in the energy range 30 MeV to about 30 GeV, there are several instrumental characteristics that limit the energy range for which time variation investigations of blazars are viable. At the low end of the energy range, below $\sim$ 70 MeV, there are systematic uncertainties that make the spectral information marginally useful. In addition, the deteriorating point spread function (PSF) and energy resolution at low energies, make analysis more difficult. At high energies, although the systematic uncertainties are reduced, and the PSF and energy resolution are more reasonable, because of the steeply falling spectra, few photons are detected above 5 GeV.

The angular resolution of EGRET is energy dependent, varying from about 8° at 60 MeV to 0.4° above 3 GeV (68% containment). The positions of sources are detected with varying accuracy: better than 0.1° for the very bright sources, or at least 0.5° for sources just above the detection threshold.

The threshold sensitivity of EGRET ($>$ 100 MeV) for a single observation is $\sim 3 \times 10^{-7}$ photons cm^{-2} s^{-1}, and is only about a factor of 50-100 below the maximum blazar flux ever observed. The dynamic range for most observations of blazar variations is, therefore, fairly small.

2.2. EGRET DATA ANALYSIS

The blazars described here were typically observed by EGRET for a period of 1 to 2 weeks; however, several of them were observed for 3 to 5.5 weeks. Following the standard EGRET processing of individual γ-ray events, summary event files were produced with γ-ray arrival times, directions and energies. For the observations reported here, photons coming from directions greater than 30° from the center of the field of view (FOV) were not used, in order to restrict the analysis to photons with the best energy and position determinations. In addition, exposure history files were produced containing information on the instrument's mode of operation and pointing. These maps were used to generate skymaps of counts and intensity for the entire field of view for each observation, using a grid of 0.5° $\times$ 0.5°. The intensity maps were derived simply by dividing the counts by the exposure. The EGRET data processing techniques are described further by Bertsch et al. (1989).

The number of source photons, distributed according to the instrument PSF in excess of the diffuse background, was optimized. An E^{-2} photon spectrum was initially assumed for the source search. The background diffuse radiation was taken to be a combination of a Galactic component caused by cosmic ray interactions in atomic and molecular hydrogen gas (Hunter et al. 1997), as well as an almost uniformly distributed component that is believed to be of extragalactic origin (Sreekumar et al. 1998).

The data were analyzed using the method of maximum likelihood as described by Mattox et al. (1996) and Esposito et al. (1998). The likelihood value, L, for a model of the number of γ-rays in each pixel of a region of the map is given by the product of the probability that the measured counts are consistent with the model counts assuming a Poisson distribution. The probability of one model with likelihood, L_1, better representing the data than another model with likelihood, L_2, is determined from twice the difference of the logarithms of the likelihoods, $2(\ln L_2 - \ln L_1)$. This difference, referred to as the test statistic TS, is distributed like χ^2 with the number of degrees of freedom being the difference in the number of free parameters in the two models. The flux of the point source and the flux of the diffuse background emission in the model are adjusted to maximize the likelihood. The significance of a source detection in sigma is given approximately by the square root of TS.

2.3. EGRET OBSERVATIONS

The 51 blazars listed in Table 1 were all detected by EGRET above 100 MeV during the period of EGRET observations from 1991 April to 1995 September (Phases 1 through 4 of CGRO) (Mukherjee et al. 1997). Some of these blazar associations are not certain; Mattox et al. (1997a) find only 42 identifications to have high confidence. Conversely, some of the unidentified high-latitude EGRET sources are likely to be blazars. In addition to the 42 considered strongest, Mattox et al. (1997a) note 16 possible associations with bright flat-spectrum, blazar-like radio sources. Typically, each blazar listed in Table 1 was seen in several different viewing periods (VPs). The maximum and minimum fluxes observed for each blazar is indicated in Table 1. A more complete list of blazar detections by EGRET may be found in the third EGRET catalog (Hartman et al. 1998).

TABLE 1
EGRET-DETECTED BLAZARS

Name	Other Names	z^a	Maximum^b Flux	Minimum^b Flux	Source Type	V^c
0202+149	4C+15.05		24.5 ± 10.7	< 9	FSRQ	2.29
0208-512		1.003	131.9 ± 24.7	15.9 ± 8.7	FSRQ	11.38
0219+428	3C 66A	0.444	25.5 ± 5.8	< 17	BL LAC	1.05
0235+164	OD160	0.94	82.8 ± 9.2	< 24	BL LAC	7.96
0336-019	CTA26	0.852	186.2 ± 7.6	< 15	FSRQ	5.48
0420-014	OA 129	0.915	51.2 ± 10.5	< 13	FSRQ	2.14
0440-003	NRAO 190	0.844	84.4 ± 12.0	< 11	FSRQ	5.05
0446+112		1.207	105.4 ± 19.2	< 12	FSRQ	2.72
0458-020		2.286	30.8 ± 9.5	< 11	FSRQ	0.47
0521-365		0.055	37.5 ± 11.2	< 13	BL LAC	2.04
0528+134		2.06	307.6 ± 34.6	< 40	FSRQ	> 15
0537-441		0.894	89.8 ± 14.5	< 18	BL LAC	4.43
0716+714			44.0 ± 11.0	< 35	BL LAC	1.21
0735+178		> .424	40.9 ± 21.3	< 17	BL LAC	0.30
0827+243		2.046	68.1 ± 14.4	< 26	FSRQ	2.00
0829+046	OJ+49	0.18	19.1 ± 10.6	< 30	BL LAC	0.11
0836+710	4C+71	2.17	31.4 ± 8.9	< 7	FSRQ	1.92
0917+449		2.18	32.1 ± 9.8	< 31	FSRQ	0.29
0954+556	4C+55	0.901	48.1 ± 15.7	< 11	FSRQ	0.61
0954+658		0.368	17.8 ± 9.4	< 5	BL LAC	1.24
1101+384	Mrk 421	0.031	26.2 ± 6.9	< 16	BL LAC	0.56
1127-145	OM-146	1.187	80.6 ± 19.9	< 16	FSRQ	1.69
1156+295	4C+29	0.729	192.2 ± 47.8	< 6	FSRQ	3.01
1219+285	ON231	0.102	12.7 ± 6.8	< 11	BL LAC	0.92
1222+216	4C 21.35	0.435	50.7 ± 15.6	< 8	FSRQ	3.42
1226+023	3C 273	0.158	55.7 ± 11.9	< 11	FSRQ	3.48
1229-021		1.0448	13.4 ± 4.0	< 8	FSRQ	1.10
1253-055	3C 279	0.538	287.8 ± 10.9	< 18	FSRQ	> 15
1331+170	OP+151	2.0838	12.8 ± 8.1	< 6	FSRQ	0.22
1406-076	OQ-010	1.494	127.6 ± 23.4	< 10	FSRQ	> 15
1424-418		1.522	31.5 ± 9.9	< 13	FSRQ	1.26
1510-089		0.361	28.6 ± 7.6	< 28	FSRQ	0.30
1604+159	4C+15.54	0.357	41.4 ± 12.2	< 18	BL LAC	0.84
1606+106	4C+10.45	1.24	62.5 ± 13.0	< 15	FSRQ	3.34
1611+343	OS+319	1.404	73.2 ± 15.6	< 10	FSRQ	7.05
1622-253		0.786	60.6 ± 22.3	< 38	FSRQ	0.34
1622-297		0.815	245.6 ± 31.8	< 15	FSRQ	> 15
1633+382	4C38.41	1.810	98.3 ± 9.3	< 88	FSRQ	4.72
1730-130		0.902	120.2 ± 36.5	< 42	FSRQ	1.79
1739+522	4C+51.37	1.375	51.0 ± 27.6	< 29	FSRQ	1.03
1908-021			43.7 ± 27.5	< 14	FSRQ	0.81
1933-400		0.966	99.4 ± 32.2	< 9	FSRQ	1.78
2022-077			72.0 ± 13.3	< 5	FSRQ	3.51
2032+107	OW+154	0.601	24.4 ± 10.2	< 12	BL LAC	0.20
2052-474		1.489	37.6 ± 21.6	< 11	FSRQ	1.58
2155-304		0.116	32.3 ± 7.8	< 17	BL LAC	1.77
2200+420	BL Lac	0.069	78.1 ± 38.3^d	< 16	BL LAC	0.69^d
2209+236			35.8 ± 18.0	< 10	FSRQ	0.43
2230+114	CTA 102	1.037	38.8 ± 13.9	< 27	FSRQ	1.79
2251+138	3C 454.3	0.859	119.3 ± 18.7	< 50	FSRQ	3.21
2356+196	OZ+193	1.066	29.7 ± 9.0	< 17	FSRQ	0.87

a Redshift.
b Units: 10^{-8} ph cm^{-2} s^{-1}. Upper limits are 2 sigma.
c Variability index (see text for definition).
d Both the maximum flux and the variability index would increase if the recent flare in BL Lac were taken into account (Bloom et al. 1997).

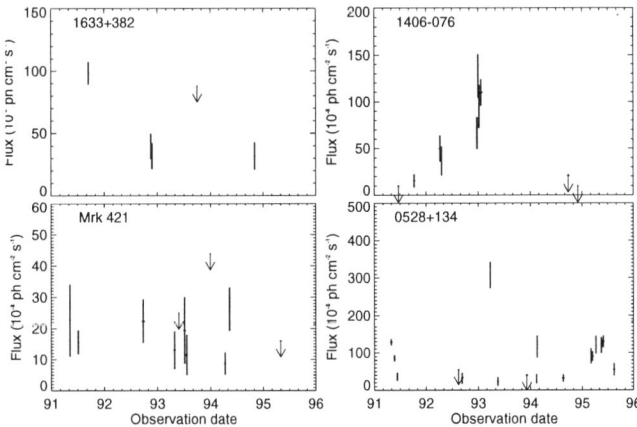

Figure 1. Flux history of four blazars (PKS 1633+382, PKS 1406-076, Mrk 421, and PKS 0528+134) from 1991 April to 1995 September, as detected by EGRET. 2σ upper limits are indicated by downward arrows.

3. Time variability

The fluxes of the blazars detected by EGRET have been found to be variable on time scales of a year or more down to well under a day. Long term variations of blazars have been addressed earlier by several authors (eg. von Montigny et al. 1995; Hartman et al. 1996a; Mukherjee et al. 1997). In some cases, many of the detected blazars have exhibited flux variations up to a factor of about 30 between different observations. Figure 1 shows the flux history of four EGRET-detected blazars. The horizontal bars on the individual data points denote the extent of the VP for that observation. Fluxes have been plotted for all detections greater than 2σ. For detections below 2σ, upper limits at the 95% confidence level are shown. A systematic uncertainty of 6% was added in quadrature with the statistical uncertainty for each flux value, consistent with the analysis of McLaughlin et al. (1996) on EGRET source variability.

In order to quantify the flux variability of the blazars in Table 1, Mukherjee et al. (1997) calculated the variability index, $V = \log Q$, as defined by McLaughlin et al. (1996), where $Q = 1 - P_\chi(\chi^2, \nu)$. Here $P_\chi(\chi^2, \nu)$ is the probability of observing χ^2 or something larger from a χ^2 distribution with ν degrees of freedom. The flux versus time data were fit to a constant flux and the reduced χ^2_ν, for ν degrees of freedom, was calculated using the least square fit method. For a nonvariable source, a constant flux is expected to fit the data well, and the mean value of the χ^2 distribution is expected to

High Energy Gamma-Ray Emission from Blazars: 221

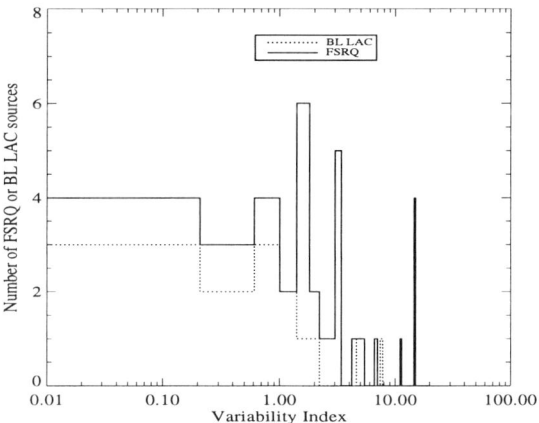

Figure 2. Distributions of the variability indices for BL Lac objects and FSRQs.

be equal to the number of degrees of freedom in the data. The quantity V is used to judge the strength of the evidence for flux variability. Following the classification of McLaughlin et al., $V < 0.5$ was taken to indicate non-variability, $V \geq 1$ to indicate variability, and $0.5 \leq V < 1$ as uncertain. Table 1 lists the value of V for each source.

Of the 51 blazars reviewed here, 35 are found to be variable ($V \geq 1$), 9 are non-variable ($V < 0.5$), and 7 fall in the range of uncertain variability. It should be noted that, although the criterion used here to gauge variability of a blazar is somewhat arbitrary, it does provide a way to compare the numbers obtained. Also, as McLaughlin et al. (1996) have noted, changing these criteria by 20% yields similar results. If the FSRQs and the BL Lac objects are considered separately, it is found that 76% of the FSRQs in the sample are variable, while 16% are non-variable. Similarly, for the BL Lac objects, 50% are variable, while 21% are definitely non-variable. It should be noted that the low intrinsic luminosity of BL Lac objects could bias observations (see discussions in §5.2). Figure 2 shows the distribution of the variability indices for the FSRQs and the BL Lac objects. The BL Lac objects in the data set are found to be less variable on the average than the FSRQs. Recent observations of flares in BL Lac objects, however, modify some of these conclusions. For example, BL Lac was detected during a γ-ray outburst with an average flux of $(171 \pm 42) \times 10^{-8}$ photons cm^{-2} s^{-1} in July 1997 (Bloom et al. 1997). BL Lac would have a high value of V in Table 1, if this information were taken into account.

Figure 3 shows a plot of the variability index as a function of the

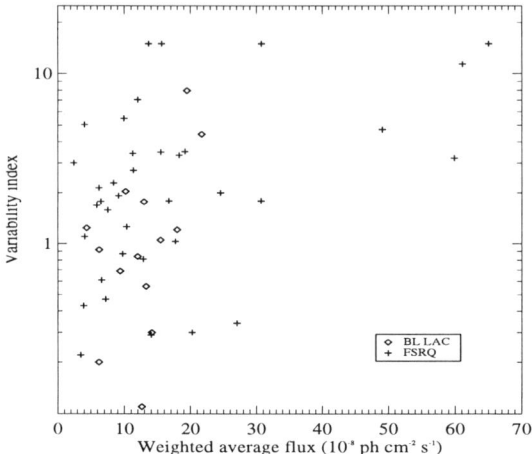

Figure 3. Variability index plotted as a function of the weighted average flux for BL Lac objects and FSRQs. The BL Lac objects are denoted by diamonds and the FSRQs by plus signs in the figure.

weighted average flux for the blazars in Table 1. Note that the sources that have the highest average fluxes all have high variability indices. Only 2 out of 18 blazars with average flux less than 1×10^{-7} photons cm^{-2} s^{-1} have a variability index greater than 2.5. In fact, there are no non-variable blazars in the sample that have high flux.

The study of short term variability in blazars is always limited by the small numbers of photons detected in the short time intervals at γ-ray energies. The shortest time-scale variations detected for blazars with EGRET are for PKS 1622-297 (Mattox et al. 1997b) and 3C 279 (Wehrle et al. 1997). For both these objects the flux was found to increase by a factor of two or more in less than 8 hours. Other objects that have shown flux variations over the period of a few days are 3C 279 (Kniffen et al. 1993), 3C 454.3 (Hartman et al. 1993), 4C 38.41 (Mattox et al. 1993), PKS 1406-076 (Wagner et al. 1995), and PKS 0528+134 (Hunter et al. 1993; Mukherjee et al. 1996). The short time-scale of γ-ray flux variability (e.g. in 3C 279 or PKS 1622-297) when combined with the large inferred γ-ray luminosities, implies that the blazar emission region is very compact. Gamma-ray tests for beaming from variability and flux measurements using the Elliott-Shapiro relation and γ-ray transparency arguments are summarized in a recent review on γ-ray blazars by Hartman et al. (1997). A factor-of-two flux variation on an observed time-scale $\delta t_{\rm obs}$ limits the size r of a stationary isotropically emitting region to be roughly $r \leq c\delta t_{\rm obs}/(1+z)$ by simple light-travel time arguments. Under the assumptions of isotropic radiation and Eddington-limited accretion, the implied minimum black hole masses of blazars are $\geq 8 \times 10^{11}$ $M_\odot$ for PKS 1622-297 (Mattox et al. 1997b) from EGRET observations and $\geq 7.5 \times 10^8$ $M_\odot$ for PKS 0528+134 from COMPTEL observations (Collmar et al. 1997).

4. Luminosity

The γ-ray luminosity can be estimated by considering the relationship between the observed differential energy flux $S_0(E_0)$, where the subscript "0" denotes the observed or present value, and Q_e the power emitted in dE, where $E = E_0(1 + z)$ in the Friedman universe.

$$Q_e[E_0] = 4\pi S_0(E_0)(1+z)^{b-1} \Theta D_L{}^2(z, q_0) \tag{1}$$

where

$$D_L = \frac{c}{H_0 q_0{}^2}[1 - q_0 + q_0 z + (q_0 - 1)(2 q_0 z + 1)^{1/2}] \equiv \frac{cz}{H_0} g(z, q_0). \tag{2}$$

H_0 is the Hubble parameter, q_0 is the deceleration parameter, b is the spectral index, z is the redshift, and Θ is the beaming factor. H_0 is chosen to be 70 and q_0 to be 0.5, although the results obtained here are not highly sensitive to these choices. The beaming factor is taken to be 1. The spectral index is obtained using the analysis described in §4. The luminosity as a function of the redshift is determined using equation (1) and is plotted in Figure 4 for the blazars detected by EGRET. The typical detection threshold for EGRET as a function of z, for relatively good conditions, is also shown in the figure. The actual threshold varies somewhat with exposure and region of the sky, and the average threshold is a little higher than the curve shown, but the shape is the same. The BL Lac objects are indicated in the figure with dark diamonds, and one sees clearly that they are predominantly closer and lower in luminosity.

Recently, Chiang & Mukherjee (1998) have calculated the evolution and luminosity function of the EGRET blazars, and have estimated the contribution of this source class to the diffuse extragalactic gamma-ray background. They find that the evolution is consistent with pure luminosity evolution. According to their estimates, only 25% of the diffuse extragalactic emission measured by SAS-2 and EGRET can be attributed to unresolved γ-ray blazars, contrary to some of the other estimates (eg. Stecker & Salamon 1996). Below 10 MeV, the average blazar spectrum suggests that only about 50% of the measured γ-ray emission could arise from blazars (Sreekumar, Stecker & Kappadath 1997). This leads to the exciting possibility that other sources of diffuse extragalactic γ-ray emission must exist.

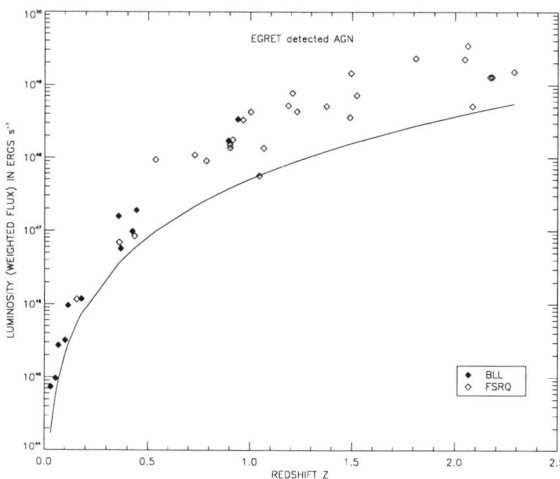

Figure 4. Luminosity vs redshift for blazars detected by EGRET. The BL Lac objects are indicated with filled symbols. The typical detection threshold for EGRET is shown as a solid curve.

5. Spectra

5.1. SPECTRA IN THE EGRET ENERGY RANGE

EGRET spectra of blazars typically covers at least two decades in energy (from 30 MeV to 10 GeV) and are well described by a simple power-law model of the form $F(E) = k(E/E_0)^{-\alpha}$ photons cm^{-2} s^{-1} MeV^{-1}, where the photon spectral index, α, and the coefficient, k, are the free parameters. The energy normalization factor, E_0, is chosen so that the statistical errors in the power law index and the overall normalization are uncorrelated.

The average blazar spectrum has a spectral index of about -2.15. Figure 5 shows the photon spectral index of the blazars plotted as a function of the redshift. There are marginal indications that suggest that the BL Lac objects have slightly harder spectrum in the EGRET energy range than the FSRQs. Mukherjee et al. (1997) find that the average spectral index of the BL Lac objects is about -2.03, compared to about -2.20 for the FSRQs. For some individual blazars there has been noted a trend for the spectrum to harden during a flare state (eg. in blazars 1222+216, 1633+382, and 0528+134; Sreekumar et al. 1996; Mukherjee et al. 1996). A spectral study of blazars as a class has been performed by Mücke et al. (1996) and Pohl et al. (1997).

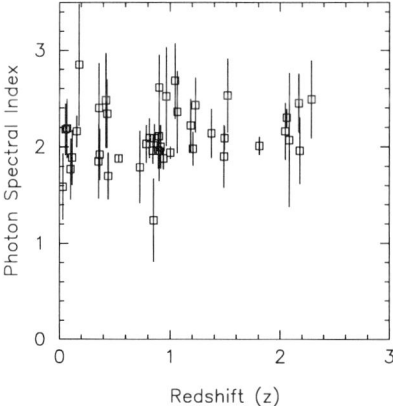

Figure 5. Photon spectral index as a function of the redshift for blazars detected by EGRET.

5.2. SPECTRAL ENERGY DISTRIBUTIONS AND GAMMA-RAY MODELS

The processes by which γ-rays are produced in blazars can be best understood by the study of the correlated multiwaveband observations of blazars extending from radio to γ-ray wavebands. One of the most significant findings of EGRET is that, in the radio to γ-ray multiwavelength spectra of blazars, the power in the γ-ray range equals or exceeds the power in the infrared-optical band. Any model of high-energy γ-ray emission in blazars needs to explain this basic observational fact. The high γ-ray luminosity of the blazars suggests that the emission is likely to be beamed and, therefore, Doppler-boosted into the line of sight. This is in agreement with the strong association of EGRET blazars with radio-loud flat-spectrum radio sources, with many of them showing superluminal motion in their jets. This information has helped to favor jet models of emission over models in which the γ-ray production is directly associated with accretion onto a massive black hole (e.g. Becker & Kafatos 1993).

The jet models explain the radio to UV continuum from blazars as synchrotron radiation from high energy electrons in a relativistically outflowing jet which has been ejected from an accreting supermassive black hole (Blandford & Königl 1979). The emission in the MeV-GeV range is believed to be due to the inverse Compton scattering of low-energy photons by the same relativistic electrons in the jet. However, two main issues remain questionable: the source of the soft photons that are inverse Compton scattered, and the structure of the inner jet, which cannot be imaged directly. The soft photons can originate as synchrotron emission either from within the jet (the synchrotron-self-Compton or SSC process: Maraschi, Ghisellini, & Celotti 1992; Bloom & Marscher 1996), or from a nearby accretion disk, or they can be disk radiation reprocessed in broad-emission-line clouds (the external radiation Compton process or the ERC process: Dermer & Schlickeiser 1994; Sikora, Begelman, & Rees 1994; Blandford & Levison 1995; Ghisellini & Madau 1996). In contrast to these leptonic jet models, the proton-initiated cascade (PIC) model (Mannheim & Biermann 1989, 1992) predicts that the high-energy emission comes from knots in jets as a consequence of diffusive shock acceleration of protons to energies so high that the threshold of secondary particle production is exceeded.

Figure 6 shows the simultaneous spectral energy distribution of 3C 279 during January-February 1996, when the source was detected at its highest state ever (Wehrle et al. 1997). The figure shows the relative amounts of energy detected in equal logarithmic frequency bands. The power output in γ-rays dominates the bolometric luminosity of the sources, as mentioned in §1. Wehrle et al. (1997) note that the γ-rays vary by more than the square of the observed IR-optical flux change, a fact that could be hard to explain by some specific blazar emission models. Although the data do not rule out SSC models, Wehrle et al. point out that the data are most likely explained by the "mirror" model of (Ghisellini & Madau 1996). In this model the flaring region in the jet photo-ionizes nearby broad-emission-line clouds, which in turn provide low energy external seed photons that are inverse Compton-scattered to high energy γ-rays.

Recently, a model combining the ERC and SSC scenarios has been used to fit the simultaneous COMPTEL and EGRET spectra of PKS 0528+134 by Böttcher & Collmar (1998). Figure 7 shows their fit to the gamma-ray spectrum during the high state of the source during March 1993. In their model Böttcher & Collmar assume a spherical blob filled with ultrarelativistic pair plasma which is moving out along an existing jet structure perpendicular to an accretion disk around a black hole of mass 5×10^{10} $M_\odot$. They argue that the observed spectral break between COMPTEL and EGRET energy ranges can plausibly be explained by a variation of the Doppler beaming factor in the framework of a relativistic jet model for AGNs.

The EGRET results have demonstrated that in order to model the spectra of blazars it is very important to get a truly simultaneous coverage across the entire electromagnetic spectrum before, during, and after a flare in the high-energy γ-ray emission. The limited data that we have on most of the blazars prevents us from being able to distinguish between the different theoretical models, on the basis of the spectra alone. For example, both the SSC and ERC models have been shown to reproduce the multiwavelength spectrum of 3C 279 rather well (Hartman et al. 1996b; Maraschi, Ghisellini

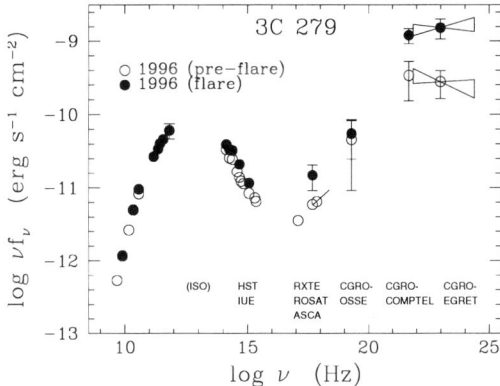

Figure 6. Radio to γ-ray energy distribution of 3C 279 in low (open circles) and flaring state (filled circles) in 1996 January-February (Wehrle et al. 1997).

& Celotti 1992; Ghisellini & Maraschi 1996). The SSC model was similarly found to fit the multiwavelength spectrum of PKS 0528+134 during the March 1993 flare reasonably well (Mukherjee et al. 1996). The low-state data of PKS 0528+134 (Aug 1994) was fit well with the ERC model, as demonstrated by Sambruna et al. (1997). The SSC, ERC, and PIC models have all been shown to fit the multiwavelength spectrum of 3C 273 well (von Montigny 1997).

The differences between the γ-ray variability properties of BL Lac objects and FSRQs can be explained in light of the model of Ghisselini and Madau (GM) (1996). In their model, soft photons from the jet are reprocessed by broad line region (BLR) clouds. Subsequently, these soft photons are emitted back into the jet where they scatter off of electrons in relativistically moving "blobs" to create high-energy γ-rays. Since BL Lac objects generally have very weak emission lines, it may be that they have much less BLR gas available for reprocessing soft photons. If there is an initial outburst of soft photons created via the synchrotron process in a jet "blob," then BL Lac objects can still create γ-rays via the SSC process, though perhaps with lower amplitude than γ-rays created via the GM model. (This effect is somewhat dependent on adjustable model parameters.) In this scenario, BL Lac objects may undergo several SSC outbursts which fail to reach the EGRET detection threshold, thus giving the appearance that BL Lac objects as a class experience less dramatic and less frequent γ-ray flares. The general properties of the low frequency outbursts of BL Lac objects,

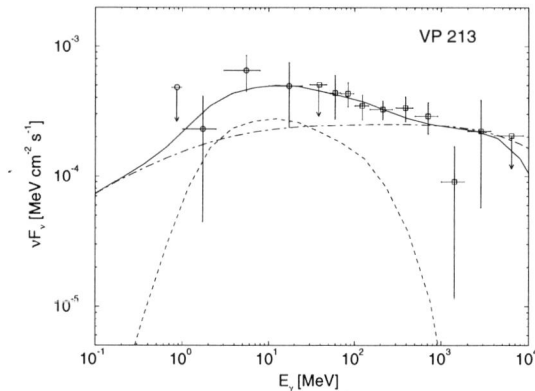

Figure 7. Fit to the γ-ray (COMPTEL and EGRET) spectrum of PKS 0528+134 in its high state in VP 213. See Böttcher & Collmar (1998) for fit parameters.

however, would be very similar to that of the FSRQs.

In order to achieve a better understanding of the emission mechanisms of γ-rays from blazars a study of the correlated short time scale ($\sim 1-3$ days) γ-ray variations with those at other frequency bands is needed. Since the predictions of time delays between the flux changes at various frequencies are different for the individual models for both the seed photons and the nature of the inner jet, this method could provide a means to discriminate between the different models. The differences in the model predictions are discussed in more detail by Marscher et al. (1995). Gamma-ray variability in the different models may have different impacts on the spectral behavior during the build-up and decline of an outburst. Studying the short-time-scale behavior and looking for spectral changes while following a complete outburst may be the key to pin down the basic emission mechanisms.

6. Summary

In conclusion, the EGRET results have shown the importance of the γ-ray window on blazars. The high luminosities and strong time variability observed have pushed theoretical models to emphasize relativistic jets of particles seen at small angles to the line of sight. The EGRET observations have established that the γ-ray window is critical for understanding the properties of blazars. Future observations with CGRO and successor γ-ray observatories like INTEGRAL and GLAST should play a key role in resolving the physics of these powerful sources.

The author presents this work on behalf of the EGRET Team and acknowledges contributions from D. L. Bertsch, S. D. Bloom, B. L. Dingus, J. A. Esposito, C. E. Fichtel, R. C. Hartman, S. D. Hunter, G. Kanbach, D. A. Kniffen, Y. C. Lin, H. A. Mayer-Hasselwander, L. M. McDonald, P. F. Michelson, C. von Montigny, A. Mücke, P. L. Nolan, M. Pohl, O. Reimer, E. Schneid, P. Sreekumar, and D. J. Thompson. The author would particularly like to thank P. Sreekumar for critical comments on the draft. The author also acknowledges support from NASA Grant NAG5-3696.

7. References

Becker, P. A. & Kafatos, M. 1993, in: Proceedings of the 2nd COMPTON Symposium, College Park, MD 1993, AIP Conference Proc. No. 304, eds: C. E. Fichtel, N. Gehrels, & J. P. Norris, pg. 620
Bertsch, D. L., et al. 1989, *Proc. of the Gamma Ray Observatory Science Workshop*, ed. W. N. Johnson, 2, 52
Blandford, R. D. & Königl, A. 1979, ApJ, 232, 34
Blandford, R. D. & Levison, A. 1995, ApJ, 441, 79
Bloom, S. D. & Marscher, A. P. 1996, ApJ, 461, 657
Bloom, S. D., et al. 1997, ApJ, 490, L145
Böttcher, M. & Collmar, W. 1998, A&A, 329, L57
Chiang, J. & Mukherjee, R. 1998, ApJ, 496, 752
Collmar, W., et al. 1997, A&A, 328, 33
Dermer, C. D. & Schlickeiser, R. 1994, ApJS, 90, 945
Esposito, J. A., et al. 1998, in preparation
Ghisellini, G. & Madau, P. 1996, MNRAS, 280, 67
Ghisellini, G. & Maraschi, L. et al. 1996, "Blazar Continuum Variability," A. S. P. Conf. Series Vol. 110, pg 436
Hartman, R. C., et al. 1993, ApJ, 407, L41
Hartman, R. C., et al. 1996a, "Blazar Continuum Variability," A. S. P. Conf. Series Vol. 110, pg 333
Hartman, R. C., et al. 1996b, ApJ, 461, 698
Hartman, R. C., et al. 1997, *Proc. of the Fourth Compton Symposium*, eds. C. D. Dermer, M. S. Strickman, & J. D. Kurfess, CP410, 307
Hartman, R. C., et al. 1998, ApJ, submitted
Hughes, E. B., et al. 1980, IEEE Trans. Nucl. Sci., NS-27, 364
Hunter, S. D., et al. 1993, ApJ, 409, 134
Hunter, S. D., et al. 1997, ApJ, 481, 205
Kanbach, G., et al. 1988, Space Sci. Rev., 49, 69
Kanbach, G., et al. 1989, *Proc. of the Gamma Ray Observatory Science Workshop*, ed. W. N. Johnson, 2, 1
Kniffen, D. A., et al. 1993, ApJ, 411, 133

Mannheim, K. & Biermann, P. L. 1989, A&A, 221, 211
Mannheim, K. & Biermann, P. L. 1992, A&A, 53, L21
Maraschi, L., Ghisellini, G., & Celotti, A. 1992, ApJ, 397, L5
Marscher, A. P., et al. 1995, PNAS, 92, 11439
Mattox, J. R., et al. 1993, ApJ, 410, 609
Mattox, J. R., et al. 1996, ApJ, 461, 396
Mattox, J. R., et al. 1997a, ApJ, 481, 95
Mattox, J. R., et al. 1997b, ApJ, 476, 692
McLaughlin, M. A., et al. 1996, ApJ, 473, 763
von Montigny, C., et al. 1995, ApJ, 440, 525
von Montigny, C., et al. 1997, ApJ, 483, 161
Mücke A., et al. 1996, IAU Symposium 175, (Dordrecht: Kluwer)
Mukherjee, R., et al. 1996, ApJ, 470, 831
Mukherjee, R., et al. 1997, ApJ, 490, 116
Pohl, M., et al. 1997, A&A, submitted
Sambruna, R. M., et al. 1997, ApJ, 474, 639
Sikora, M., Begelman, M. C., & Rees, M. J. 1994, ApJ, 421, 153
Sreekumar, P., et al. 1996, ApJ, 464, 628
Sreekumar, P., Stecker, F. W., & Kappadath, S. C. 1997, *Proc. of the Fourth Compton Symposium*, eds. C. D. Dermer, M. S. Strickman, & J. D. Kurfess, CP410, 307
Sreekumar, P., et al. 1998, ApJ, 494, 523
Stecker, F. W., & Salamon, M. H. 1996, ApJ, 464, 600
Swanenburg, B. N., et al. 1978, Nature, 275, 298
Thompson, D. J., et al. 1993a, ApJS, 86, 629
Thompson, D. J., et al. 1995, ApJ, 101, 259
Thompson, D. J., et al. 1996, ApJS, Vol. 107, 227
Wagner, S. et al. 1995, ApJ, 454, L97
Wehrle, A. et al. 1997, ApJ, submitted

STRONG GRAVITY AND X-RAY SPECTROSCOPY

A. MACIOŁEK-NIEDŹWIECKI
Łódź University, Department of Physics
Pomorska 149/153, 90-236 Łódź, Poland

AND

P. MAGDZIARZ
University of Durham, Department of Physics
South Road, Durham DH1 3LE, UK

Abstract. This paper reviews effects of general relativity in an X-ray spectrum reflected from a cold matter accreting onto a black hole. The spectrum consists of the iron Kα line and the Compton reflection. We sketch the overall picture of radiative processes in the central parts of the accretion flow with relation to the relativistic effects derived from the discrete features in the X-ray spectrum. We discuss implications for detection of relativistic effects and computational tools of spectral analysis.

1. Introduction

Among various observational signatures of accreting black holes (e.g., Madejski 1998; Paradijs 1998) perhaps the most convincing source of information on the strong gravitational field in the vicinity of the black hole comes from the X-ray spectroscopy. In the central part of the accreting flow the emitting matter exists in at least two phases: a hot, mildly relativistic, moderately thick plasma emitting hard X-ray continuum (e.g., Poutanen 1998) and a cold, Compton-thick, thermal plasma of temperatures characteristic for optical/UV/soft X-ray energy range (e.g., Blaes 1998). Such two emitting phases of matter are radiatively coupled to each other, leading to a complex variable geometry of the accreting flow (e.g., Życki, Done & Smith 1998; Magdziarz et al. 1998). However, both the multi-temperature thermal continuum emitted by the cold matter and the hard X-ray continuum produced by Comptonization of the soft photons by the hot, thermal plasma are less useful for probing the geometry of the emitting source: the first one, due to complex atomic physics and superposition of emission from mostly more distant components, and the second one due to basic featureless continuum character. On the other hand, illumination of the cold matter by the hard X-rays produces Compton reflection component with superimposed discrete signatures coming from bound-free absorption and fluorescence lines (e.g., George & Fabian 1991). The most pronounced of these are the iron absorption edge and the iron Kα fluorescence line, at 7.1 keV and 6.4 keV, respectively (for the neutral matter), both discovered in the X-ray spectra of black hole systems by *Ginga* (e.g., Pounds et al. 1990). Modeling the energy shifts of those spectral features (with respect to the known rest energies) we can investigate physical conditions in the vicinity of the black hole, where the generation of the X-ray luminosity takes place. *ASCA* observations have confirmed that the X-ray spectra contain significant contribution from the matter emitting in the immediate vicinity of the black hole, however we need a much higher quality data to investigate the velocity field of that matter, its physical state, and the gravitational potential in the emission region.

In this paper we review the theoretical models developed for the analysis of relativistic effects in the spectra generated by black hole accretion (section 2) and summarize the observational data indicating presence of such effects (section 3). We also discuss the possibility of the estimation of the black hole spin by modelling the X-ray spectral features (section 4).

2. Physical outline

In order to investigate the relativistic distortion affecting the spectrum of the radiation coming from the region with the strong gravitational field, in particular to obtain the precise profiles of the discrete features, one has to solve directly the geodesic equation. No analytical solutions are available apart from the Fabian et al. (1989) formulae for the emission line profile from a disk around a Schwarzschild black hole. However, these formulae fail in the strong field limit. Therefore, the calculation of the relativistic effects requires quite complex numerical procedures.

Since the publication of the paper by Bardeen (1970) it became clear that black holes in accretion powered systems should be described by the Kerr metric rather than the less general Schwarzschild metric, as the accretion is likely to spin up the black hole. The Kerr black hole can be characterized by two parameters, the mass, M, and the dimensionless spin parameter, a. The maximum value of $a = 0.998$ cannot be exceeded in accretion systems (Thorne 1974) due to a torque exerted on the hole by radiation from the accreting matter (cf. Moderski & Sikora 1996 for the recent calculation of the black hole spin evolution).

All the relativistic effects affecting the spectra, except the variability time-scale, do not depend on the absolute distance from the black hole, and are functions of the distance in units of the gravitational radius, $r_\mathrm{g} = GM/c^2$, only. Therefore, the same quantitative effects would occur for both the stellar mass ($10 M_\odot$) and supermassive ($10^6 - 10^9 M_\odot$) black holes. The value of the spin influences the trajectories close to the black hole yielding the distance of the last (quasi) stable circular orbit $r_\mathrm{ms} = 6 r_\mathrm{g}$ for $a = 0$ and $r_\mathrm{ms} = 1.23 r_\mathrm{g}$ for $a = 0.998$.

The velocity field of the accreting matter is usually treated in the approximated way in models of relativistic effects: it is assumed that the matter accretes in a form of a geometrically thin accretion disk (cf. Jaroszyński & Kurpiewski 1997 for the observational effects in geometrically thick disks expected in low-efficiency systems) and, for $a > 0$, it is assumed that the disk rotates in the planar plane of the black hole (cf. Bardeen & Petterson 1975 for frame dragging effects on disks not in the equatorial plane). The accreting material is assumed to flow along circular equatorial geodetics with a superimposed small radial inflow (neglected in calculations) for $r > r_\mathrm{ms}$, and to be in free fall for $r < r_\mathrm{ms}$.

The first calculation of the spectrum of the matter accreting onto a Kerr black hole has been presented by Cunningham (1975). His results, however, do not have the form suitable for data analysis, as they are tabulated only for a few values of the relevant parameters. So far, the most efficient procedure for the data analysis comes from the concept of the photon transfer function, which is constructed by calculating the trajectories of a large number of photons (e.g., Laor 1991).

Further issue is an effect of the relativistic transfer within the source. As pointed out by Cunningham (1976) the returning radiation may give additional contribution to illumination of the accretion disk. This effect was studied by Dabrowski et al. (1997) who found that the returning disk radiation in the Kerr metric gives no significant effect, and particularly it cannot enhance equivalent width of the line by more than 20 %. On the other hand, the light bending may become very important when one considers the transfer of the radiation from the X-ray source to the disk, leading to strong enhancement of the equivalent width of the iron line (Martocchia & Matt 1996) and the amount of the Compton reflected radiation.

Figure 1 shows the emission line profiles from the accretion flows. The lines are typically broad and skewed rather than the double-peaked structures derived in the Newtonian approximation (cf. Fabian 1997 for the basic discussion of the effects in the line profiles). In general the characteristics of the lines allow to distinguish them from the line shapes produced by other physical processes (Fabian et al. 1995). The precisely measured line

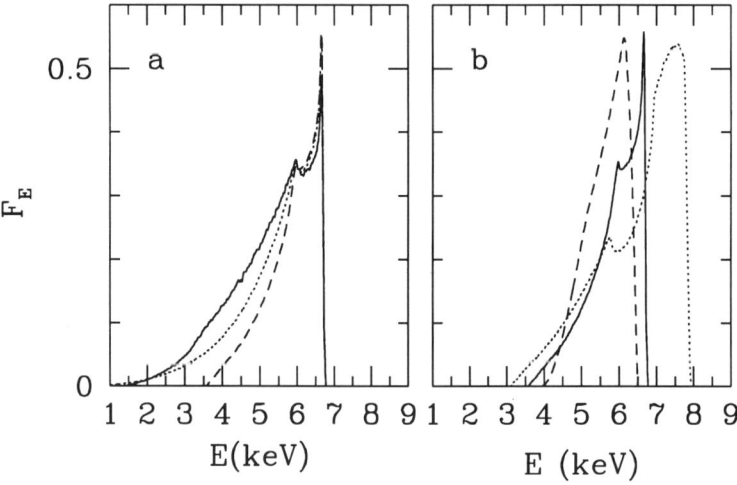

Figure 1. a) The dependence of the line profile on the geometry of the innermost part of the accretion flow. The rest energy of the line is assumed to be 6.4 keV (neutral iron Kα line). All the curves give the line profiles for the inclination angle 30°, the outer radius of the emission region, $r_{out} = 100 r_g$, and the radial emission law favoring the emission from the central parts (the local intensity of the emission of the line $I(r) \sim r^{-3}$). The dashed and dotted curves correspond to the Schwarzschild geometry with the inner radii, $r_{in} = 6 r_g$ and $2 r_g$, respectively (the latter includes the contribution from the free-falling material). The solid curve corresponds to the Kerr geometry ($a = 0.998$) with $r_{in} = 1.3 r_g$. The Kerr metric disk gives more redshifted photons due to the location of the inner edge of the disk. Then, the high-a line has an excess in the red tail, with respect to the low-a line, if only the fluorescence from the circularly flowing disk material is taken into account. However, the contribution from the free-falling matter in the Schwarzschild metric removes this robust difference between the line profiles. b) The dependence of the line profile on the inclination angle. The curves correspond to the disk in the Schwarzshild metric extending between $6 r_g$ and $100 r_g$ with $I(r) \sim r^{-3}$, observed at 60° (dotted curve), 30° (solid curve) and 10° (dashed curve). The position of the high energy peak of the line and the extent of the red tail can be used for the precise determination of the inclination and the distance of the emission region.

profile can be used for the determination of such system parameters as the inclination or the location of the emission region. However, the analysis of relativistic effects may likely lead to unreliable results, since modelling of the iron line and the iron edge is strongly dependent of the continuum model in most of X-ray data (e.g., Zdziarski, Johnson & Magdziarz 1996). The X-ray spectral deconvolution is a non-linear, complex procedure and it cannot be simplified by any indirect method of the analysis of the line profile. E.g., it is possible to constrain the parameters of the system using the extreme frequency shifts of the line profile (Bromley, Miller & Pariev 1998), however, determination of these extreme shifts requires prior deconvolution of the line profile from the data, which yields by itself the best

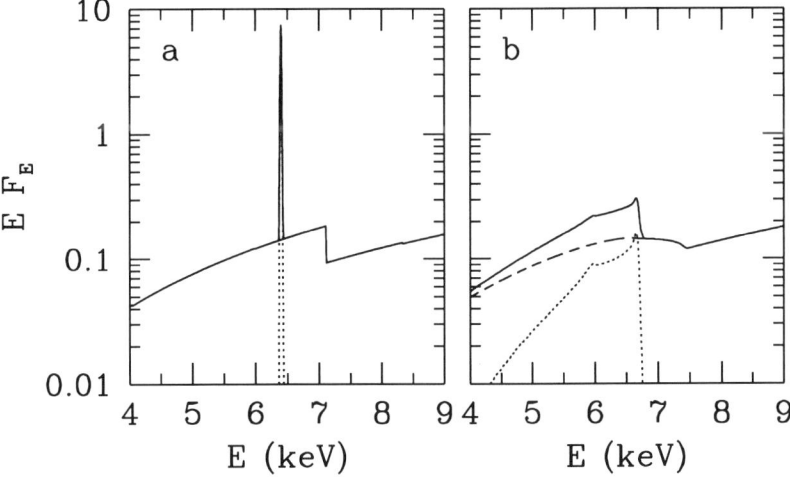

Figure 2. The figure compares the shape of the iron Kα line and absorption edge in the reflection spectrum arising from the X-ray irradiation of the slab at rest (a) with those in the reflection from the innermost region of the accretion disk in the Schwarzschild geometry (b) with $r_{in} = 6r_g$, $r_{out} = 100r_g$, $I(r) \sim r^{-3}$ and $\cos\theta_{obs} = 30°$. The solid curve shows the sum of the line (dotted curve; $EW = 150$ eV in both geometries) and the Compton reflection (dashed curve). The relativistic transfer effects smear both spectral components, which add up giving rise to the complex spectral feature around 7 keV.

fitting set of the parameters of a given assumed model.

The particularly important, but usually neglected, effect is due to the relativistic distortion of the Compton reflected radiation (cf. Figure 2). The reflected continuum is affected similarly to the fluorescent line since it originates from the same matter (cf. Matt, Perola & Piro 1991), although the dependence of reflection and fluorescence on ionization is different, and they are not necessarily smeared in the same manner. A further shortcoming of the usually applied models is the simplified treatment of the fluorescent line, which is approximated as a δ-function. Due to Compton scattering of fluorescent photons in the disk, the line flux emerges locally as a broadened spectral feature, which effect is especially pronounced in highly ionized disks (e.g., Matt, Fabian & Ross 1996). The above problems will be addressed in detail in our forthcoming paper (Maciołek-Niedźwiecki & Magdziarz 1998). The spectral components, relevant to the determination of the iron line profile, are shown in the best fit of our model to the *Ginga* data of the Seyfert 1 galaxy MCG-6-30-15.

Finally, the innermost region of the accretion flow appears to be extremaly complex. It is well known, that solutions of the standard accretion disk with dissipation rate proportional to pressure are unstable (Shakura &

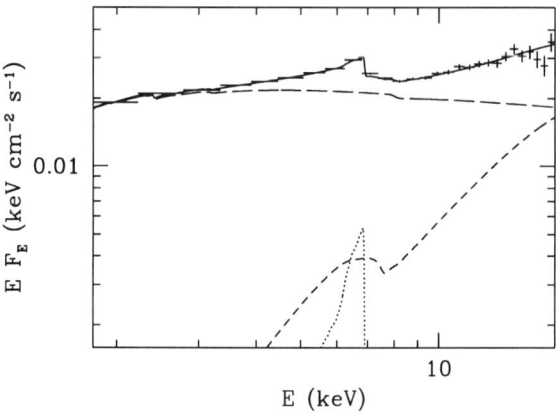

Figure 3. The average *Ginga* spectrum of MCG-6-30-15. The fitted model (solid curve) is a power-law (the energy spectral index $\alpha = 1.1$), attenuated by the ionized absorber (long-dashed curve), the Compton reflection with the amplitude corresponding to the solid angle of the reprocessor as seen from the X-ray source of 3π (for isotropic X-ray illumination; short dashed curve) and the Fe Kα line with $EW = 300$eV (dotted curve). All the components are self-consistently transferred through the Schwarzschild metric field with the assumption that the emission comes from the region between $6r_g$ and $100r_g$. The fitted radial emission law, $I(r) \sim r^{-3.9}$, and the inclination $\theta_{obs} = 40°$. The fit weakly prefers the relativistically smeared spectrum over a reflection from a distant reprocessor. Furthermore, the distant reprocessor requires nearly edge-on inclination (in contradiction with its Seyfert 1 classification), as well as the solid angle subtended by the reflector $\sim 10\pi$, which would require a complex model with strongly obscured X-ray source. On the other hand, the parameters of the relativistic model would be compatible with a nearly flat disk, provided that the isotropic X-ray source is close to the black hole horizon. In this case the light bending effects can account for both the very steep illumination of the disk (as implied by the fitted radial emissivity) and the enhanced amplitude of the iron line and Compton reflection.

Sunyaev 1976) likely leading to the disk fragmentation (e.g., Krolik 1998). Such a complex structure of the central region has been indeed invoked to explain both the nature of EUV emission in AGNs in general (e.g., Kuncic, Celotti & Rees 1997) and the overall spectral phenomenology in some particular sources (e.g., Magdziarz & Blaes 1998; for NGC 5548). The complex radiative coupling of the multi-phase medium convolved with relativistic dynamics seems to be plausible alternative to a simple disk geometry. However, the profiles of the discrete features in such geometry need more complex treatment, and have not been calculated as yet.

3. Observational outline

The presence of the broad iron lines, with profiles indicating strong relativistic effects, has been found in a number of Seyfert 1s observed by *ASCA* (e.g., Nandra et al. 1997a). Such objects show no spectral contamination from jet activity and have relatively low absorption in the direction of the central part of the accretion flow. The iron line has typically $EW \sim 200$ eV and the profile consistent with that expected from an accretion disk with a bulk of emission coming from a region within $\sim 20 r_g$. The observed peak energy at 6.4 keV suggests that the emission comes mainly from nearly neutral matter, although the observed anticorrelation of the equivalent width of the broad line with the source luminosity (Nandra et al. 1997b) indicates that the reflecting matter tends to be more ionized in more luminous sources. It is also likely that the profiles contain a small ($EW \sim 25$ eV) constant and narrow component from a more distant reprocessor, e.g., a molecular torus. Orientation of the disk surface inferred for the sample of Seyfert 1 galaxies seems to be consistent with the unification scheme (Antonucci & Miller 1985), however, an analysis of a sample of Seyfert 2s suggests that either a population of the obscured objects or geometry of the central part of the accretion flow is more complex (Turner et al. 1998; Elvis et al. 1998). We note, however, that the above results are likely to be subject to the modelling problems indicated in section 2.

The most convincing example of relativistic broadening of the Fe Kα line comes from observations of the Seyfert 1 galaxy MCG-6-30-15 (Tanaka et al. 1995; cf. Figure 4). The profile seems to vary with the source luminosity in the sense that it shows stronger relativistic effects in fainter states of the source (Iwasawa et al. 1996). At the minimum state the profile seems to require either models with fluorescence from matter free-falling below the last stable orbit or models with maximally rotating black hole (cf. section 4). The signatures of the relativistic smearing of both the iron line and absorption edge have been also found in *Ginga* observations of Galactic black hole binaries: V404 Cyg (Życki, Done & Smith 1997) and Nova Muscae (Życki, Done & Smith 1998).

4. The value of the black hole spin

The spin of the black hole appears to be crucial for the properties of accretion powered objects, according to the unification scheme of AGNs, in which one expects that radio-loud jet dominated active galaxies host rapidly rotating black holes (cf. Blandford 1990). To test this unification model, one needs an independent measurement of the spin value. The mass of the black hole can be directly determined from the orbital motion in some X-ray binaries. A lower limit on M can be derived for AGNs from the requirement that their luminosity cannot exceed the corresponding Eddington limit. On the other hand, the estimation of the spin is based on more sophisticated arguments. The spectrum of the accretion disk depends on a (Novikov &

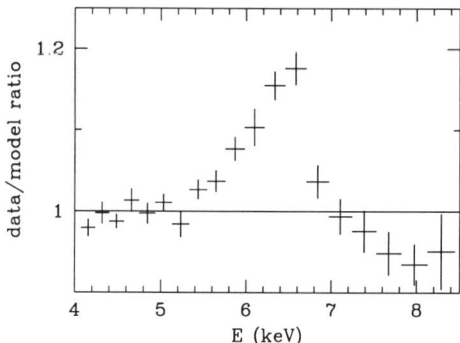

Figure 4. A ratio of the average data over power-law model for the *ASCA* observation of MCG-6-30-15 (July 1994). All four *ASCA* detectors where taken into account to achieve high signal-to-noise ratio which is sufficient to reveal the skewed profile of the line, with the maximum at 6.6 keV and the broad tail at lower energies, independently of the assumed model of the continuum. The iron absorption edge above 7 keV appears consistent with relativistic smearing.

Thorne 1973), however, we do not understand the reprocessing of the disk radiation sufficiently well to be able to constrain the spin value from the shape of the accretion disk spectrum, although the shape of the ultrasoft component in Galactic sources with superluminal jets indicates that they are powered by rapidly rotating black holes (Zhang, Cui & Chen 1997).

The effect of the black hole rotation on the line profile has been investigated, e.g., by Kojima (1991) who did not find any significant dependence of the line profile on a if fluorescence occurs at a distance $r > 6r_g$. As for the rapidly rotating black hole the disk extends well below $6r_g$, the emission from that region gives the extra flux in the red wing of the line, which makes the low-a and high-a lines clearly distinguishable (see Figure 1). Reynolds & Begelman (1997) have pointed out, however, that fluorescence by the material inside the radius of marginal stability can have an observable influence on the iron line profile causing the red wing to be much wider. In fact, Jaroszyński (1997) finds that the difference in the velocity field, in the innermost region, between low-a and high-a cases does not yield any robust difference in the line profiles. Given the uncertainty in such parameters of the system as the radial emissivity law or inclination (which is not independently constrained, except for some black hole binaries, and must be found in the fitting procedure) we cannot expect to constrain the value of a from the line profile with the present or even forthcoming X-ray data quality, unless we better understand the properties and geometry of the matter in the region inside $6r_g$. In fact, the model by Reynolds & Begelman (1997) implies, that the free-falling matter gets completely ionized and cannot contribute to the line profile, except for very inefficient X-ray sources. The efficiency as low as $\eta \sim 10^{-5}$ is needed (where $L_X \sim \eta \dot{m} c^2$, L_X is the X-ray luminosity and $\dot{m}$ is the accretion rate) to avoid the complete ionization of that region. Such a low efficiency is very unlikely in the luminous accretion powered systems. Furthermore, the model by Reynolds & Begelman (1997) relies on the central location of the X-ray source, which could be the case if the X-ray emission was related to the formation of the jet activity. However, if the emission is connected locally with the dissipation of the energy in the disk (as in the model with magnetic flares above the disk surface), one may expect that the line will be dominated by the flourescence occurring in the disk. There is virtually no dissipation of energy once the material has passed the last stable orbit, so in this case the contribution from the free-falling matter should not be very significant, independently of its ionization stage.

If any of the above arguments allows to exclude the contribution from the free-falling matter, the observation of the very extended red tail of the line can be used as an unambiguous evidence for the rapid rotation of the central black hole.

5. Summary

X-ray spectroscopy offers, for the first time, the unique opportunity to investigate the geometry of the region in the immediate vicinity of the black hole horizon. Currently available data suggest that the observed X-ray spectra contain a significant contribution from matter emitting in the vicinity of the black hole, however, results are still uncertain mostly due to complex response of X-ray instruments, which confuses the iron line profile with both the iron edge and the underlying continuum. The progress in determination of the iron line and, consequently, the innermost geometry of the accretion flow, is expected to result from new X-ray observatories of high spectral resolution, to be launched in the near future.

Acknowledgements We thank Greg Madejski for his comments and hospitality during our stay in NASA/GSFC where this paper has been written.

References

Antonucci, R. R. & Miller, J. S., 1985, ApJ, 297, 621
Bardeen, J. M., 1970, Nature, 226, 64
Bardeen, J. M. & Petterson, J. A., 1975, ApJ, 195, L65
Blaes, O. M., 1998, in: *Accretion Processes in Astrophysical Systems: Some Like It Hot*, eds. Holt, S. S., & Kallman T. R.
Blandford, R. D., 1990, in: *Active Galactic Nuclei*, ed. J. L. Courvoisier, M. Mayor (Berlin: Springer)
Bromley, B. C., Miller, W. A. & Pariev, V. I., 1998, Nature, 391, 54

Cunningham, C. T., 1975, ApJ, 202, 788
Cunningham, C. T., 1976, ApJ, 208, 534
Dabrowski, Y., Fabian, A. C., Iwasawa, K., Lasenby, A. N. & Reynolds, C. S., 1997, MNRAS, 288, L11
Elvis, M., Fiore, F., Giommi, P. & Padovani, P., 1998, ApJ, 492, 79
Fabian, A. C., 1997, in *X-ray Imaging and Spectroscopy of Cosmic Hot Plasmas*, ed. F. Makino, K. Mitsuda (Tokyo: Universal Academy Press)
Fabian, A. C., Rees, M. J., Stella, L. & White, N. E., 1989, 238, 729
Fabian, A. C., et al., 1995, MNRAS, 277, L11
George, I. M. & Fabain, A. C., 1991, MNRAS, 249, 352
Iwasawa, K., et al., 1996, MNRAS, 282, 1038
Jaroszyński, M., 1997, Acta Astronomica, 47,339
Jaroszyński, M. & Kurpiewski, A., 1997, AA, 326, 419
Kojima, Y., 1991, MNRAS, 250, 629
Krolik, J. H., 1998, ApJL, in press
Kuncic, Z., Celotti, A. & Rees, M. J., 1997, MNRAS, 284, 717
Laor, A., 1991, ApJ, 376, 90
Maciołek-Niedźwiecki, A. & Magdziarz, P., 1998, in preparation
Madejski, G. M., 1998, in *Theory of Black Hole Accretion Disks*, eds. M. A. Abramowicz, G. Björnsson & J. E. Pringle (Cambridge University Press), in press
Magdziarz, P. & Blaes, O., 1998, in *Proc. of IAU Symp. 188*, Kyoto, Japan, in press
Magdziarz, P., Blaes, O., Zdziarski, A. A., Johnson, W. N. & Smith, D. A., 1998, MNRAS, in press
Martocchia, A. & Matt, G., 1996, MNRAS, 282, L53
Matt, G., Fabian, A. C. & Ross, R. R., 1996, MNRAS, 278, 1111
Matt, G., Perola, G. C. & Piro, L., 1991, AA, 247, 25
Moderski, R. & Sikora, M., 1996, MNRAS, 283, 854
Nandra, K., George, I. M., Mushotzky, R. F., Turner, T. J. & Yaqoob, T., 1997a, ApJ, 477, 602
Nandra, K., George, I. M., Mushotzky, R. F., Turner, T. J. & Yaqoob, T., 1997b, ApJ, 488, L91
Novikov, I. D. & Thorne, K. S., 1973, in *Black Holes*, ed. C. DeWitt and B. DeWitt (New York: Gordon & Breach)
van Paradijs, J., 1998, in *The Many Faces of Neutron Stars*, ed. R. Buccheri, J. van Paradijs, M. A. Alpar (Kluwer Academic Publishers), in press
Pounds, K. A., Nandra, K., Stewart, G. C., George, I. M. & Fabian, A. C., 1990, Nature, 344, 132
Poutanen, J., 1998, in *Theory of Black Hole Accretion Disks*, eds. M. A. Abramowicz, G. Björnsson & J. E. Pringle (Cambridge University Press), in press
Reynolds, C. S. & Begelman, M. C., 1997, MNRAS, 488, 109
Shakura, N. I. & Sunyaev, R. A., 1976, MNRAS, 175, 613
Tanaka, Y., et al., 1995, Nature, 375, 659
Thorne, K. S., 1974, ApJ, 191, 507
Turner, T. J., George, I. M., Nandra, K. & Mushotzky, R. F., 1998, ApJ, 493, 91
Zhang, S. N., Cui, Wei & Chen, Wan, 1997, ApJ, 482, L155
Zdziarski, A. A., Johnson, W. N. & Magdziarz, P.,1996, MNRAS, 283, 193
Życki, P. T., Done, C. & Smith, D. A., 1997, ApJ, 488, L113
Życki, P. T., Done, C. & Smith, D. A., 1998, ApJ, 496, L25

MY INVOLVEMENT IN THE EARLY YEARS OF RADIO ASTRONOMY

M.K. DAS GUPTA[1]
*P 282, CIT Scheme VI M, Kankurgachi,
Calcutta- 700054*

1. Emerging Years

Radio Astronomy probably began with rather accidental detection and at least partial identification of cosmic radio noise by Karl Jansky in 1932 (Jansky, 1932). While trying to find the cause of radio noise of the transatlantic telephonic conversations, Jansky found that most of the radio static noise were from thunderstorms. But even when there were no storm, the faint noise persisted and it was found to be mostly coming from the direction of the center of Our Galaxy, the Milky Way. From repeated observations he concluded that the principal source showed a periodicity of 23 hours and 56 minutes, characteristic of sidereal period of objects like stars. Jansky's discovery was subsequently corroborated by Grote Reber, another radio engineer from Illinois, USA (Reber, 1940). He constructed a radio telescope, first of its kind, of a diameter 31 feet and tuned to a wavelength of 1.87 meters. He also observed that the radio waves were mostly coming from the center of the Milky Way. His radio map also shows significant radio waves from the directions of the constellations of Cygnus and Casseopeia. World War II intervened and further researches in this new field were discontinued. During the closing years of the war, a remarkable discovery was made by another radio engineer, named J.S. Hey in England. Almost every afternoon coastal Radar receivers all around the British Isles detected strong radio signals. These were first thought to be due to a special type of jamming of radio signals (perhaps introduced by Germans!). However, Hey by his remarkable intuition traced the origin of these signals in the setting sun which was carrying a large sunspot group at that time. He was thus

---

[1]Former Professor and Head, Centre of Advanced Study in Radio Physics and Electronics, Calcutta University, Calcutta 700009

credited with the discovery of radio emission from the disturbed sun. Hey also discovered a couple of discrete sources of radio emission. Radiation from the quite sun at an wavelength of 3.2 cm was first detected by Southworth in 1942. These radiations were all of continuum type. In mid-forties H.C. Van de Hulst of Leiden Observatory, Holland, predicted from theoretical considerations that a line radiation at a wavelength of 21 cm might be observable from a hyperfine transition in neutral hydrogen atom in the interstaller space. Consequently, in the years to follow, this line radiation was simultaneously detected by three groups of scientists from Holland, USA, and Australia. This opened one of the most exciting chapters in radio astronomy and helped radio astronomers to determine for the first time, the spiral structure of our local milky-way galaxy.

Immediately after the war a band of experienced radio and electronic engineers who were responsible for the development of radar, directional antenna system, sensitive receivers and other special types of electronic instruments, went back to the universities and research institutes to pursue radio astronomical investigations. By the early fifties, the new science of radio astronomy – a queer combination of the old with the new – gained the general acceptance of astronomers, astrophysicists and cosmologists. Radio astronomical observatories sprung up in many countries, including India. Colossal radio telescopes, radio interferometers, dynamic spectrographs were constructed and commissioned in these observatories with the aim of collecting as much of radio energy as possible.

2. Overcoming the Fundamental Limitations of Radio Astronomy

In the early years of radio astronomy, observations were severely limited by (a) poor angular resolution of the radio telescopes and (b) limited sensitivity of the receiving system. The history of radio astronomy has been one of steadily increasing angular resolution and sensitivity. These are achieved by novel techniques over the last four decades. In order to match the resolving power even of an unaided human eye a radio telescope must have a reflector dish several miles in diameter. Bigger and better reflector dishes were made but the prohibitive cost, engineering difficulties and the imperfections of the surface made 100 meters aperture about the ultimate limit, until the development of radio interferometry which is the radio analogue of optical interferometry.

In this technique, signals from two separated radio telescopes are carried by coaxial cables simultaneously to a central point receiving system for comparison. For practical limitations, connecting cables were not suitable for base lines exceeding a few Kilometers. Cables were soon replaced by radio or microwave relay for bringing the received signal at one station to the other for comparison. 'Long Baseline Interferometry', thus developed could successfully resolve many of the discrete radio sources. Meanwhile electronic computers and magnetic tapes became commercially available and data could be transferred for comparison by storing them in tapes. This was an important breakthrough since it removed all the limitations to the size of the effective aperture. This 'tape recorded interferometry' or 'very long base interferometry' ushered in a revolution in radio astronomy. The USA-USSR link produced an effective aperture of 10,000 kms giving a resolution of 0.0003 seconds of arc at a wavelength of 3.5 cms. Radio sources even at the farthest reaches of the observable universe could thus be detected and mostly resolved.

3. A Radio Astronomer by Luck

The concluding phase of the freedom movement of India, particularly the period of 1942-1947 was a vital one for me so far as my later career was concerned. As a student of Physics at the University of Dacca (then in undivided India), I was particularly fortunate to receive affectionate blessings from Prof. S.N. Bose, Head of the Physics Department. Our result was announced in May, 1947, and I was expecting to be absorbed in the department to continue my research work on atmospherics (natural radio waves, generated in lightning discharges). But that was not to be!

Immediately after the partition of India, proclaiming independence on August 15th, 1947, we decided to come to Calcutta from Dacca. For a couple of years, I worked at the Physics department, University of Calcutta, as a research assistant under Professor S.K. Mitra, a pioneer of upper atmospheric researches. Sometime in 1949, after I was pursued by Prof. Mitra, I applied for one of the Overseas Scholarships jointly sponsored by the Govt. of India, State Govt. and the University of Calcutta. He wrote a personal letter to Prof. P.M.S. Blackett of the University of Manchester exploring the possibility of my entry into the field of Radio Astronomical research at Manchester. The department ran the Jodrell Bank Experimental Station (now called Nuffield Radio Astronomy Laboratory) under the dynamic leadership of Dr. A.C.B. Lovell, one of the founding fathers of Radio Astronomy. Professor Blackett readily agreed to take me in for research in Radio Astronomy. That was a 'turning point' of my career.

On October 24, 1950, I first met Prof. Blackett in his chamber at the University of Manchester. After a warm welcome, Prof. Blackett requested Dr. Lovell to take me to Jodrell Bank so that I could start the work from the next day. Such a quick decision and immediate action revealed to me that time was so precious that it should not be unnecessarily wasted. On our arrival at Jodrell, Dr. Lovell showed around the laboratories and introduced me to research workers, particularly to Hanbury Brown and Roger Jennison. I was told that Hanbury Brown had conceived a novel type of radio interferometer of much improved resolving power which will have to be designed and fabricated by Jennison and myself. The theory of this novel interferometer (intensity interferometer) was worked out by Hanbury Brown and R.Q. Twiss (1954). In this system the detected signal will have to be transmitted from the distant station over radio or microwave links to the home station for correlation studies. Jennison and myself were given the task of designing and constructing the intensity interferometer for our proposed studies on the apparent angular structures of Cygnus A and Cassiopeia A, two of the most intense radio sources in the sky.

It took a few months to complete a smaller version of the proposed interferometer just to check whether the new interferometer worked in principle. Our trial run on the radio sun worked as expected and we could measure its apparent angular diameter at 125 MHz. Our activities had to be accelerated to complete the construction of the much larger interferometer (two sets of huge antenna arrays together with highly sensitive receiver systems, correlator, relay transmitter and other necessary equipments) in the shortest possible time. Dr. Lovell used to visit our laboratory almost twice a day constantly urging upon us that we must win the race of resolving Cygnus A and Cassiopeia A. It took almost a year to complete the related constructions. I was subsequently awarded a DSIR (UK) research fellowship so that I could continue to be at Manchester. Thus, by and by, I became a radio astronomer.

4. Hints of Success: Discovery of Double Radio Sources in Cygnus A

Meanwhile, by the summer of 1951, Martin Ryle's team at Cambridge increased the resolution of the radio telescope (Ryle, 1952). Graham Smith, a colleague of Ryle used this instrument to produce a 1 arc minute error box for Cygnus A. Waltar Baade used the newly completed Palomar 5 meter telescope to find the nature of the optical counterpart of the radio source Cygnus A. The galaxy looked partitioned (by dust cloud, not known during those days) and the radio source was wrongly interpreted to be the result of a collision of two galaxies. Jennison and myself started using the new interferometer (Fig. 1), consisting of two telescopes, one fixed to the ground at the home station, and the other which could be dismantled and taken on a moving truck to be erected at different locations, several Kms away. We could, for the first time achieve a baseline of as large as 20 kms. We found that Cygnus A radio waves were not at all coming from the "Colliding

Fig. 1 : R. Jennison (facing the camera) and M.K. Das Gupta in the home station.

Galaxies". They seem to be coming from two *distinct regions of space*, each about 200,000 light years in size (Fig. 2). They are located at the opposite sides of the so-called colliding galaxies discovered by Baade and Minkowski (1952). These results were published in a series of papers (Hanbury Brown, Jennison & Das Gupta, 1952; Jennison & Das Gupta, 1953; Jennison & Das Gupta, 1956ab). At last, a fundamental nature of the radio sources were revealed. Eventually, with higher resolution instruments the radio maps of the two lobes were made by Mitton and Ryle (1969) as shown in Fig. 2. Since then further researches revealed that most of the radio galaxies discovered, if not all, are double in structure with an enigmatic optical source at the Centre, a fundamental discovers which we first made in the early fifties. It may be noted that W.T. Sullivan III (1982) in his book 'Classics in Radio Astronomy', reprinted our original papers with a note – "... This first double radio source immediately raised the question of why the optical

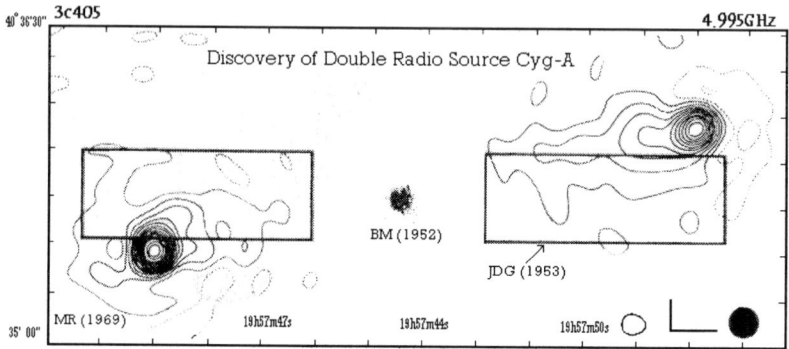

Fig. 2 : First identification of the Cygnus A (3C405) as a double radio source (rectangular boxes). Also superposed on the figure are the observations of Baade & Minkowski (1952) and Mitton & Ryle (1969). The figure is adapted from Thorne (1994).

image and radio brightness distribution should be so different. And to-day we are still asking the same question."

Subsequent development of the subject involves in modeling these double radio sources. The revelation that the enormous energy emitted by these sources are supplied by the central region of the galaxy through *radio jets* is also equally exciting. It is now widely believed that these radio jets come out as results of accretion of matter on supermassive black holes located at the galactic centers (see, Begelman, Blandford & Rees, 1984 for an early review and Chakrabarti, 1996 for a recent review). Our humble discovery thus perhaps was the first indirect evidence for astrophysical black holes much before astronomers accepted the concept of black hole in the first place!

References

BEGELMAN, M.C., BLANDFORD, R.D. & REES, M.J. 1984, *Rev. Mod. Phys.*, **56**, 255.
CHAKRABARTI, S.K. 1996, *Phys. Rep*, **266**, 5-6, 229.
HANBURY BROWN, R. & TWISS, R.Q. 1954, *Phil Mag. Ser*, 7, **45**, 633.
HANBURY BROWN, R., JENNISON, R.C. & DAS GUPTA, M.K. 1952, *Nature*, **170**, 1061.
JANSKY, K.G. 1932, *Proc IRE*, **20**, 1920.
JENNISON, R.C. & DAS GUPTA, M.K. 1953, *Nature*, **172** 996.
JENNISON, R.C. & DAS GUPTA, M.K. 1956, *Phil Mag. Ser.*, 8, **1**, 55.
JENNISON, R.C. & DAS GUPTA, M.K. 1956, *Phil Mag. Ser.*, 8, **1**, 65.
MITTON, S. & RYLE, M. 1969, *M.N.R.A.S.*, **146**, 221.
REBER, G. 1940, *ApJ*, **91**, 621.
RYLE, M. 1952, *Proc. Roy Soc.*, **A211**, 351.
SULLIVAN III, W.T. 1982, *Classics in Radio Astronomy* (Dordrecht: Reidel), Paper Nos. 30 and 33.
THORNE, K.S. 1994, *Black Holes and Time Warps: Einstein's Outrageous Legacy*, (W.W. Norton & Co.: New York).

RADIO OBSERVATIONS OF ACTIVE GALACTIC NUCLEI: EVIDENCE FOR DISKS AND BLACK HOLES

D.J. SAIKIA
National Centre for Radio Astrophysics
TIFR, Pune University Campus, Pune 411007, India

1. Introduction

Radio galaxies and quasars are amongst the most luminous of active galaxies with radio continuum luminosities of $\approx 10^{45}$ ergs s^{-1}, which is about 10^6 times more luminous than the continuum emission from nearby spiral galaxies. The total energy content in relativistic particles and magnetic fields is at least about 10^{60} ergs. The radio structure of a typical radio source shows the following features: two lobes of radio emission, with steep non-thermal spectra ($\alpha > 0.5$ where flux density $S \propto \nu^{-\alpha}$) and located on opposite sides of the parent optical galaxy or quasar. The separations between the two lobes range from less than a few tens of parsecs in the compact symmetric objects or CSPs to the largest sources where the overall linear size is a few Mpc (q_o=0.5 and H_o=100h km s^{-1} Mpc^{-1}). In addition, there is a core or nuclear component which is of high surface brightness and has a flat and complex radio spectrum. The core is the stationary unresolved base of the jet. The observed radio jets, which are seen on scales of parsecs to hundreds of kpc, are the signatures of the beams which are supplying energy from the central nuclear powerhouse to the outer extended lobes of emission (Begelman, Blandford & Rees 1984; Hardee, Bridle & Zensus 1996).

The structure of the extended radio emission depends on the luminosity of the lobes. For the high-luminosity or Fanaroff-Riley class II sources with a luminosity at 178 MHz, $P_{178} > 10^{25}$ W Hz^{-1} sr^{-1} the extended lobes are characterised by bright hotspots usually at the outer edges of the lobes. These hotspots are identified with the regions where the beams interact with the external medium. The jets in these sources traverse outwards with high Mach numbers dissipating only a small fraction of their energy along their paths, and terminate in a strong shock at the region of interaction with the external medium. In this region, the bulk kinetic energy of the jet is converted to highly relativistic particles and there is generation of radio emission via synchrotron emission. The relativistic material expands outwards to generate the extended lobes of radio emission (cf. Scheuer 1974; Blandford & Rees 1974). The radio jets are highly asymmetric or usually one-sided in these objects due to relativistic speeds on the scale of parsecs to tens of kpc. The jets in the high-luminosity sources transfer energy efficiently from the nucleus to the outer lobes. In the lower luminosity Fanaroff-Riley class I sources, the extended lobes are diffuse with no prominent hotspots at the outer edges. Here the jets are not well collimated, highly dissipative, have low Mach numbers and entrain more mass per kpc than the jets in FRII sources (Bridle & Perley 1984; Hardee, Bridle & Zensus 1996). Although the radio jets in FRI sources are reasonably symmetric on large scales of tens of kpc, they are very asymmetric or one-sided on VLBI or parsec-scales. Similar collimated structures or radio jets but of much lower luminosity have also been seen in many nearby Seyfert and mildly active galaxies, such as Mkn 3 (Kukula et al. 1993), NGC3079 (Irwin & Seaquist 1988), NGC4151 (Pedlar et al. 1993) and NGC4388 (Hummel & Saikia 1991). In this review, I concentrate largely on the powerful radio galaxies and quasars and examine the possible evidence for supermassive objects such as black holes and disks in the nuclear regions from radio observations.

In the unified scheme for the FRII radio galaxies and quasars, they are believed to be intrinsically similar objects but appearing to be different because they are inclined at different angles to the line of sight. Here, the core-dominated quasars are seen at small angles to the line-of-sight, the lobe-dominated quasars at intermediate angles while the radio galaxies lie close to the plane of the sky (Barthel 1989; Antonucci 1993). The flux density of the core is Doppler boosted when the source is inclined close to the line-of-sight. In radio galaxies, the quasar nucleus is hidden from our view by a putative torus, but may be seen in scattered light as in the radio galaxy 3C234 (Young et al. 1998). For the FRI radio galaxies, the beamed counterparts are the BL Lac objects (Urry & Padovani 1995). In this review I summarise briefly some of the radio observations of large- and small-scale jets, and discuss the evidence from radio observations in favour of the canonical scenario of a black hole and disks, which are intimately related to the source of energy. For parts of this review I have relied extensively on recent excellent reviews by Pearson (1996) and Wilson (1996).

2. The compact radio cores

Observations of the nuclei of radio cores by VLBI techniques have led to significant advances in our understanding of active galactic nuclei (Zensus

& Pearson 1987, 1990; Zensus & Kellermann 1994; Pearson 1996). These observations of cores have shown that the nuclear radio jets on the scale of parsecs are well-collimated and that the jets are established on the scales of a few light years or so. The VLBI-scale jets are usually very asymmetric or one-sided although the extended emission is often two-sided, consistent with the ideas that the apparent asymmetry of the jets is due to bulk relativistic motion. Assuming that the oppositely directed jets are intrinsically symmetric and the flow speed is given by v=βc, the ratio of brightness of the approaching jet to the receding one is given by (cf. Blandford & Königl 1979) $R_s = \{(1+\beta cos\phi)/(1-\beta cos\phi)\}^{n+\alpha}$, where ϕ is the angle of inclination of the source axis to the line of sight. Here n is 3+α for a blob of plasma and 2+α for a piece of continuous jet. For a jet to counter-jet brightness ratio, R_s, of about >50, as in the jet in NGC315, $\beta cos\phi$ <0.54, and the angle between the source and the line-of-sight is <58° (Venturi et al. 1993).

The measurements of proper motions of knots in the parsec scale jets show these often exhibit superluminal speeds of advancement from a stationary core. The apparent speed β_{app} is given by $\beta_{app} = \beta sin\phi/(1-\beta \cos\phi)$ where v=βc is the true velocity (Pearson & Zensus 1987). Some of the strongly core-dominated sources such as the well-known quasars 3C273 (Davis, Unwin & Muxlow 1991; Unwin, Davis & Muxlow 1994; Bahcall et al.1995) and 3C345 (Wardle et al. 1994; Brown, Roberts & Wardle 1994; Zensus, Cohen & Unwin 1995) have been studied extensively over many years and provide us with the most detailed picture of conditions in parsec scale radio jets. The sources exhibit a stationary flat-spectrum core component and a steep-spectrum one-sided radio jet. The knots or brightness peaks in the outer components traverse outwards with apparent velocities in the range of 2-10c. The quasars 4C39.25 (Marcaide et al. 1994) and 3C395 (Lara et al. 1997) show evidence of stationary components in addition to the moving ones. The stationary components possibly correspond to jet positions close to the line-of-sight in a twisted or helical jet.

Detailed studies of individual sources show that the jets have complex substructure. For example consider the 16 mag quasars 3C345 at a redshift of 0.595 which is one of the best-studied superluminal radio sources (Zensus, Cohen & Unwin 1995 and references therein). The components or knots which are regions of enhanced emission in an underlying jet flow, follow curved trajectories rather than ballistic radial trajectories from the nucleus. The kinks and wiggles are more pronounced close to the nucleus, different components have different speeds and the speeds also change with time. The apparent change in speed may be intrinsic or due to geometrical effects such as a change in orientation. These changes are most rapid close to the core, while farther down the length of the jet, the trajectories are more or less parallel to the ridge of the jet. Complex intrinsic substructure suggest that components might also have undergone a split while passing through a recollimation shock (Gómez et al. 1995, 1996). The nuclear jet in 3C345 seems to undergo a transition at a distance of about 1.5 mas from the core. The magnetic field orientations also change from being predominantly orthogonal to the jet axis on the small scales to being longitudinal on larger scales (Leppänen, Zensus & Diamond 1995; Lobanov & Zensus 1996). This may be due to strong shocks on the smallest scales, while at larger distances the shearing of field lines possibly produce a more longitudinal field. The strong jet features also appear to decelerate close to the core reaching a minimum speed at about 1.5 mas and then accelerates up to about 0.9 mas yr^{-1}. The motions in 3C345 have been modeled assuming that the intrinsic pattern speed is constant with a Lorentz factor of about 12, and that the trajectory starts at a small angle to the line of sight of ($\phi < 2°$) and then curves away ($\phi \approx 4°$). There is a suggestion that the tracks could be helical.

While individual sources provide valuable insights into the physical conditions in the object, a statistical study reveal the properties of the population as a whole (Vermeulen & Cohen 1994; Vermeulen 1996). A comparison of the observations with simulations of different effects such as precession, bending, acceleration, distribution of bulk Lorentz factors enable us to examine these effects in a large sample. Apparent velocities or upper limits for a sample of sources from the Pearson-Readhead and Caltech-Jodrell Bank VLBI surveys (Polatidis et al. 1995; Thakkar et al. 1995; Taylor et al. 1994) show that the β_{app} distribution has a somewhat uniform distribution between 1 and 5h^{-1} and then tapers off to about 10h^{-1} in a q$_o$=0.5 Universe. The distribution can be reproduced if there is a large range in γ_b from about 2 to 18, with a peak at about 4 in a skewed distribution when h=0.55. The mean values of β_{app} are 2.1 for the 7 galaxies, 2.3 for the 8 BL Lacs and 3.2 for the 44 quasars, after ignoring the upper limits. The trend is broadly consistent with the unified scheme (Vermeulen 1996).

Another significant result from the statistical study of a large sample is the dependence of β_{app} on radio core luminosity (cf. Vermeulen 1996). While low values of β_{app} may occur at any luminosity, the high values are found only at high radio luminosities. The high radio luminosity objects are beamed towards us and their luminosity is high because of relativistic beaming. However, the absence of similar high values of β_{app} at lower luminosities suggest that objects which are about 2 to 3 orders of magnitude lower in luminosity are incapable of producing highly relativistic jets. There is a correlation between 5 GHz radio luminosity and β_{app} or Lorentz factor, suggesting that the highly relativistic jets can be produced in a restricted range of luminosity.

2.1. THE CORES OF FRII RADIO GALAXIES AND QUASARS

The core-dominated radio sources have strong cores and have been studied extensively since the early days of VLBI. On the other hand the cores of lobe-dominated radio galaxies and quasars are weak, because of their larger angles to the line of sight in the relativistic beaming scenario, and are being studied more extensively in more recent years (Hough 1994; Vermeulen et al. 1993). Their proper motions are also expected to be smaller because of their larger angles of inclination. A sample of 25 lobe-dominated 3CR quasars are being studied by Hough (1994) while a sample of quasars from the Jodrell Bank 966 MHz survey is being studied by Zensus & Porcas (1987). The parsec-scale jets are on the same side as that of the larger-scale jets. Most of the objects exhibit superluminal speeds which range from slightly over 1 to about 4, consistent with the correlation of β_{app} with core prominence as expected in the unified schemes (cf. Browne 1987). The properties of the parsec-scale jets as well as the bases of the kpc-scale jets can be explained with Lorentz factors of about 5.

One of the best studied radio galaxies is the powerful radio source Cygnus A (Carilli & Barthel 1996). It has a well-collimated nuclear jet on the same side as the larger-scale one, and a weak counter jet. The brightness ratio of the nuclear jet to the counter jet one is about 5±3. Two features in the jet yield proper motions of $\beta_{app} = (0.35\pm0.15)h^{-1}$ and $(0.55\pm0.15)h^{-1}$ (Carilli, Bartel & Diamond 1994; Bartel et al. 1995), while closer to nucleus the estimates of β_{app} are lower (Krichbaum et al. 1993, 1996). ¿From the observed values of Cygnus A, Bartel et al. (1995) estimate β to be in the range of about 0.6 and 1 and $55° < \phi < 85°$ for h=0.5.

2.2. THE LARGE-SCALE JETS

The tendency for extended radio jets to be detected in sources with strong cores in a sample of lobe-dominated quasars led Saikia (1984) to suggest that the extended jets also traverse at relativistic speeds and their apparent asymmetry or one-sidedness is due to relativistic beaming. More recently Hough (1994) has noted the prominence of the cores and the bases of the kiloparsec scale radio jets, and have suggested that perhaps both are enhanced due to relativistic motion with Lorentz factors of ~5 for nuclear jets and between about 2 and 5 for the extended jet segments. Bridle et al. (1994) have made very deep images of a sample of 12 3CR lobe-dominated quasars. All 12 sources are known to have radio jets, and counter-jet candidates were detected in 7 of them although these were not contiguous and well-defined like the radio jets. The flux density ratio of the jet to counter-jet brightness range from about 1.2:1 to >175:1 with a median value of about 12:1. Using these VLA observations along with VLBI data on the nuclear jets in these objects, Wardle & Aaron (1995) suggest jet speeds in the range of 0.6−0.8c and inclination angles of ≤ 70°. The extended radio jets also remain significantly relativistic over length scales of several tens of kpc. One of the strongest pieces of evidence that the extended jets are beamed comes from the correlation of depolarization asymmetry with jet sidedness, also known as the Laing-Garrington effect (Laing 1988; Garrington et al. 1988). Here, the counter-jet lobe depolarises more rapidly seen it is being seen through the magnetoionic medium associated with the host galaxy, establishing that the jet is on the approaching side.

A small but significant fraction of sources appear to be completely one-sided (cf. Saikia et al. 1990). The apparent asymmetry appears to be due to relativistic beaming and requires speeds up to about 0.8c for the extended emission. One of the most asymmetric objects is the quasar 3C273 which has been mapped to a very dynamic range, and the limit to the brightness ratio is about 4000. Saikia et al. (1990) suggested that the lobe south of the main jet could be the counter-lobe seen in projection. This will decrease the degree of asymmetry to about 200, and could be understood in the relativistic beaming scenario. Depolarization observations of this component should help test this possibility.

2.3. MISALIGNMENTS BETWEEN THE LARGE AND SMALL-SCALE STRUCTURE

In sources inclined at small angles to the line of sight an intrinsic asymmetry can appear amplified due to projection effects. The apparent misalignment between the nuclear and large-scale jets in core-dominated radio sources was interpreted to be due to such a projection effect nearly two decades ago. This was extended to the apparent non-collinearity observed in the double-lobed structure of quasars by Kapahi & Saikia (1982) where they reported a strong correlation between the apparent non-collinearity and the predominance of the core component which is taken as a statistical measure of orientation of the source axis to the line of sight.

In samples of radio sources complete to a certain flux density limit, there is a population of radio sources where the nuclear and large-scale radio jets are well-aligned or slightly misaligned and a population where the nuclear jets tend to be orthogonal to the large-scale ones (Xu et al. 1994). The sources with large misalignments tend to be sources with strong radio cores, have high optical polarization and strong radio variability. These are all signatures of jets which are aligned close to our line of sight and the jets are relativistically beamed. The jets in many of these sources can be interpreted to be due to helical jets with low pitch angles (Conway & Murphy 1993) seen at a small angle of inclination to the line of sight. These helical jets could be due to precession of binary black hole systems in the centres of active nuclei. On the other hand, the close alignment of the nuclear and large-scale jets in most lobe-dominated radio sources suggest that the ejection axes are usually steady over time scales of about 10^8 yr or so, which might be due to a jet anchored to a black hole in the centre.

2.4. THE CORES OF FRI GALAXIES

On the large scales, the jets in FRI radio galaxies are not well-collimated and appear to terminate in diffuse extended lobes of radio emission. It is important to understand whether this is due to low Mach number jets which are highly dissipative and entrain matter as they traverse outwards or whether there is a fundamental difference in their collimation close to the nucleus. One significant way of addressing the problem is to examine the parsec-scale or nuclear jets in these objects and study their collimation on these small scales. The cores are generally weaker and they are believed to be the inclined at large angles to the line of sight, the beamed counterparts being the BL Lac objects. Many of them are however relatively nearby and could be studied with high linear resolution. The main studies have been reported by the Italian group (Giovannini et al. 1995, and references therein). These observations show that the jets in FRI sources are usually one-sided although one and possibly two of the sources are known to have two-sided radio jets. The jets are more asymmetric on the small scales than the larger ones, the flux density ratio often being larger than about 30.

One of the best studied FRI radio galaxies is the nearby radio galaxy Virgo A (M87, 3C274). This galaxy, has been studied on the smallest scales where Junor & Biretta (1995) have seen a well-collimated jet down to a distance of 10^{16} cm of the central compact variable core, which is identified with the central nuclear power house. Parsec scale features have subluminal speeds with $\beta_{app} = 0.28 \pm 0.08$ (Reid et al. 1989) while Biretta, Zhou & Owen (1995) have found speeds of about $2.5 \pm 0.3c$ in small components in knot D, about 3 arcsec from the nucleus, suggesting that the jet accelerates. Another well-known galaxy is Centaurus A where the SHEVE team found the source to be subluminal with $\beta_{app} = 0.08^{+0.09}_{-0.07}$, with the parsec-scale jet aligned with the large-scale one. There is also evidence of a weaker counter-jet (Tingay et al. 1994). The radio galaxy 3C120 has an FRI structure but the core exhibits superluminal motion. It is a core-dominated radio galaxy, exhibits significant variability at radio frequencies (Walker, Benson & Unwin 1987) and there has been some debate on evidence of motion in an arcsec scale knot (Walker 1997).

These observations basically show that the jets of FRI sources close to the nucleus are well-collimated, usually one-sided and exhibit subluminal speeds for those with the weaker cores, and are basically similar to those of the FRII sources in total intensity. Polarization studies of the parsec-scale jets are required to show if there are differences between FRI and FRII sources, which might provide some further clues on their collimation.

2.5. BLACK HOLES AND THE FRI-FRII DICHOTOMY

A fundamental question in our understanding of radio galaxies and quasars is what triggers the formation of a radio loud AGN. In studies of elliptical galaxies, the fraction of sources which have a radio luminosity above a certain value is a strong function of the optical luminosity of the elliptical galaxy. In nearby Es and S0s of low radio luminosity, the radio power appears to be related to the bulge luminosity of the parent galaxies (Auriemma et al. 1977; Sadler, Jenkins & Kotanyi 1989; Wilson 1996 for a review). For a sample of galaxies for which black hole masses have been estimated, Kormendy & Richstone (1995) have shown that the black hole mass is correlated with the bulge luminosity. Although the total number of galaxies is small, there is no evidence of a massive black hole with a low bulge luminosity. These trends suggest that high radio luminosity might be correlated with high black hole mass. It would be interesting to establish whether the bulge masses of FRI and FRII galaxies are different, and as more galaxies with black hole masses are estimated it would be important to plot the radio luminosity of radio galaxies and quasars with the black hole masses.

Baum and her collaborators (Zirbel & Baum 1995; Baum et al. 1995) have reported several important differences between FRI and FRII radio galaxies. For similar host galaxy magnitude or radio luminosity, the FRII galaxies produce more optical line emission by about an order of magnitude compared to the FRI sources. They attribute the strong line emission in FRIIs to a strong UV continuum source, while the line emission in FRIs is due to processes in the host galaxy itself. They speculate that this might be due to differences in the central engine itself, and suggest that while the central engines in FRIs are fed at a low accretion rate while in FRIIs the accretion rate is high leading to the central engine depositing a higher fraction of its energy as radiant energy.

3. Energy supply and the origin of powerful radio galaxies

There is a growing school of thought that the powerful radio galaxies are the result of relatively recent mergers which involve at least one disk galaxy. The evidence in favour of mergers include the disturbed or peculiar optical morphology of about 50 per cent of high-luminosity radio galaxies compared to only about 7 per cent for the low-luminosity ones (cf. Heckman et al. 1986; Smith & Heckman 1989; Owen & Laing 1989). These include bridges, tails and evidences of tidal distortions. Powerful radio galaxies with strong emission lines are found to have unusually blue colours which are spatially extended. These have been interpreted to be due to merger induced star formation. Powerful radio galaxies with peculiar optical morphologies are often dynamically different (Smith & Heckman 1989, 1990). Deep imaging of quasars have also shown them to be often tidally interacting with companions. This appears to be more common in radio loud quasars than the radio quiet ones (Hutchings & Neff 1990; references in Clements & Pérez-Fournon 1997).

Assuming that the formation of the powerful radio source is related to mergers of galaxies, one still needs to understand the fueling of the central massive object in the nuclear region while the interactions are on the scales of the galaxy. Extensive studies of interactions of disk galaxies over the last decade have shown that gas can accumulate within about the central 200 pc or so (Barnes & Hernquist 1991). The infall of gas is also sensitive to the presence of bulges. In the absence of bulges, bars form at the central region leading to infall of gas and a burst of star formation leaving less gas to be fed to the central black hole. On the other hand, when bulges are present there is only a weak starburst with more gas being available for fueling the central black hole (Mihos & Hernquist 1994). This may partly account for the relationship for sources with strong bulges to harbour active galaxies. At closer distances, the gas might fragment and fall deeper into the nucleus, leading to fueling of the AGN. The detailed flow from the scale of parsecs to an accretion disk associated with a black hole is far from clear.

4. The central engine and the formation of jets

The important questions in the study of radio galaxies and quasars are an understanding of the source of energy and the formation of two oppositely-directed collimated jets. Soon after the discovery of radio galaxies and quasars, it was realised that the source of energy was gravitational in origin. The models of energy generation in active nuclei involved accretion on to a black hole in the centre of the galaxy. A strong argument in favour of such a scenario was that it is an efficient way of converting mass into energy to explain the enormous luminosities of these objects. The efficiencies for conversion of mass to energy is about 6 % for Schwarzschild black holes, about 32 % for maximally rotating Kerr black holes while it is less than about 1 % for nuclear processes (cf. Wiita 1991).

In the twin-beam model (Blandford & Rees 1974; Scheuer 1974), hot, buoyant plasma is channelised into twin jets traversing outwards in opposite directions by a flattened cloud of cooler gas. The gas escapes along the minor axis of the flattened gas cloud, forming de Laval nozzles with a minimum radius at the transonic point (cf. Norman et al. 1981). Such a scenario was difficult to sustain after the imaging of jets with VLBI techniques which showed that they are collimated on scales of parsecs. In the application of their model to Cygnus A Blandford & Rees (1974) found the nozzle to form at a distance of about 200 pc. Models where nozzles form at distances within a few pc of the black hole have been discussed by Norman et al. (1981) and Smith et al. (1983), but these short nozzles are also fat and have opening angles larger than what are seen in VLBI observations of such sources. Further, if the collimation occurs on scales much smaller than a pc, the X-ray emission from the hot dense gas would violate the observed limits or luminosities of these objects.

In the black hole scenario, the models for the collimation of jets are due to funnels or magnetically focussed beams. The collimation would depend on the structure of the disk or torus and hence also on the accretion rate. The critical accretion rate is set by the Eddington limit, where L_E is the standard Eddington luminosity $L_E = 1.3 \times 10^{38} M_{BH}$ ergs s^{-1} where M_{BH} is the mass of the black hole in $M_\odot$. For subcritical accretion rates, the energy can be released from the surface of the disk. However, such outflows are likely to be very poorly collimated and not applicable to radio jets.

Quasars and radio galaxies are highly luminous objects, and at the critical accretion rate, the photons heat the disk substantially forming a geometrically thick accretion disk, with narrow funnels formed at the polar regions. Such funnels are the candidates for formation of collimated outflows. Radiation pressure driven outflow collimated by the funnels are limited to Lorentz factors of 1.6 for electron-proton plasma and 4 electron-positron plasma. These Lorentz factors are less than what is typically inferred for superluminal radio sources. However, even in these cases, the opening angles are far larger than what is inferred either from VLBI observations of jets, or sizes of hotspots in compact steep spectrum radio sources (Wiita 1991).

Models where magnetic fields are responsible for the ejection and collimation of radio jets seem promising (cf. Wiita 1996; Camenzind 1998) and there seems to be a consensus on this scenario although the details remain unclear. Since the early work in this field, (Blandford & Znajek 1977; Blandford & Payne 1982), there have been different MHD schemes which depend on the assumptions of the boundary conditions (cf. Wiita 1996; Begelman 1996; Massaglia & Bodo 1998) Ouyed, Pudritz & Stone (1997) have reported numerical solutions to the time dependent, non-linear equations of ideal MHD to study outflows driven from the surfaces of an accretion disk. They have attempted different initial magnetic field configurations, and find that the outflows either self collimate or remain episodic for the duration of their calculations. They also find that the internal structure of the disks may be crucial in understanding the formation of outflows.

5. Evidence for black holes and disks

There has been striking advances in finding evidence for tori or disks associated with supermassive objects in the nuclear regions of active galaxies. At radio frequencies one of the most spectacular ones is the water vapour masers around NGC4258 where there are three groups of maser lines. One is at the systemic velocity of the galaxy, while the others are redshifted and blueshifted. The disk is thin, seen nearly edge on and slightly warped. The velocities show a remarkably good fit to Keplerian motion within the observed range of radii from 0.13−0.26 pc (Miyoshi et al. 1995; Moran et al. 1995). The mass which must lie within the inner radius is 3.5×10^7 $M_\odot$, leading to a mass density of $> 4 \times 10^9$ $M_\odot$ pc^{-3}. If this were to be a star cluster, its life time would be short compared to the age of the galaxy and the object is likely to be a black hole (Maoz 1995). At other wavelengths, the studies of proper motions of stars in our Galaxy indicate a mass of 2 .45±0.4×10^6 $M_\odot$ (Eckart & Genzel 1997), while HST observations provide evidence of disk extending from about 20 to 100 pc in M87 (Ford et al. 1994; Harms et al. 1994). The implied mass from the measured radial velocities is about 2.4×10^9 $M_\odot$. At X-ray wavelengths the observations of the broad 6 keV Fe line in MCG-6-30-15 whose profile shows the features expected from a relativistically rotating disk. The inner radius is only about 3-5 Schwarzschild radius from the black hole (Tanaka et al. 1995).

In radio galaxies and quasars the highly relativistic jets suggest deep potential wells, which could be associated with black holes. The steadiness of the axis of ejection in many sources over time scales of 10^7 yr and in some cases up to about 10^9 yr suggest a channel which might be anchored to a single massive object. The helical jets seen in several VLBI scale jets, and also evidence of precession in larger-scale jets such as in 2300−189 could be understood in terms of precession of a black hole to which the collimating channel or jet is anchored. In addition to these suggestions of supermassive objects, there is perhaps somewhat more direct observational evidence of disks from radio observations which we describe briefly here.

5.1. THE VLBI SCALE JETS AND A DISK IN NGC1275

The core-dominated radio source 3C84 is associated with the nearby radio galaxy NGC1275 in the Perseus cluster and there is an X-ray detected cooling flow (Böhringer et al. 1993). The large-scale structure of the source (Pedlar et al. 1990) shows diffuse extended halo or lobes of emission similar to those seen in FRI radio galaxies. The VLBI observations show a prominent jet towards the south with evidence of subluminal motion, and a weak counterjet. The spectra of the counterjet show that it has an abrupt cutoff at low frequencies which is not consistent with synchrotron self absorption but could be understood in terms of free-free absorption by a disk which has been ionized by the central continuum source. The disk affects the spectrum of the counter jet which is seen through the disk without affecting the spectrum of the prominent jet (Vermeulen, Readhead & Backer 1994; Walker, Romney & Benson 1994; Levinson, Laor & Vermeulen 1995).

5.2. THE COMPACT SYMMETRIC OBJECT 4C31.04

The compact symmetric objects or CSOs are a subset of the compact steep spectrum or CSS objects and are believed to be at an early stage of evolution of the radio source. In the CSOs and also towards cores which have a steep radio spectrum such as 3C236, there is high prevalence of broad HI lines in absorption with widths >100 km s^{-1} (Conway 1996). One of the attractive models for these broad HI features is that these are due to circumnuclear disks or tori which are common in AGN on scales of < 100pc from the central engine. The core-dominated radio sources which are believed to be inclined close to the line of sight have no detected HI absorption features and this can be understood in this simple geometric model. The approaching jet in these sources is boosted by relativistic beaming while the receding one is usually undetected due to relativistic dimming, and there is no strong feature against which the HI disk may be seen in absorption. The CSOs are not beamed and the two minilobes which are seen on scales of about a 100 pc or so, could be occulted by the HI disk. In these sources the minilobe farther from us is likely to be seen through the disk for a large range of orientations. The disk model also provides an explanation for the width of the lines of several hundred km s^{-1}. In these thick disks the expected velocity dispersion of the clouds at a distance of 100 pc is about 100 km s^{-1} for a 10^8 M$_\odot$ black hole (cf. Conway 1996).

VLBA HI and continuum observations of this object show a sharp absorption edge and also absorption features due to individual clouds. The strong absorption towards the eastern lobe suggests that this is the counter lobe seen through the obscuring material in the disk, while the sharp edge in the western lobe would require a finite thickness for the disk which is seen nearly edge on. The radio axis is along the disk axis, and there is also evidence of absorption due to individual clouds which are evaporating off the inner edge of the disk (Conway 1996). The size and geometry of the disk is only slightly larger than the one imaged in NGC4261 by the HST (Jaffe et al. 1993).

5.3. HI ABSORPTION IN THE SEYFERT GALAXY NGC4151

The radio continuum structure of this well-known Seyfert galaxy shows elongated structure extended over about 10 arcsec with several knots of emission, and suggests collimated ejection along a PA 77° (Pedlar et al. 1993). High-resolution neutral hydrogen absorption measurements with an angular resolution of 0.15 arcsec using MERLIN show that the component close to the nucleus shows significant absorption while no absorption is seen against any of the other components. The observations suggest a disk no thicker than 50 pc. Absorption is seen against the component moving away from the nucleus while no absorption is seen towards the components approaching us (Mundell et al. 1995).

5.4. CO EMISSION OBSERVATIONS IN CEN A

Cen A (NGC5128) at a distance of 3 Mpc is the nearest giant elliptical galaxy associated with powerful radio source. It has been studied extensively at a number of wavebands. Observations of $^{12}CO(2-1)$ emission in the central region of Cen A suggests that the emission originates from an edge-on circumnuclear ring with a radius of 100 pc and a velocity of 220 km s^{-1}. The central dynamical mass has been inferred to be 1.4×10^9 M$_\odot$. The ring is orthogonal to the radio jet, suggesting that it might be related to an inner accretion disk associated with a supermassive object (Rydbeck et al. 1993).

5.5. TORUS OR DISK FROM POLARIZATION OBSERVATIONS

In addition to the molecular and atomic component in the disks, there is also likely to be a significant ionized component due to X-ray or UV heating from the AGN (cf. Levison et al. 1995). The emission line clouds in AGN may also be distributed in the form of a disk. There is evidence for free-free absorption in the CSO object 4C31.04 discussed earlier. If there is a significant ionized component with a magnetic field, it could also affect the polarization properties of the radio components seen through this obscuring disk.

In the unified scheme for high-luminosity radio galaxies and quasars, the quasar nucleus is hidden from our view by a putative torus disk which is also partially ionized. Such a scenario should also affect the observed polarization properties of the cores, with the lobe-dominated sources having lower core polarization at a given frequency and higher rotation measure than the core-dominated sources due to Faraday effects. Saikia & Kulkarni (1998) report that the core polarization of weak-cored quasars has a median value of less than about 0.4% and is indeed much smaller than for core-dominated quasars where the median value is about 2.5%. They suggest that this might be due to the depolarization caused by the edge of the obscuring torus or disk. They also examine the relative orientation of the core-polarization E-vector at λ6cm and the radio axis for the weak-cored quasars. The sample is small but does not show a significant trend reported earlier for cores of moderate strength. This could also be due to Faraday rotation by the material in the edge of torus or disk. These results are consistent with the basic ideas of the unified scheme, and also lend evidence for an ionized material in the form of a disk (Saikia & Kulkarni 1998).

6. Concluding remarks

In this article a few of the possible pieces of evidence from radio observations for the existence of disks and black holes in the centre of radio galaxies and quasars have been presented. There is an increasing body of evidence in favour of the canonical picture of an AGN with a supermassive black hole at the centre with an accretion disk. However there are still theoretical challenges in understanding the formation of jets, although magnetic processes seem promising, and the infall of matter from the scale of galaxy interactions to the scale of accretion disks in order to feed the black hole.

References

Antonucci, R.R.J. (1993) Unified models for active galactic nuclei and quasars, *Ann. Rev. A&A* **31**, 473–521

Auriemma, C. et al. (1977) A determination of the local radio luminosity function of elliptical galaxies, *A&A* **57**, 41–50

Bahcall, J. et al. (1995) HST and MERLIN observations of the jet in 3C273, *ApJ* **452**, L91–93

Barnes, J.E. & Hernquist, L.E. (1991) Fueling starburst galaxies with gas-rich mergers, *ApJ* **370**, L65–L68

Bartel, N. et al. (1995) The nuclear jet and counter-jet region of the radio galaxy Cygnus A, *Proc. Natl. Acad. Sci. USA* **92**, 11371–11373

Barthel, P.D. (1989) Is every quasar beamed? *ApJ* **336**, 606–611

Baum, S.A., Zirbel, E.L. & O'Dea, C.P. (1995) Toward understanding the Fanaroff-Riley dichotomy in radio source morphology and power, *ApJ* **451**, 88–99

Begelman, M.C. (1996) The acceleration and collimation of jets, in *Proc. Natl. Acad. Sci. USA* **92**, 11442–11446

Begelman, M.C., Blandford, R.D. & Rees, M.J. (1984) Theory of extragalactic radio sources, *Rev. Mod. Phys.* **56**, 255–351

Biretta, J.A., Zhou, F. & Owen, F.N. (1995) Detection of proper motions in the M87 jet, *ApJ* **447**, 582–596

Blandford, R.D. & Königl, A. (1979) Relativistic jets as compact radio sources, *ApJ* **232**, 34–38

Blandford, R.D. & Payne, D.G. (1982) Hydromagnetic flows from accretion discs and the production of radio jets, *MNRAS* **199**, 883–903

Blandford, R.D. & Rees, M.J. (1974) A twin-exhaust model for double radio sources, *MNRAS* **169**, 395–415

Blandford, R.D. & Znajek, R.L. (1977) Electromagnetic extraction of energy from Kerr black holes, *MNRAS* **179**, 433–456
Böhringer, H. et al. (1993) A ROSAT HRI study of the interaction of the X ray-emitting gas and radio lobes of NGC1275, *MNRAS* **264**, L25–L28
Bridle, A.H. et al. (1994) Deep VLA imaging of twelve extended 3CR quasars, *AJ* **108**, 766–820
Bridle, A.H. & Perley, R.A. (1984) Extragalactic radio jets, *ARA&A* **22**, 319–358
Brown, L.F., Roberts, D.H. & Wardle, J.F.C. (1994) Evolution of the parsec-scale linear polarization structure of the superluminal quasar 3C345, *ApJ* **437**, 108–121
Browne, I.W.A. (1987) Extended structure of superluminal radio sources, in *Superluminal radio sources* eds. Zensus, J.A. & Pearson, T.J., Cambridge University Press, 129-147
Camenzind, M. (1998) Origin, acceleration and flaring of jets, in *Astrophysical Jets open problems* eds. Massaglia, S & Bodo, G., Gordon & Breach, 3–29
Carilli, C.L. & Barthel, P.D. (1996) Cygnus A, *A&A Revs.* **7**, 1–54
Carilli, C.L., Bartel, N. & Diamond, P. (1994) Second epoch VLBI observations of the nuclear jet in Cygnus A: subluminal jet proper motion measured, *AJ* **108**, 64–75
Clements, D.L. & Pérez-Fournon, I. (1997) *Quasar Hosts*, ESO Astrophysics Symposia, Springer
Conway, J.E. (1996) VLBI observations of HI absorption in CSO/SSCs, in *The second workshop on Gigahertz peaked spectrum and compact steep spectrum radio sources*, eds Snellen, I.A.G. et al., Sterrewacht, Leiden, 198–207
Conway, J.E. & Murphy, D.W. (1993) Helical jets and the misalignment distribution for core-dominated radio sources, *ApJ* **411**, 89–102
Davis, R.J., Unwin, S.C. & Muxlow, T.W.B. (1991) Large-scale superluminal motion in the quasar 3C273, *Nature* **354**, 374–376
Eckart, A. & Genzel, R. (1997) Stellar proper motions in the central 0.1 pc of the Galaxy, *MNRAS* **284**, 576–598
Ford, H.C. et al. (1994) Narrow-band HST images of M87: Evidence for a disk of ionized gas around a massive black hole, *ApJ* **435**, L27–L30
Garrington, S. T. et al. (1988) A systematic asymmetry in the polarization properties of double radio sources with one jet, *Nature* **331**, 147-149
Giovannini, G. et al. (1995) Very-long-baseline radio interferometry observations of low power radio galaxies, in *Proc. Natl. Acad. Sci. USA* **92**, 11356–11359
Gómez, J.L. et al. (1995) Parsec-scale synchrotron emission from hydrodynamic relativistic jets in active galactic nuclei, *ApJ* **449** L19–21
Gómez, J.L. et al. (1996) Time variable synchrotron emission from hydrodynamic relativistic jets, in *Energy transport in radio galaxies and quasars*, eds Hardee, P.E., Bridle, A.H. and Zensus, J.A. (Astron. Soc. Pacific), 159–164
Hardee, P.E., Bridle, A.H. & Zensus, J.A. (1996) Energy transport in radio galaxies and quasars, ASP Conf. Series 100
Harms, R.J. et al. (1994) HST FOS spectroscopy of M87: Evidence for a disk of ionized gas around a massive black hole, *ApJ* **435**, L35–L38
Heckman, T.M. et al. (1986) Galaxy collisions and mergers –The genesis of very powerful radio sources, *ApJ* **311**, 526–547
Hough, D.H. (1994) Relativistic motion in lobe-dominated quasars, in *Compact Extragalactic Radio Sources*, eds. Zensus, J.A. & Kellermann, K.I., NRAO, 169–174
Hummel, E. & Saikia, D.J. (1991) The anomalous radio features in NGC4388 and NGC4438, *A&A* **249**, 43–56
Hutchings, J.B. & Neff, S.G. (1990) Optical environments of radio quasars, *AJ* **99**, 1715–1721
Irwin, J.A. & Seaquist, E.R. (1988) Nuclear jets in the radio lobe spiral galaxy NGC3079, *ApJ* **335**, 658–667
Jaffe, W. et al. (1993) A large nuclear accretion disk in the active galaxy NGC4261, *Nature* **364**, 213–215
Junor, W. & Biretta, J.A. (1995) The radio jet in 3C274 at 0.01 pc resolution, *AJ* **109**,

500–506
Kapahi, V.K. & Saikia, D.J. (1982) Relativistic beaming in the central components of double radio quasars, JA&A **3** 465–483
Kormendy, J. & Richstone, D. (1995) Inward bound – the search for supermassive black holes in galactic nuclei, Ann. Rev. A&A **33**, 581–624
Krichbaum, T.P. et al. (1993) First 43 GHz VLBI observations with the 30m radio telescope at Pico Veleta, A&A **275**, 375–389
Krichbaum, T.P., Alef, W. & Witzel, A. (1996) High-resolution VLBI imaging of the inner jet of Cygnus A, in Cygnus A – a study of a radio galaxy eds Carilli, C.L. & Harris, D.E., Cambridge University Press, 92–97
Kukula, M.J. et al. (1993) High-resolution radio observations of Markarian 3, MNRAS **264**, 893–899
Laing, R.A. (1988) The sidedness of jets and depolarization of jets in powerful extragalactic radio sources, Nature **331**, 149-151
Lara, L. et al. (1997) Radio observations of the quasar 3C395 from parsec to kiloparsec scales, A&A **319**, 405–412
Leppänen, K.J., Zensus, J.A. & Diamond, P.J. (1995) Linear polarization imaging with very long baseline interferometry at high frequencies, AJ **110**, 2479–2492
Levinson, A., Laor, A. & Vermeulen, R.C. (1995) Constraints on the parsec-scale environment in NGC1275, ApJ **448**, 589–599
Lobanov, A.P. & Zensus, J.A. (1996) Physics of parsec-scale regions in 3C345, in Energy transport in radio galaxies and quasars, eds Hardee, P.E., Bridle, A.H. and Zensus, J.A. (Astron. Soc. Pacific), 109–114
Marcaide, J.M. et al. (1994) Peculiar quasar 4C39.25 deciphered, in Compact Extragalactic Radio Sources, eds. Zensus, J.A. & Kellermann, K.I., NRAO, 141–148
Maoz, E. (1995) A stringent constraint on alternatives to a massive black hole at the center of NGC4258, ApJ **447**, L91–L94
Massaglia, S & Bodo, G. (1998) Astrophysical Jets open problems Gordon & Breach
Mihos, J.C. & Hernquist, L.E. (1994) Ultraluminous starbursts in major mergers, ApJ **431**, L9–L12
Miyoshi, M. et al. (1995) Evidence for a black hole from high rotation velocities in a subparsec region of NGC4258, Nature **373**, 127–129
Moran, J. et al. (1995) Probing active galactic nuclei with H_2O megamasers, in Proc. Natl. Acad. Sci. USA **92**, 11427–11433
Mundell, C.G. et al. (1995) MERLIN observations of neutral hydrogen absorption in the Seyfert nucleus of NGC4151, MNRAS **272**, 355–362
Norman, C. et al. (1981) Hydrodynamic formation of twin-exhaust jets, ApJ **247**, 52–58
Ouyed, R., Pudritz, R.E. & Stone, J.M. (1997) Episodic jets from black holes and protostars, Nature **385**, 409–414
Owen, F.N. & Laing, R.A. (1989) CCD surface photometry of radio galaxies –I. FR class I and II sources, MNRAS **238**, 357–378
Pearson, T.J. (1996) Observations of parsec-scale jets, in Energy transport in radio galaxies and quasars, eds Hardee, P.E., Bridle, A.H. and Zensus, J.A. (Astron. Soc. Pacific), 97–108
Pearson, T.J. & Zensus, J.A. (1987) Superluminal radio sources: introduction, in Superluminal radio sources eds. Zensus, J.A. & Pearson, T.J., Cambridge University Press, 1–11
Pedlar, A. et al. (1990) The radio structure of NGC1275, MNRAS **246**, 477–489
Pedlar, A. et al. (1993) The radio nucleus of NGC4151 at 5 and 8 GHz, MNRAS **263**, 471–480
Polatidis, A.G. et al. (1995) The first Caltech Jodrell-Bank VLBI survey. I. λ=18cm observations of 87 sources, ApJS **98**, 1–32
Reid, M.J. et al. (1989) Superluminal motion and limb brightening in the nuclear jet of M87, ApJ **336** 112–120
Rydbeck, G. et al. (1993) High resolution $^{12}CO(2-1)$ observations of molecular gas in

Centaurus A, *A&A* **270**, L13-L16
Sadler, E.M., Jenkins, C.R. & Kotanyi, C.G. (1989) Low luminosity radio sources in early-type galaxies, *MNRAS* **240**, 591-635
Saikia, D.J. (1984) Extended radio jets and compact cores in quasars, *MNRAS* **208**, 231–238
Saikia, D.J. & Kulkarni, A.R. (1998) Polarization of cores and the unified scheme, *MNRAS*, in press
Saikia, D.J. et al. (1990) A VLA and MERLIN study of extragalactic radio sources with one-sided structure, *MNRAS* **245**, 408-426
Scheuer, P.A.G. (1974) Models of extragalactic radio sources with a continuous energy supply, *MNRAS* **166**, 513-528
Smith, M.D. et al. (1983) Bubbles, jets and clouds in active galactic nuclei, *ApJ* **264**, 432–445
Smith, E.P. & Heckman, T.M. (1989) Multicolor surface photometry of powerful radio galaxies. II Morphology and stellar content, *ApJ* **341**, 658–678
Smith, E.P. & Heckman, T.M. (1990) Stellar dynamics of powerful radio galaxies, *ApJ* **356**, 399–415
Tanaka, Y. et al. (1995) Gravitationally redshifted emission implying an accretion disk and massive black hole in the active galaxy MCG-6-30-15, *Nature* **375**, 659–661
Taylor, G.B. et al. (1994) The 2nd Caltech Jodrell-Bank VLBI survey. I. observations of 91 of 193 sources, *ApJS* **95**, 345–369
Thakkar, D.D. et al. (1995) The first Caltech Jodrell-Bank VLBI survey. II. λ=18cm observations of 25 sources, *ApJS* **98**, 33–40
Tingay, S.J. et al. (1994) Centaurus A, the core of the problem, *Aust.J. Phys.* **47**, 619–624
Unwin, S.C., Davis, R.J. & Muxlow, T.W.B. (1994) The whole earth array: 3C273 at 18cm, in *Compact Extragalactic Radio Sources*, eds. Zensus, J.A. & Kellermann, K.I., NRAO, 81–86
Urry, C.M. & Padovani, P. (1995) Unified schemes for radio-loud active galactic nuclei, *PASP* **107**, 803-845
Venturi, T. et al. (1993) VLBI observations of a complete sample of radio galaxies II. The parsec-scale structure of NGC315, *ApJ* **408**, 81–91
Vermeulen, R.C. (1996) Superluminals: when have we seen them all? in *Energy transport in radio galaxies and quasars*, eds Hardee, P.E., Bridle, A.H. and Zensus, J.A. (Astron. Soc. Pacific), 117–122
Vermeulen, R.C. et al. (1993) Superluminal motion in the large, strongly lobe-dominated quasar 3C47, *ApJ* **417**, 541–546
Vermeulen, R.C. & Cohen, M.H. (1994), Superluminal motion statistics and cosmology, *ApJ* **430**, 467–494
Vermeulen, R.C., Readhead, A.C.S & Backer, D.C. (1994) Discovery of a nuclear counter jet in NGC1275: A new way to probe the parsec-scale environment, *ApJ* **430**, L41–L44
Walker, R.C. (1997) Kiloparsec-scale motions in 3C120 - revisited, *ApJ* **488**, 675-681
Walker, R.C., Benson, J.M. & Unwin, S.C. (1987) The radio morphology of 3C120 on scales from 0.5 parsecs to 400 kiloparsecs, *ApJ* **316** 546-572
Walker, R.C., Romney, J.D.& Benson, J.M (1994) Detection of a VLBI counter jet in NGC1275: A possible probe of the parsec-scale accretion region, *ApJ* **430**, L45–L48
Wardle, J.F.C. & Aaron, S.E. (1996) How fast are the large-scale jets in quasars? in *Energy transport in radio galaxies and quasars*, eds Hardee, P.E., Bridle, A.H. and Zensus, J.A. (Astron. Soc. Pacific), 123–128
Wardle, J.F.C. et al. (1994) Interpretation of VLBI kinematic and polarization data: application to 3C345, *ApJ* **437**, 122-135
Wiita, P.J. (1991) in *Beams and jets in astrophysics*, eds Hughes, P.A., Cembridge University Press, 379–427
Wiita, P.J. (1996) in *Energy transport in radio galaxies and quasars*, eds Hardee, P.E., Bridle, A.H. and Zensus, J.A. (Astron. Soc. Pacific), 395–403
Wilson, A.S. (1996) On the nature of radio galaxies, in *Energy transport in radio galaxies*

and quasars, eds Hardee, P.E., Bridle, A.H. and Zensus, J.A. (Astron. Soc. Pacific), 9-24

Xu, W. et al. (1994) The bimodal distribution of misalignment angle in powerful extragalactic radio sources, in *Compact Extragalactic Radio Sources*, eds. Zensus, J.A. & Kellermann, K.I., NRAO, 7–10

Young, S. et al. (1998) The obscured BLR in the radio galaxy 3C234, *MNRAS* **294**, 478–484

Zensus, J.A., Cohen, M.H. & Unwin, S.C. (1995) The parsec-scale jet in quasar 3C345, *ApJ* **443**, 35–53

Zensus, J.A. & Kellermann, K.I. (1994) Compact extragalactic radio sources, National Radio Astronomy Observatory

Zensus, J.A. & Pearson, T.J. (1987) Superluminal radio sources, CUP, Cambridge

Zensus, J.A. & Pearson, T.J. (1990) Parsec-scale radio jets, CUP, Cambridge

Zensus, J.A. & Porcas, R.W. (1987) Superluminal motion in a randomly oriented quasar sample, in *Superluminal radio sources* eds. Zensus, J.A. & Pearson, T.J., Cambridge University Press, 126–128

Zirbel, E.L. & Baum, S.A. (1995) On the FRI/FRII dichotomy in powerful radio sources: analysis of their emission-line and radio luminosities, *ApJ* **448**, 521–547

HIGH PROPER MOTION STARS IN THE VICINITY OF SGR A*: EVIDENCE FOR A SUPERMASSIVE BLACK HOLE AT THE CENTER OF OUR GALAXY

A. M. GHEZ, B. L. KLEIN, M. MORRIS, E. E. BECKLIN
Department of Physics and Astronomy
UCLA
Los Angeles, CA 90095-1562

1. Introduction

Although the notion that the Milky Way galaxy contains a supermassive central black hole has been around for more than two decades, it has been difficult to prove that one exists. The challenge is to assess the distribution of matter in the few central parsecs of the Galaxy. Assuming that gravity is the dominant force, the motion of the stars and gas in the vicinity of the putative black hole offers a robust method for accomplishing this task, by revealing the mass interior to the radius of the objects studied. Thus objects located closest to the Galactic Center provide the strongest constraints on the black hole hypothesis.

To probe the inner region of the Galaxy, it is crucial to attain the highest resolution possible. However, turbulence in the Earth's atmosphere distorts astronomical images and typically limits the angular resolution of long-exposures to ∼0.5 - 1 arcsec, an order of magnitude worse than the theoretical limit for large ground-based telescopes. With a distance near 8 kpc to the Galactic center (Reid 1993), these types of observations have previously been limited to estimating a central mass constrained only to a volume of radius greater than or equal to ∼ 0.1 pc (e.g., Lacy et al. 1980, McGinn et al. 1989, Haller et al. 1996; Genzel et al. 1996). In contrast to long exposures, short exposures such as the one shown in Figure 1, although distorted, preserve high spatial resolution information which can be used to recover diffraction limited images via a number of different techniques, such as the relatively simple and straight forward method of "Shift-and-Add" (Christou 1991). Two groups have utilized this technique to derive proper motion measurements for stars within the central clusters: Eckart & Genzel

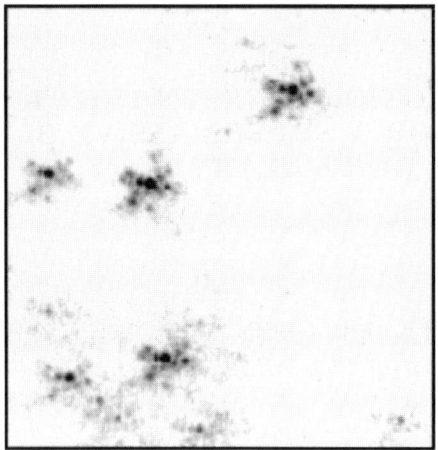

Fig. 1: One of the many short exposures ($t_{exp} = 0.13$ sec) obtained with the Keck telescope. The field of view is $\sim 5'' \times 5''$ centered roughly on SgrA* and oriented such that North is up and East is to the left. Each star has the same speckle pattern, which is dominated by one bright speckle.

(1996, 1997), who used the ESO 3-m NTT to achieve an angular resolution of 0″15 in a 4 year (1992-1996) study and Ghez et al. (1998), who collected data with the W. M. Keck 10-m telescope to obtain a resolution of 0″05 in a 2 year (1995-1997) study.

2. Shift-and-Add

One method for obtaining diffraction limited images from a series of short exposures (specklegrams) is "shift-and-add" (eg., Christou 1991). In a short exposure, the image of a star breaks up into a number of speckles, each of which can be thought of as a noisy diffraction limited images of the source, with one bright speckle dominating the overall pattern. Furthermore, each star in the short exposure shown in Figure 1 is distorted by the atmosphere in the same way, indicating that the field of view is well within the near infrared isoplanatic patch. By adding together the specklegrams, shifted to align the brightest speckle of a reference source (IRS 16 C for the Keck data and IRS 16 NE or IRS 7 for the ESO data), a shift-and-add image is generated with a point spread function composed of a diffraction-limited core atop of a broad seeing halo (see Figure 2&3).

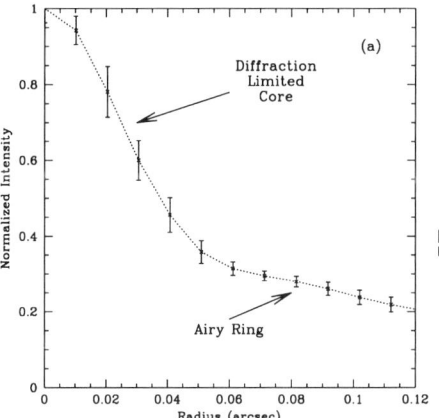

Fig. 2a: The Keck shift-and-add point spread function. The diffraction limited core, which contains ~10% of the flux, is built up from the brightest speckle in each contributed frame, while the seeing halo results from the fainter surrounding speckles. Airy rings encircle each core indicating that the diffraction-limit has truly been achieved.

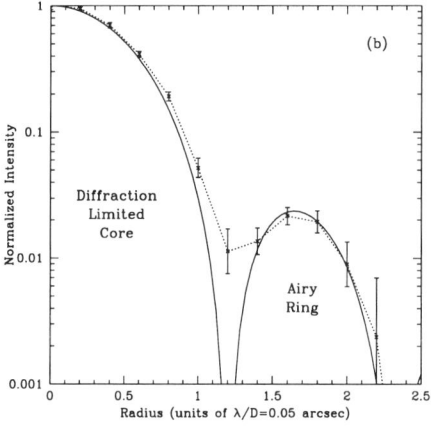

Fig. 2b: A Gaussian fit to the seeing halo has been subtracted from the data (dotted line) to emphasize the Airy ring contribution and the theoretical point spread function of Keck (solid line) is plotted for comparison.

3. Stellar Kinematics

Proper motion measurements have been reported for 104 stars located in the central stellar cluster (see Figure 4), this . Figure 5 shows the Keck positions measurements for stars within a 1 $arcsec^2$ region centered on the nominal position of Sgr A*, where motion for several of the stars can be easily seen. The distribution of velocities is nonuniform, with the highest velocity sources clustered toward the field center (see Figure 6). The coincidence

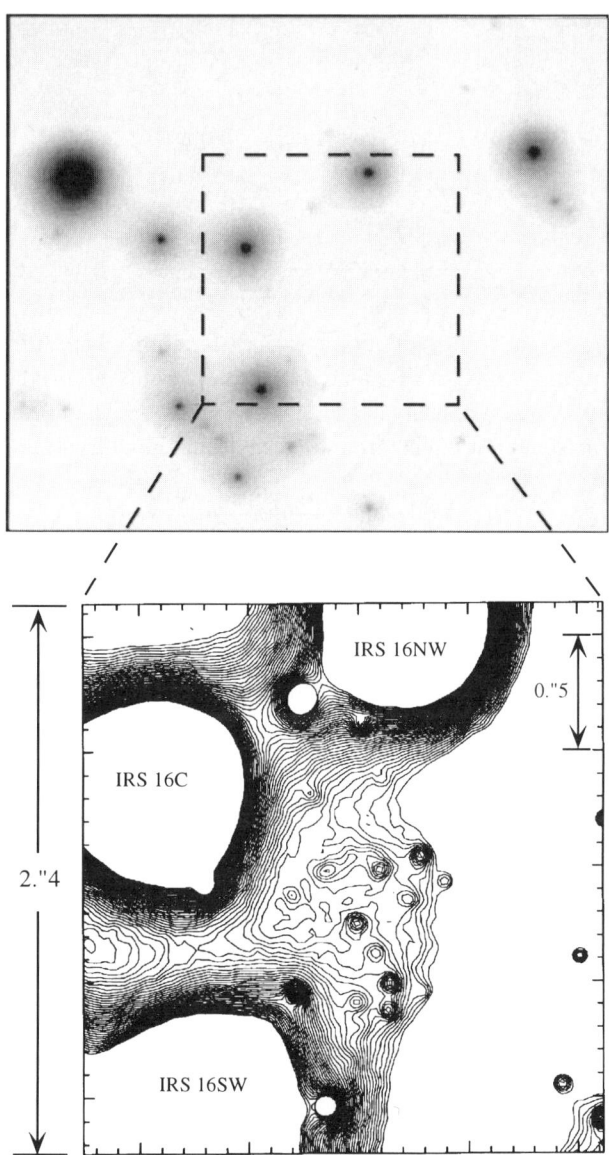

Fig. 3: The raw shift-and-add image constructed from ∼3,000 specklegrams obtained with the Keck telescope in June 1995. The top panel is scaled to the cores of the bright (K = 9-10 mag) IRS 16 stars, whereas the bottom panel highlight the cluster of faint K = 14-15 mag stars, which is roughly centered on the position of Sgr A*.

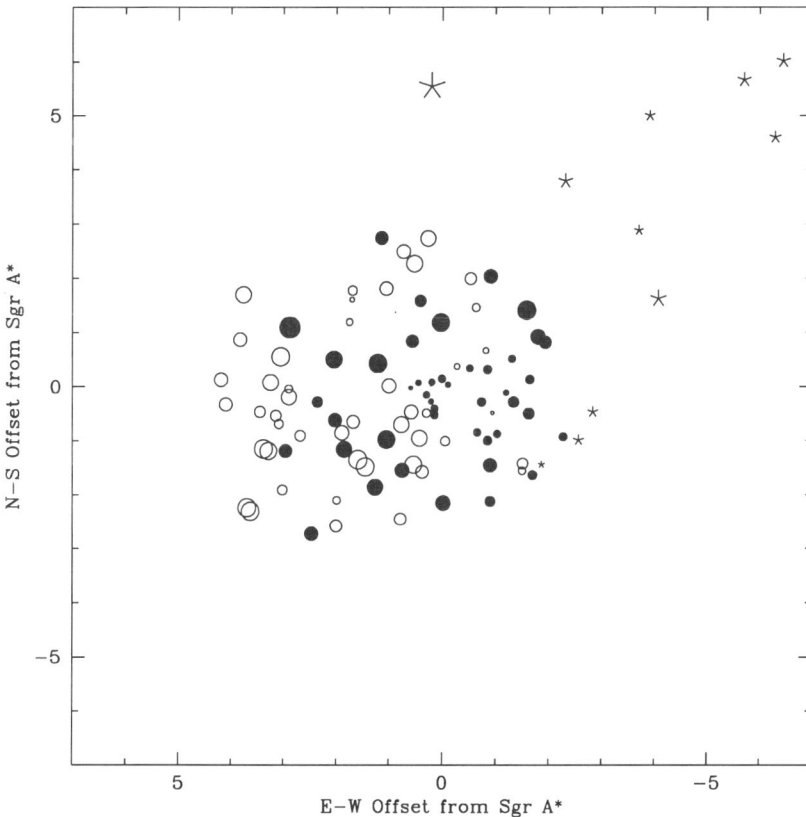

Fig. 4: The positions of the 104 stars with reported proper motions with point sizes scaled to their K magnitudes, which range from K = 6.5 to 17 mag. The filled points are the 44 stars measured by both the Keck (Ghez et al. 1998) and ESO (Eckart & Genzel 1997; Genzel et al. 1997) studies. The starred points are the 14 stars measured by only the ESO study and the unfilled points are the 46 stars reported only by the Keck study.

between the nominal position of Sgr A*, the peak of the stellar surface density, and the peak of the velocity dispersion, suggests that Sgr A* is indeed at the dynamical center of our galaxy. The position of Sgr A* is therefore assumed to be the center for the analysis that follows.

The velocity dispersion as a function of projected radius is shown in Figure 7 for the Keck measurements. The velocity dispersion at small radii is clearly much higher than the 50 km/sec dispersion observed at larger radii. Furthermore, fitting these data to a power law, $v(r) \sim r^\alpha$, results in best fit α of -0.53 ± 0.1, an excellent match with to that expected from

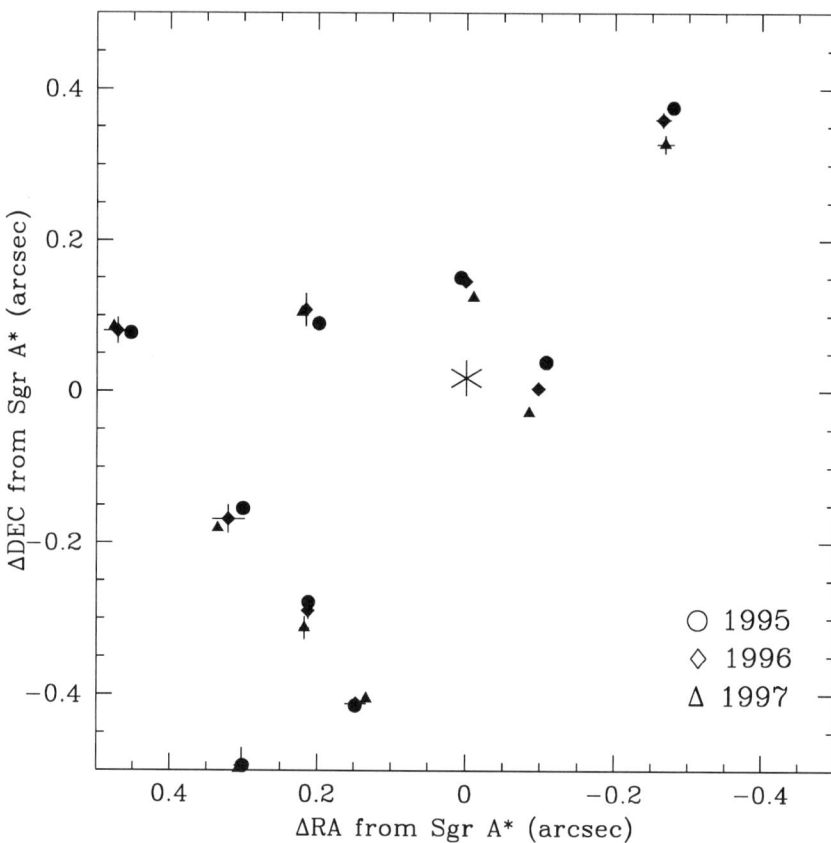

Fig. 5: The Keck position measurements of stars a 1 $arcsec^2$ centered on the position of Sgr A* (starred point, which depicts the location of this radio source). Significant velocities, which reach 1400 ± 100 km/sec, are easily detected in this region. Each year's measurement is represented by a different symbol: 1995 by triangles, 1996 by squares, and 1997 by circles.

Keplerian orbits ($\alpha = -0.5$). This behavior suggests that the stars' motions are dominated by the gravitational force of a large central mass confined to a radius less than the smallest radial bin - 0.015 pc. It also suggest that this central mass completely dominates the mass distribution out to the radius of the outermost bin - 0.1 pc.

4. The Central Dark Mass

The two-dimensional positions and velocities measured for stars in the central stellar cluster provide excellent constraints on the distribution of mat-

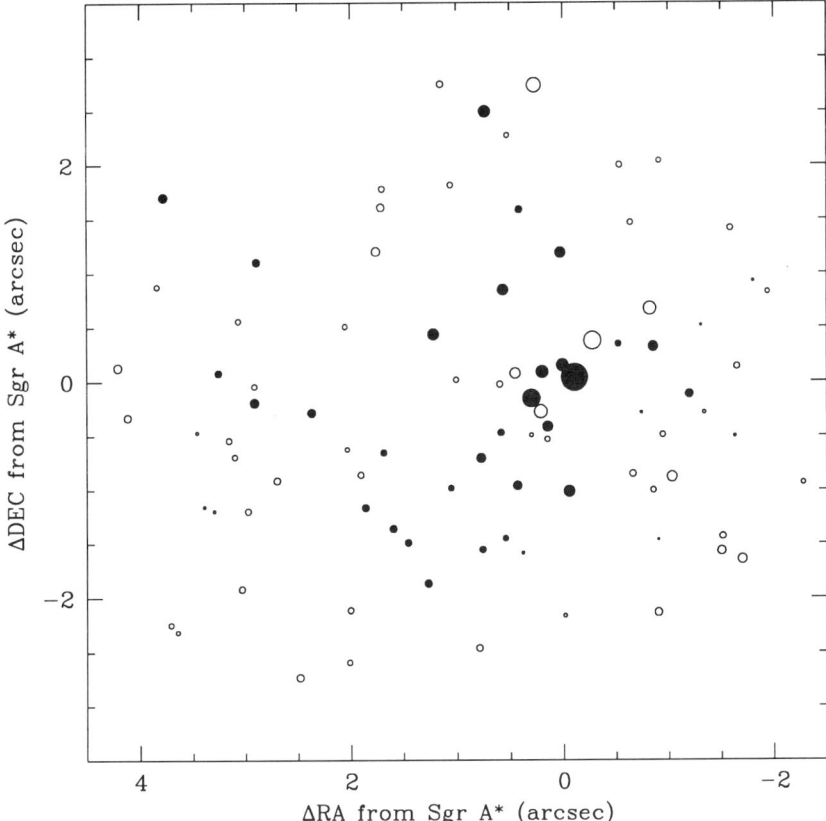

Fig. 6: The positions of the 90 stars measured with Keck are displayed with pointsizes scaled to the stars' velocities. Proper motion measurements with SNR of at least 4 are plotted as filled points. A clear increase in the velocities is visible at the field center, where stars reach velocities of 1,400 km/sec.

ter at the center of the Galaxy. In principle, if all six components of the position and velocity vectors could be observed, each star would yield an estimate of the mass enclosed within its radius. With the two-dimensional projections, the individual stars provide only lower limits on the enclosed mass, M_{min}, under the assumption that the stars are gravitationally bound in which case

$$M_{min} = \frac{v^2 R}{2G}.$$

Every star imposes a minimum mass that exceeds the enclosed mass of luminous matter extrapolated from the power law relationship derived at larger radii by Genzel et al. (1996). Considering only stars with $v/\sigma_v \geq 4$,

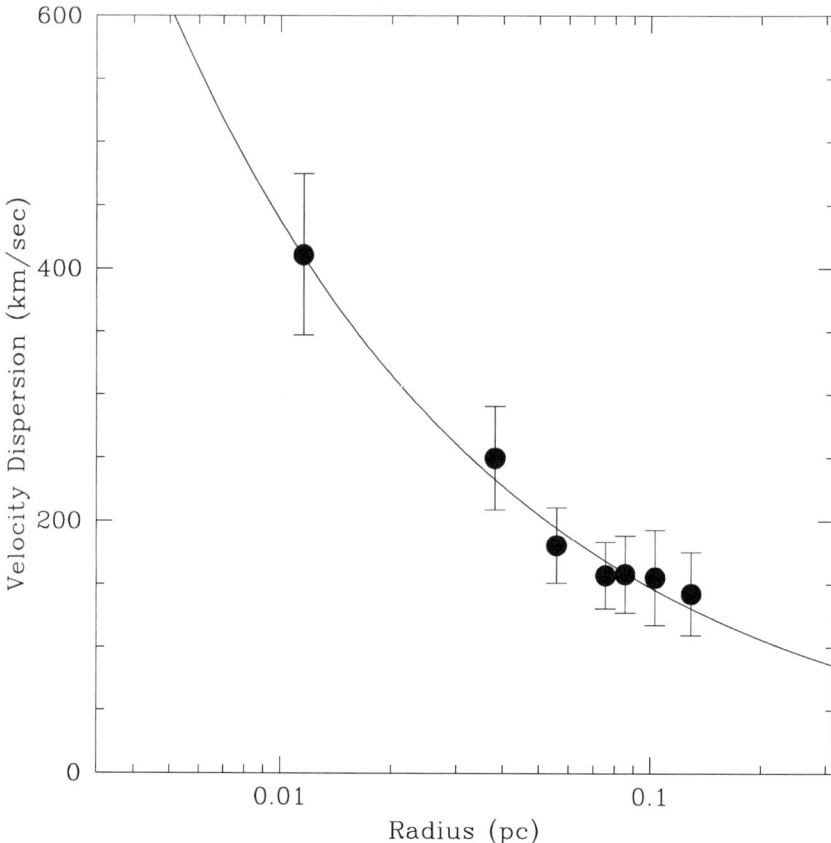

Fig. 7: The projected stellar velocity dispersion as a function of projected distance from Sgr A* is consistent with Keplerian motion, which implies the gravitational field is dominated by mass within 0.1 pc.

the minimum enclosed mass estimates reach values of $2-3 \times 10^6 M_\odot$, with the apparent members of the Sgr A* cluster having M_{min}'s ranging from $0.2-1 \times 10^6 M_\odot$ (see Figure 8). Thus the stars appear to be moving under the influence of a gravitational potential generated by at least a few million solar masses of dark matter.

Projected mass estimators analyze stars grouped in concentric annuli around the dynamical center to account for projection effects and produce estimates of the true enclosed mass. The well-known and frequently used virial mass estimator, M_{virial}, has the form

$$M_{virial} = \frac{3\pi}{2G} <v^2> / <1/R> .$$

Applied to the Galactic center data set, this mass estimator suggests that

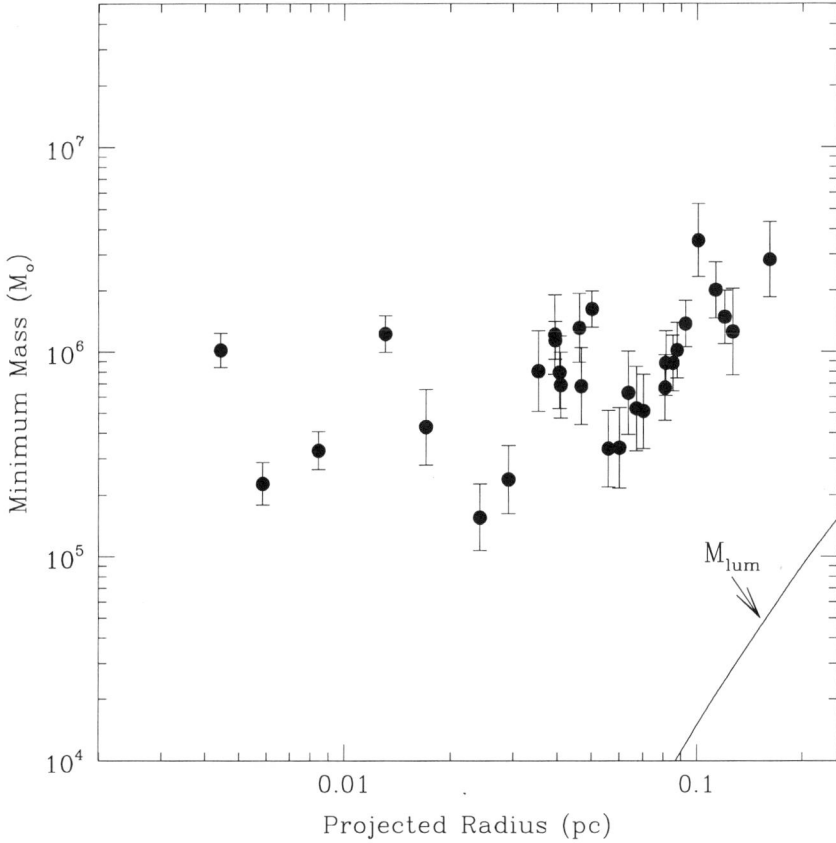

Fig. 8: The minimum enclosed mass as a function of projected radius inferred from the 30 stars measured using Keck with $v/\sigma_v \geq 4$. The solid curve in the lower right corner is an extrapolation of the enclosed luminous matter curve from Genzel et al. (1996). Each star suggests the presence of a central dark mass of roughly a million $M_\odot$.

$2.5\pm0.2\times10^6 M_\odot$ of dark matter is located predominantly, if not exclusively, at radii smaller than 0.015 pc. Bahcall & Tremaine (1981), however, pointed out that the virial mass is a biased, inefficient, and, in some cases, an inconsistent mass estimator; they proposed a set of new projected mass estimators for a tracer population moving under the influence of a central potential, M_{BT}. For the case of isotropic orbits this estimator is given by

$$M_{BT} = \frac{16}{\pi G} < v^2 R > .$$

The assumption of a dominating central potential is well justified, given the Keplerian fall-off of the velocity dispersion as a function of radius (Figure

6). For the Galactic center data set the two methods produce very similar results, with the M_{BT} values being only slightly larger, $2.6 \pm 0.2 \times 10^6 M_\odot$. These values agree very well with those obtained over similar radial distances by Genzel et al. (1997) and match well with those obtained at larger radii (see Figure 9). Overall, the enclosed mass results suggest, consistent with both the functional form of the velocity dispersion vs. radius and the minimum mass estimates, that the majority of stars observed are moving in a potential dominated by $2.6 \times 10^6 M_\odot$ of matter contained within 0.015 pc of Sgr A*. Since the total luminosity within 0.015 pc of Sgr A* observed in our maps is a meager L_K of 40 $L_\odot$, the implied mass to light ratio is $M/L_K \sim 6 \times 10^5$. As the K-band covers only a small range of wavelengths, it is useful to compare this value to that observed for the Sun, which has a M/L_K of ~ 40. Given the high mass to light ratio observed, the central mass concentration is certainly composed primarily of dark matter.

Strong constraints on the distribution of central dark matter arise from the enclosed mass measurements. Confining the density distribution of dark matter to radii smaller than 0.015 pc implies a minimum density of $10^{12} M_\odot pc^{-3}$, surpassing the volume averaged mass densities inferred for dark matter at the center of any other galaxy by at least two orders of magnitude. One intriguing possibility is that the dark compact object we are observing is a single supermassive black hole, as has been inferred for several other galaxies such as M87 (Ford et al. 1994; Harms et al. 1994) and NGC 4258 (Greenhill et al. 1995; Myoshi et al. 1995). This would be a unique solution if the minimum radius of the enclosed mass measurements corresponded to the Schwarzchild radius for a $2.6 \times 10^6 M_\odot$ black hole ($R_{sh} = 2GM_{bh}/c^2$, $R_{sh,2.6\times10^6 M_\odot} = 2.5 \times 10^{-7} pc = 11 R_\odot$), in which case the central "density" would be $\rho_{o,bh} = 4 \times 10^{25} M_\odot pc^{-3}$; however the minimum radius is still a factor of 40,000 larger than R_{sh} and thus other scenarios still need to be explored.

One alternative to the single black hole scenario is a cluster of dark matter in the form of stellar remnants, brown dwarfs, or even elementary particles. In general, astrophysical clusters can often be approximated by a Plummer model, $\rho(r) = \rho_o \left(1 + \frac{r^2}{r_c^2}\right)^{-\alpha/2}$, which requires the specification of two parameters in addition to the central density: a characteristic size scale, r_c, and the power law, α. Although an α of 2 holds for the visible stellar cluster with $\rho_o = 4 \times 10^6 M_\odot/pc^3$ and $r_c = 0.2$ pc, such a profile produces an enclosed mass which increases linearly with radius, much steeper than that observed within the central 0.2 pc. To match the observed flat enclosed mass as a function of radius with a pure cluster model requires α to be at least 3 and a very small r_c. Since astrophysical systems have been observed with α's as large as 5, we explored the viability of clusters with α

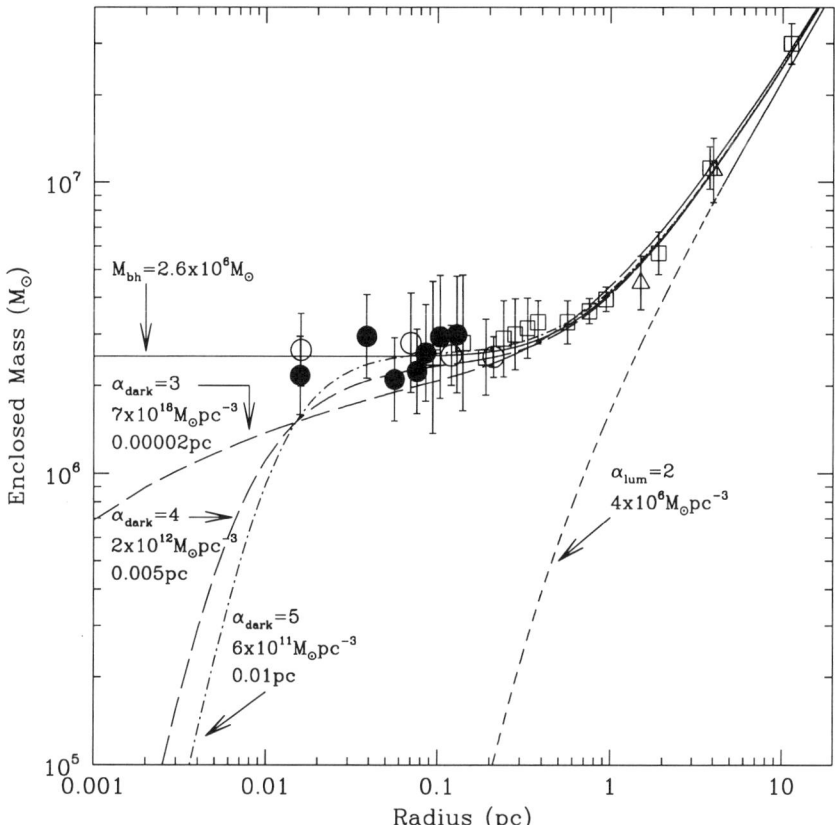

Fig. 9: The enclosed mass as a function of projected distance from Sgr A* are shown for the results of the Keck study (7 filled circles), the ESO's proper motion study (4 unfilled circles), Genzel et al. (1996) radial velocity study (13 unfilled squares), and Guesten et al. (1987) measurement of the rotating gas disk (2 unfilled triangles). From 0.1 to 0.015 the enclosed mass appears to be constant with a value of $2.6 \times 10^6 M_\odot$. Mathematically, power law dark clusters with $\alpha \geq 3$ fit the observed distributions, however they are not physically tenable (see text). The high density of the central dark mass, which exceeds $10^{12} M_\odot/pc^3$, is indicative of a single supermassive black hole.

ranging from 3 to 5. Mathematically dark cluster models can be made to fit the observed data; α of 5 requires a r_c of 0.01 pc and ρ_o of $6 \times 10^{11} M_\odot/pc^3$, α of 4 is fit by r_c of 0.005 pc and ρ_o of $2 \times 10^{12} M_\odot/pc^3$, and α of 3 demands r_c of 0.00002 pc and ρ_o of $7 \times 10^{18} M_\odot/pc^3$ (see Figure 9). Physically, however, such dark cluster models are highly improbable (cf. Maoz 1995, 1998). A viable cluster must have both evaporation and collision timescales that are greater than the lifetime of the Galaxy ~ 10 Gyr. Clusters of objects having a single mass have evaporation timescales shorter than the age of the Galaxy for masses larger than 0.02 $M_\odot$, ruling them out from consideration. Among possible nonluminous cluster members with masses less than 0.02 $M_\odot$, white dwarfs, brown dwarfs, and very low mass objects with cosmic composition are ruled out by their short collisional timescales, which are at most 10^7 years. What cannot be eliminated with timescale considerations alone are clusters of elementary particles and very low mass (M < 0.02 $M_\odot$) black holes, however such clusters are theoretically unmotivated. Thus, the observed mass distribution is not likely to be due to a pure cluster of dark objects.

Another alternative is for only a fraction of the mass to be in a central black hole with the remaining mass contained in a cluster of dark objects as might be found in a post core-collapsed cluster. Fitting the measured enclosed mass as a function of radius with a black hole plus an $\alpha \sim 2$ cluster model, we find that only 1% of the total mass interior to 0.015 pc can be in the cluster due to rapid rise of the mass enclosed by an $\alpha \sim 2$ cluster. Although larger α clusters relax this criterion, $\alpha \sim 2$ is the expected form for a cluster surrounding a black hole (e.g., Binney & Tremaine 1987). Thus the dynamical evidence, independent of the presence of Sgr A*, leads us to the conclusion that our Galaxy harbors a $2.6 \times 10^6 M_\odot$ black hole.

Our Galaxy was neither the first nor an obvious candidate for a central supermassive black hole; however it, along with NGC 4258, has become one of the strongest cases for a $10^6 M_\odot$ black hole. The significance of a central black hole in our normal inactive Galaxy is the implication that massive black holes might be found at the centers of almost all galaxies.

References

Backer, D.C. 1994, in IAU Symp. No. 169: Unsolved Problems in the Milky Way
Bahcall, J. N., and Tremaine, S. 1981, ApJ, 244, 805
Bahcall, J. N., and Tremaine, S. 1987, Galactic Dynamics (Princeton: Princeton Univ. Press)
Beckert, T., Cuschl, W. J., Mezger, P. G., & Zylka, R. 1996, A&A, 307, 450
Christou, J. C. 1991, PASP, 103, 1040
Eckart, A. & Genzel, R. 1996, Nature, 383, 415
Eckart, A. & Genzel, R. 1997, MNRAS, 284, 576
Ford, H. C., Harms, R. J., Tsvetanov, Z. I., Hartig, G. F., Dressel, L. L., Kriss, G. A., Bohlin, R. C., Davidsen, A. F., Margon, B., Kochhar, A. K. 1994, ApJ, 435, L27
Ghez, A. M., Klein, B. L., Morris, M., Becklin, E. E. 1998, ApJ, submitted
Genzel, R., Eckart, A., Ott, T. and Eisenhauer, F. 1997, MNRAS, 291, 219
Genzel, R., Thatte, M., Krabbe, Kroker, H., & Tacconi-Garman, L. E. 1996, ApJ, 472, 153
Greenhill, L. J., Jiang, D. R., Moran, J. M., Reid, M. J., Lo, K. Y., Claussen, M. J. 1995, ApJ, 440, 619
Guesten, R., Genzel, R., Wright, M. C. H., Jaffe, D. T., Stutski, J., and Harris, A. I.

1987, ApJ, 318, 124
Haller, J. W., Rieke, M. J., Rieke, G. H., Tamblyn, P., Close, L., & Melia, F. 1996, ApJ, 456, 194
Harms, R. J., Ford, H. C., Tsvetanov, Z. I., Hartig, G. F., Dressel, L. L., Kriss, G. A., Bohlin, R., Davidsen, A. F., Margon, B., Kochhar, A. K. 1994, ApJ, 435, L35
Lacy, J. H., Townes, C. H., Geballe, T. R., & Hollenbach, D. J., 1980, ApJ, 241, 132
Lynden-Bell, D. & Rees, M. J. 1971, MNRAS, 152, 461
Maoz, E. 1995, ApJ, 447, L91
Maoz, E. 1998, ApJ, 494, L181
Matthews, K., Ghez, A. M., Weinberger, A. J., and Neugebauer, G. 1996, PASP, 108, 615
Matthews, K. and Soifer, B. T. 1994, Astronomy with Infrared Arrays: The Next Generation, ed. I. McLean, Kluwer Academic Publications (Astrophysics and Space Science, v. 190, p. 239)
McGinn, M. T., Sellgren, K., Becklin, E. E., & Hall, D. N. B., 1989, ApJ, 338, 824
Menten, K. M., Reid, M. J., Eckart, A., & Genzel, R. 1997, ApJ, 475, L111
Myoshi, M., Moran, J. M., Hernstein, J., Greenhill, L., Nakai, N., Diamond, P., & Inoue, M. 1995, Nature 373, 127
Reid, M. ARA&A, 31, 345
Rogers, A. E. E., Doeleman, S., Wright, M. C. H., Bower, G. C., Backer, D. C., Padin, S., Philips, J. A., Emerson, D. T., Greenhill, L., Moran, J. M., & Kellermann, K. I. 1994, ApJ, 434, L59
Serabyn, E, Carlstrom, J., Lay, O., Lis, D. C., Hunter, T. R., & Lacy, J. H. 1997, ApJ, 490, L77

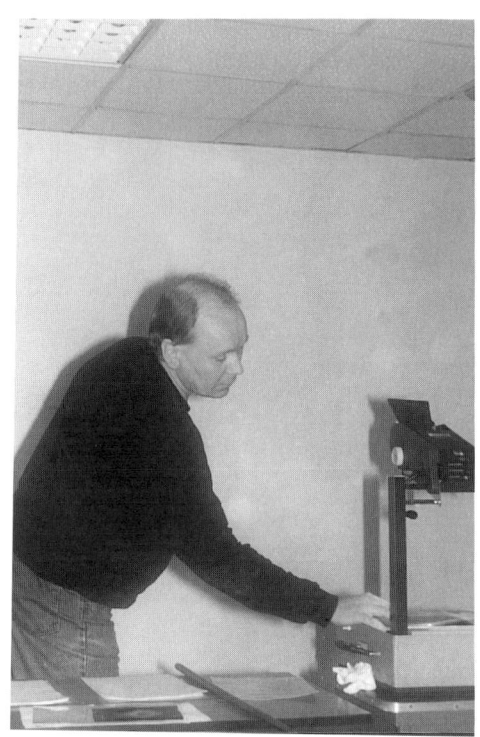

A. Sillanpää

R. Mukherjee

BLACK HOLES IN OUR GALAXY

Dynamical Evidence

PHIL CHARLES
Oxford University
Dept of Astrophysics, Keble Road, Oxford OX1 3RH, UK

Abstract. While X-ray observatories continue to identify black-hole candidates on the basis of specific characteristics (such as X-ray spectrum and fast temporal variability), it is necessary to undertake optical spectroscopic dynamical studies of these objects in order to unequivocally show that they could not be (conventional) neutron stars. Such work has expanded dramatically in recent years as a result of the discovery of ~20 *soft X-ray transients* whose extended periods of X-ray quiescence provide an ideal opportunity for ground-based studies of the mass-losing star in these close interacting binaries. Mass functions have now been derived for eight of these systems that, combined with ellipsoidal studies to constrain the orbital inclinations, imply they significantly exceed the canonical maximum neutron star mass of $3M_\odot$. They therefore represent the most secure dynamically determined black hole candidates in our Galaxy.

1. Introduction

It has been more than 30 years since the first X-ray binary was optically identified (Sco X-1), but a detailed knowledge of their binary parameters only started to come in the 1970s with the identification and study of Cyg X-1, Her X-1 and Cen X-3. But these all have early-type companions, observable in spite of the presence in the binary of luminous X-ray emission (for a review see van Paradijs & McClintock 1995). Apart from Cyg X-1, most of these high-mass X-ray binaries (HMXRBs) had pulsating neutron star compact objects, thereby providing the potential for a full solution of the binary parameters since they were essentially double-lined spectroscopic binaries. From this has come the detailed dynamical mass measurements of neutron star systems which have recently been collated by Thorsett &

Chakrabarty (1998), showing that they are all consistent with a mass of $1.35\pm0.04M_\odot$.

However, when HMXRBs are suspected of harbouring much more massive compact objects (as is the case for Cyg X-1), the mass measurement process runs into difficulties. By definition, there will be no dynamic features (such as pulsations) associated with the compact object that can be observed. Hence all mass information must come from the mass-losing companion, and the mass of the compact object cannot be determined unless the companion's mass is accurately known.

The situation for low-mass X-ray binaries (LMXBs), such as Sco X-1, is completely different, in that their short orbital periods require their companion stars to be of low mass. This can be demonstrated quite simply as follows (see King 1988). Since these are interacting binaries in which the companion fills its Roche lobe, then we may employ the useful Paczynski (1971) relation

$$R_2/a = 0.46(1+q)^{-1/3} \qquad (1)$$

where the mass ratio $q = M_X/M_2$. Combining this with Kepler's 3rd Law yields the well-known result that the mean density, ρ (in g cm^{-3}) of the secondary,

$$\rho = 110/P_{hr}^2 \qquad (2)$$

And if these stars are on or close to the lower main sequence, then $M_2 = R_2$ and hence $M_2 = 0.11 P_{hr}$. Therefore short period X-ray binaries must be LMXBs and so the companion star will be faint. The major observational problem with this is that the optical light will then be dominated by reprocessed X-radiation from the disc (or heated face of the companion star; see van Paradijs & McClintock 1994). This is why the optical spectra of LMXBs are hot, blue continua (U–B typically -1) with superposed broad hydrogen and helium emission lines, the velocities of which indicate that they largely arise in the inner disc region, thereby denying us access to the dynamical information that is essential if accurate masses are to be determined. Hence, the evidence for the nature of the compact object in most bright LMXBs has come from indirect means, usually X-ray bursting behaviour (as few are X-ray pulsars) or the fast flickering first seen in Cyg X-1 (and hence used as a suggestion of the presence of a black hole). [Note, however, that while it is useful to employ the Paczynski relation in this way, it is only valid for $q > 1$, and there is a more accurate algorithm due to Eggleton (1983) which is valid for all q.]

To make real progress in determining the nature of compact objects in our Galaxy requires dynamical mass measurements of the type hitherto employed on neutron stars in HMXRBs. But without velocity information associated with the compact object, all that can be measured (in the case of Cyg X-1, and the other two HMXRBs suspected of harbouring black holes, LMC X-1 and LMC X-3) is the mass function

$$f(M) = (M_X \sin i)^3 / (M_X + M_2)^2 \qquad (3)$$

And since $M_2 \geq M_X$, then M_X is not accurately known because M_2 has a wide range of uncertainty ($\sim$12–20M$_\odot$) given the unusual evolutionary history of the binary. The compact object in Cyg X-1 almost certainly is a black hole, but an accurate mass determination is not possible from the available data which simply constrain it to be >3.8M$_\odot$ (Herrero et al 1995). This is close to the canonical maximum mass of a neutron star, based on the oft-quoted Rhoades-Ruffini Theorem (1974). However, there are a number of assumptions built into this which need careful examination in the light of the masses of the compact objects reviewed here (see e.g. Miller 1998 and Miller et al 1998).

For the LMXBs we clearly need to find systems in which the companion star *is* visible, which requires sources where the X-ray emission switches off for some reason. This is the basis of the new field of study of the *soft X-ray transients*, hereafter SXTs (and sometimes referred to as *X-ray novae*). Remarkably, of the $\sim$23 currently known, only 6 (i.e. $\sim$25%) are confirmed neutron star systems (they display type I X-ray bursts), the remainder are all black-hole candidates, the highest fraction of any class of X-ray source.

2. X-ray Behaviour

2.1. OUTBURST

The SXTs typically outburst every $\sim$10 years. The first one (Cen X-2) was found by early rocket flights (Harries et al 1967), but the prototype of the class is widely recognised to be A0620-00 (Nova Mon 1975), for several months the brightest X-ray source in the sky and peaking at 11th mag in the optical (Elvis et al 1975; see figure 1, taken from Kuulkers 1998). Their light curves tend to show a fast rise followed by an exponential decay (see Chen et al 1997 for a compendium of all SXT light curves), the amplitude of which has been shown by Shahbaz & Kuulkers (1998) to be related to the orbital period, and the precise form of the decay is related to the peak X-ray luminosity at outburst (King & Ritter 1997; Shahbaz et al 1998a).

It takes $\sim$1 year for SXTs to reach optical/X-ray quiescence after an outburst, but note that there have been subsequent *mini*-outbursts in some systems (e.g. GRO J0422+32, see figure 1) and erratic re-brightenings in others (e.g. GRO J1655-40). The observed properties of the 9 SXTs for

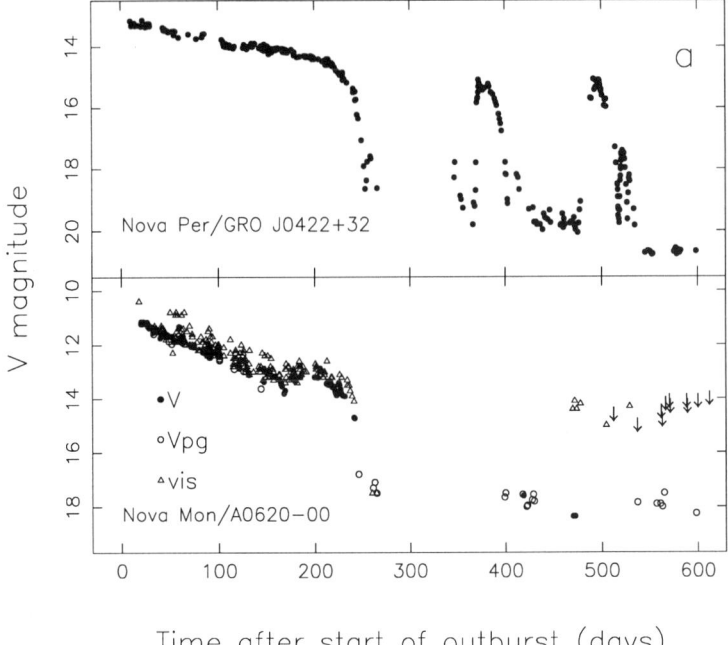

Figure 1. Optical outburst light curves for (top) GRO J0422+32 (=Nova Per 1992) and the prototype SXT A0620-00 (=Nova Mon 1975), from Kuulkers (1998). J0422+32 was the first SXT in which subsequent 'mini'-outbursts were seen, starting approximately 4 months after the end of the main outburst. The inclusion of new archival data on Nova Mon shows that it too may have exhibited this property.

which full dynamical analyses have been performed are summarised in table 1 and listed in order of orbital period.

At the time of outburst most (but not all) exhibit *ultra-soft* X-ray spectra with black-body colour temperatures of $kT\sim0.5 - 1$keV superposed on a hard power-law extending to much higher energies (see Tanaka & Lewin 1995). It is this characteristic that gives the SXTs their name, and effectively distinguishes them from the much harder Be X-ray transients that appear to be almost exclusively long-period neutron star systems. [Note also that the ultra-soft X-ray spectrum of SXTs is very different from the *super-soft* designation applied (mostly) to the (very) much cooler accreting white dwarf systems in the LMC and SMC (see Kahabka & van den Heuvel 1997).]

Additionally the SXTs (e.g. GS2023+338) can show extremely erratic variability which is very similar to that displayed by Cyg X-1. Hence the X-ray spectrum and variability are used as key discriminators to hunt for black holes. However, it must be noted that, in certain circumstances, neutron star systems can mimic these properties (e.g. Cir X-1 and X0331+53), and so we must use only dynamical evidence in the final analysis as to the nature of the compact objects (McClintock 1991).

TABLE 1. Optical/IR Properties of Soft X-ray Transients

Source	P (hrs)	Sp. Type	E_{B-V}	V	K	$v_{rot}\sin i$ (km s^{-1})	K_2 (km s^{-1})
J0422+32	5.1	M2V	0.3	22	16.2	≤ 80	381
A0620-00	7.8	K5V	0.35	18.3	6	83	433
GS2000+25	8.3	K5V	1.5	21.5	17	86	518
N Mus 1991	10.4	K0-4V	0.29	20.5	16.9	106	399
N Oph 1977	12.5	K	0.5	21.5	-	≤ 79	448
Cen X-4	15.1	K7IV	0.1	18.4	15.0	45	146
4U1543-47	27.0	A2V	0.5	16.6	-	-	124
N Sco 1994	62.9	F3-6IV	1.3	17.2	-	-	228
V404 Cyg	155.3	K0IV	1	18.4	12.5	39	208.5

2.2. QUIESCENCE

Even in quiescence, optical studies (see section 3) show that mass transfer continues in the SXTs, and indeed many have been detected by X-ray observatories (Einstein, EXOSAT, ROSAT) as very weak sources (e.g. Verbunt et al 1994). However, the observed luminosities are substantially lower than expected for the continuing accretion rate, and this has led various groups (see e.g. Abramowicz et al 1995; Narayan et al 1997) to propose that *advective accretion* is taking place. The inner disc at low accretion rates evaporates due to the X-radiation into a very hot low density corona. Such hot gas cannot radiate efficiently and transports most of its thermal energy onto the compact object (the advection process). (Such models can also account for the spectral shapes during outburst, see e.g. Chen et al 1995; Chakrabarti & Titarchuk 1995 and Chakrabarti 1998.) If it is a black hole, then that energy is lost! But if it is a neutron star then the energy will be radiated from the neutron star's surface. The model therefore predicts that black-hole SXTs will be X-ray fainter in quiescence (relative to outburst) than neutron-star systems, and there is some evidence for this (see discussion in McClintock 1998).

3. Mass Measurements

3.1. RADIAL VELOCITY CURVES

It is when they reach quiescence that the SXTs become such valuable resources for research into the nature of LMXBs. Their optical brightness has typically declined by a factor of 100 or more, with all the known SXTs having quiescent magnitudes in the range 17–23. The quiescent light is now dominated by the companion star and, while technically challenging, presents us with the opportunity to determine its spectral type, period and radial velocity curve (whose amplitude is the K-velocity). From the latter two we can calculate the mass function

$$f(M) = \frac{PK^3}{2\pi G} = \frac{M_X^3 \sin^3 i}{(M_X + M_2)^2} \qquad (4)$$

and the results for the same 9 systems (this time listed in order of their mass functions) are summarised in table 2. Hence the enormous importance of the SXTs, since all are LMXBs which have $M_X > M_2$. The mass functions in table 2 represent the *absolute minimum* values for M_X since (for all of them) $i < 90°$ and $M_2 > 0$, both of which serve only to *increase* the implied value of M_X. That is why the work of McClintock & Remillard (1986) on A0620-00 and Casares et al (1992) on V404 Cyg has generated so much interest.

It should also be noted that it can be possible to derive some dynamical information about the system even during outburst, providing spectroscopic data of sufficient resolution is obtained. Casares et al (1995) observed GRO J0422+32 during one of its subsequent "mini-outbursts" and found intense Balmer and HeIIλ4686 emission that was modulated on what was subsequently shown to be the orbital period. Furthermore, a sharp component of HeII displayed an S-wave that was likely associated with the hotspot.

However, to determine the actual value of M_X we need additional constraints that will allow us to infer values for M_2 and i.

3.2. ROTATIONAL BROADENING

Since the secondary is constrained to corotate with the primary in short period interacting binaries, we can exploit our knowledge of its size by making *assumption 1* that R_2 is given by equation (1). Hence the result (Wade and Horne, 1988)

$$v_{rot} \sin i = \frac{2\pi R_2}{P} \sin i = K_2 \times 0.46 \frac{(1+q)^{2/3}}{q} \qquad (5)$$

TABLE 2. Derived Parameters and Dynamical Mass Measurements of SXTs

Source	f(M) ($M_\odot$)	ρ (g cm^{-3})	q (=M_X/M_2)	i	M_X ($M_\odot$)	M_2 ($M_\odot$)	Ref.
V404 Cyg	6.08±0.06	0.005	17±1	55±4	12±2	0.6	[1–2]
G2000+25	5.01±0.12	1.6	24±10	56±15	10±4	0.5	[3–5]
N Oph 77	4.86±0.13	0.7	>19	60±10	6±2	0.3	[6–9]
J1655-40	3.24±0.14	0.03	3.6±0.9	67±3	6.9±1	2.1	[10–12]
N Mus 91	3.01±0.15	1.0	8±2	54^{+20}_{-15}	6^{+5}_{-2}	0.8	[13–15]
A0620-00	2.91±0.08	1.8	15±1	37±5	10±5	0.6	[16–18]
J0422+32	1.21±0.06	4.2	>12	20–40	10±5	0.3	[19–20]
4U1543-47	0.22±0.02	0.2	-	20–40	5.0±2.5	2.5	[21]
Cen X-4	0.21±0.08	0.5	5±1	43±11	1.3±0.6	0.4	[22–23]

[1] Casares & Charles 1994; [2] Shahbaz et al 1994b; [3] Filippenko et al 1995a; [4] Beekman et al 1996; [5] Harlaftis et al 1996; [6] Filippenko et al 1997; [7] Remillard et al 1996; [8] Martin et al 1995; [9] Harlaftis et al 1997; [10] Orosz & Bailyn 1997; [11, 12] van der Hooft 1997, 1998; [13] Orosz et al 1996; [14] Casares et al 1997; [15] Shahbaz et al 1997; [16] Orosz et al 1994; [17] Marsh et al 1994; [18] Shahbaz et al 1994a; [19] Filippenko et al 1995b; [20] Beekman et al 1997; [21] Orosz et al 1998; [22] McClintock & Remillard, 1990; [23] Shahbaz et al 1993.

from which q can be derived if v_{rot} is measurable. Typical values are in the range 40–100 km s^{-1} and clearly require high resolution and high signal-to-noise spectra of the secondary.

Figure 2 (second from top) shows the Casares & Charles (1994) WHT summed spectrum of V404 Cyg after doppler correcting all individual spectra into the rest-frame of the secondary. The bottom spectrum is a very high S/N spectrum of a K0IV star which was used as a template, and which clearly has much narrower (actually they are unresolved) absorption lines. The template is broadened by different velocities (together with the effects of limb darkening), subtracted from that of V404 Cyg and the residuals χ^2 tested. This gave $v_{rot} \sin i = 39\pm1$ km s^{-1} and hence $q = 16.7\pm1.4$. The full details can be found in Casares & Charles, and Marsh et al (1994). It should also be noted that while the accretion disc around the compact object might be expected to provide some velocity information, there are serious difficulties with this. The Hα line in figure 2 is extremely broad ($\geq$1000 km s^{-1}) and yet the compact object's motion in such high q systems will be very small (typically $\leq$30 km s^{-1}). Nevertheless such motions have been seen (e.g. Orosz et al 1994), but their interpretation is not straightforward as there is a small phase offset relative to the motion of the companion star,

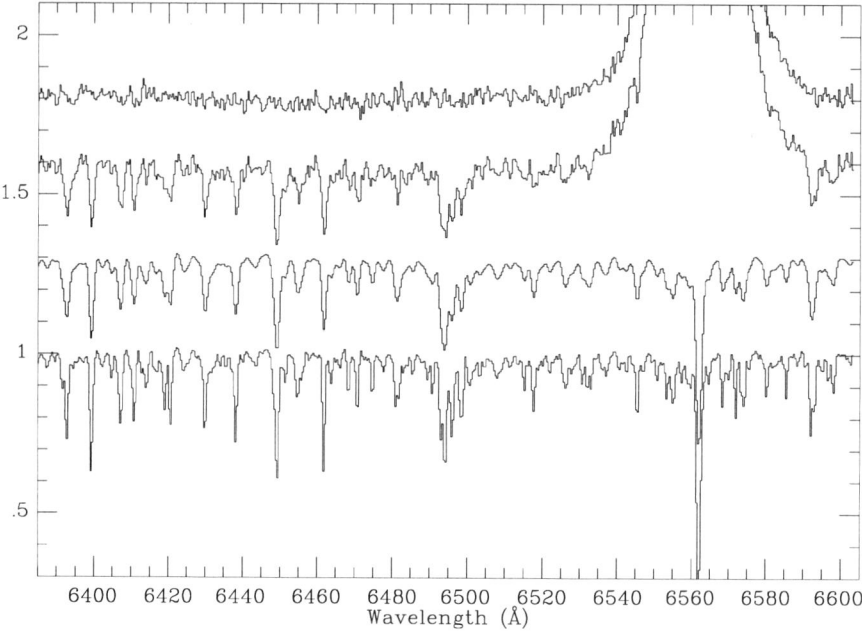

Figure 2. Determining the rotational broadening in V404 Cyg. From bottom to top: the K0IV template star (HR8857); the same spectrum broadened by 39 km s^{-1}; doppler corrected sum of V404 Cyg (dominated by intense Hα emission from the disc); residual spectrum after subtraction of the broadened template (from Casares and Charles, 1994).

and so they cannot be used as part of the dynamical study.

Having determined q, the range of masses consistent with the observed $f(M)$ is plotted in figure 3, where the only remaining unknown is the orbital inclination i. To date, none of the SXTs is eclipsing (although GRO J1655-40 shows evidence for a grazing eclipse), and so it is the determination of i that leads to the greatest uncertainty in the final mass measurement. Nevertheless there are methods by which i can be estimated.

3.3. ELLIPSOIDAL MODULATION

We exploit one more property of the secondary, it's peculiarly distorted shape responsible for the so-called *ellipsoidal modulation* as we view the varying projected area of the secondary around the orbit. This leads to the classical double-humped light curve, as shown for A0620-00 in figure 4. If the secondary's shape is sufficiently well-determined by theory (i.e. the form of the Roche lobe) then the observed light curve depends on only 2 parameters, q and i. In several cases (as described in the previous section) q

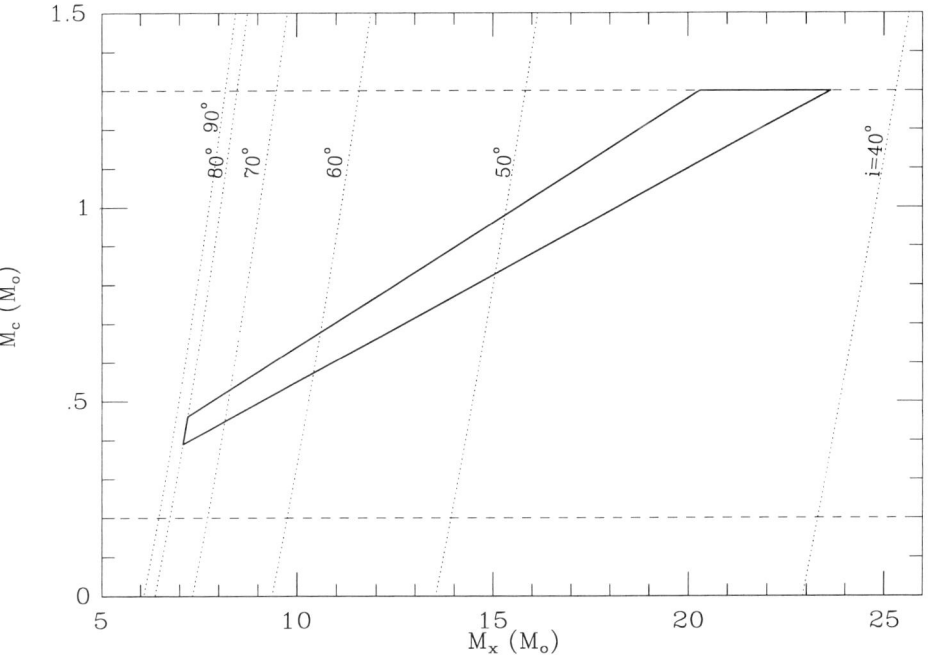

Figure 3. Constraints on M_X and M_2 for a range of values of i in V404 Cyg based on the radial velocity curve $(f(M))$ and determination of q (from the rotational broadening). It is the limited constraint on i (absence of eclipses) that leads to a wide range of M_X (from Casares and Charles, 1994).

is already determined, but in practice the ellipsoidal modulation is largely insensitive to q for values $q > 5$. Details of the light curve modelling can be found in Shahbaz et al (1993), and the collected results are in table 2.

This final stage in the SXT orbital solutions has made 2 key assumptions: *assumption 2*, that the secondary in quiescence fills its Roche lobe; *assumption 3*, that the light curve is not contaminated by any other light sources. It is felt that the former is reasonable since there is strong evidence through doppler tomography (e.g. Marsh et al 1994) for continued mass transfer in quiescence from the secondary. However, the principal (and potentially significant) uncertainty is the problem of any other contaminating light sources. This would mainly be the accretion disk, but residual X-ray heating and starspots on the surface of the secondary might also be present. It is for this reason that this work has been performed in the K band whenever possible. The disc contamination has been measured in the optical around Hα (as a by-product of the spectral type determination by searching for excess continuum light) and is typically $\leq 10\%$. It should

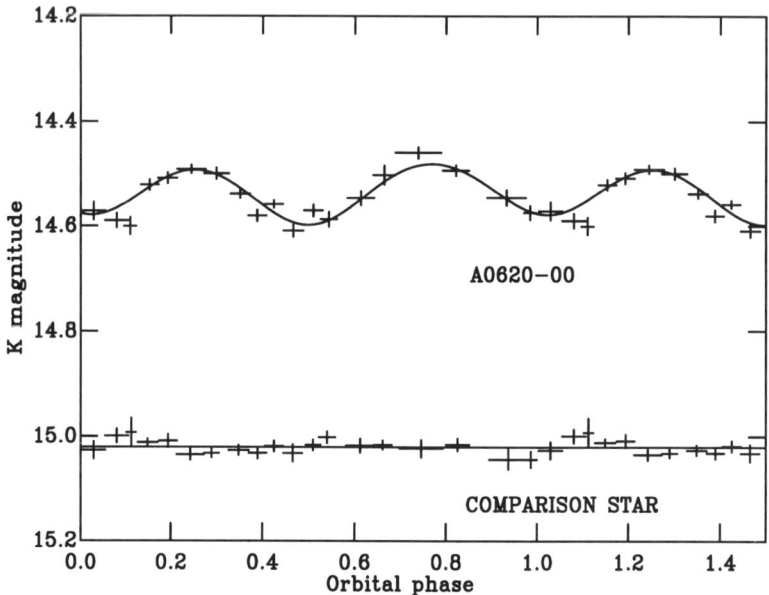

Figure 4. IR light curve of A0620-00 showing the classical double-humped modulation (from Shahbaz et al, 1994a).

therefore be even less in the IR given the blue colour of the disc. However, the outer disc edge has been found to be an IR emitter in CVs (Berriman et al 1985) and the light curves might be contaminated, as was suggested in the case of V404 Cyg by Sanwal et al (1996). This is potentially an important effect, since a contaminating (and presumably steady) contribution will reduce the amplitude of the ellipsoidal modulation, which will lead to a lower value of i being inferred, and hence a higher mass for the compact object.

For this reason Shahbaz et al (1996, 1998b) undertook IR K-band spectroscopy of the two brightest and best studied SXTs, V404 Cyg and A0620-00. Only upper limits were derived in both cases, showing that any contamination must be small and hence the masses derived can (at most) be reduced by only small amounts.

The results of these analyses are collected together in figure 5, which contains all neutron star and black hole mass measurements. It should also be noted that the value for Cen X-4 (one of only two neutron star SXTs, identified on the basis of its type I X-ray bursts) has been derived exactly by the method outlined here (Shahbaz et al 1993) and yields a value of

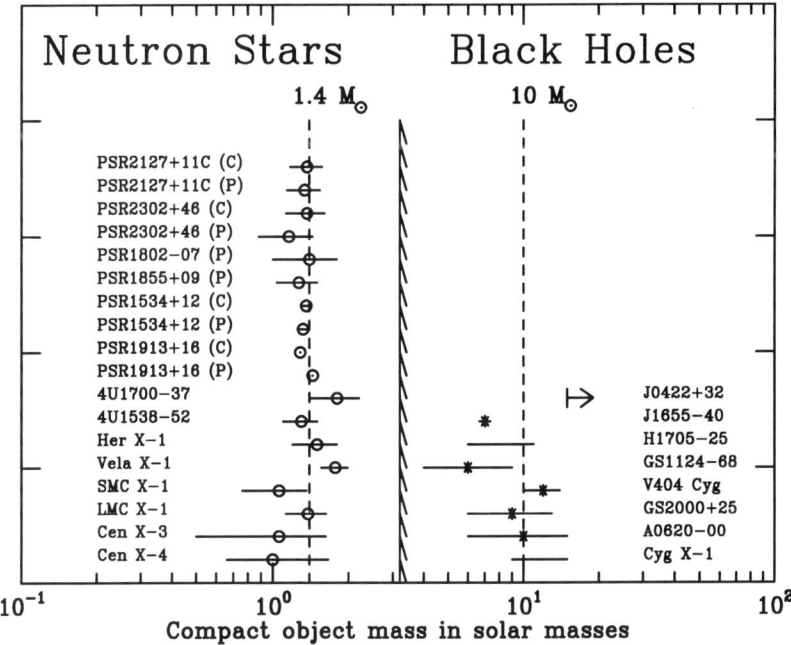

Figure 5. The mass distribution of neutron stars and black holes. Note the remarkably narrow spread of neutron star masses, and the large factor by which the BHXRB masses exceed the (canonical) maximum mass of $3.2M_\odot$.

$1.3M_\odot$, in excellent accord with that expected for a neutron star.

4. The Superluminal Transient GRO J1655-40

In 1994 there were two new X-ray transients discovered, GRS1915+105 and GRO J1655-40, that brought an entirely new type of behaviour to this field. As with many of the transient outbursts, they also emitted strongly in the radio, but VLA and VLBI observations showed that these objects also exhibited ejection events that were "superluminal" (Mirabel & Rodríguez 1994; Hjellming & Rupen 1995), the first time that such phenomena had been observed within the Galaxy. Further dynamical studies of GRS1915+105 are severely hampered by (a) its extremely high interstellar extinction ($A_V \sim 26$), leaving only a variable, $K \sim 14$ IR counterpart, and (b) its continuing and extremely variable X-ray activity that is totally unlike any of the "classical" SXTs. It is not even clear that GRS1915+105 is an LMXB (see Mirabel et al 1997), and it demonstrates an extraordinarily rich variety of X-ray variability (e.g. Morgan & Remillard 1996, Belloni et al

1997).

GRO J1655-40 (N Sco 1994), on the other hand, is optically the brightest in quiescence of all the SXTs, and so has extremely well-determined photometric light-curves, and is an excellent candidate for a dynamical study. The companion also has one of the earliest (confirmed) spectral types (mid-F) of the SXTs which means that, in quiescence, the effects of the accretion disk are very small, almost negligible. And the high γ-velocity led to the suggestion (Brandt et al 1995) that J1655-40 could be an example of a NS system that had suffered accretion-induced collapse.

The detailed system parameters have been derived from several photometric studies of J1655-40 by van der Hooft et al (1997), Orosz & Bailyn (1997) and van der Hooft et al (1998). The values recorded in table 2 are those from the latter paper due to their more conservative error analysis. J1655-40 is unusual in this class in that it has a low mass ratio of $q\sim3$ (but this has not yet been obtained from a rotational broadening study, due to its return to activity shortly after its initial outburst). At such a value, the ellipsoidal modulation is sensitive to both q and i (the latter also being tightly constrained here as a result of its grazing eclipse). Hence, once it becomes possible to spectroscopically determine the rotational broadening of the F star (when it re-enters an extended period of quiescence) it will then be possible to perform a check of the entire basis on which the quiescent SXT light curves have been modelled and used to determine i. It has also been suggested (Kolb et al 1997) that the secondary star is in a very interesting evolutionary state in which it is crossing the Hertzsprung gap and about to ascend the giant branch. This is what is driving the much higher mass transfer rate than in the other SXTs, but temporary drops in $\dot{M}$ return it to the transient domain.

5. Conclusions

The dramatic advances in the field of galactic black-hole studies have come about during this decade for 2 main reasons: (i) the almost continuous monitoring of the X-ray sky that is now provided by all-sky monitors such as those on CGRO and RXTE has provided a steady stream of new X-ray transients for subsequent ground-based observations once they reach quiescence, and (ii) the availability of high performance optical spectrographs with good red sensitivity on 4m and larger telescopes. With these facilities we have obtained more detailed information about the nature of both components of LMXBs than was hitherto possible. In particular, the discovery of the mass function of V404 Cyg has revolutionised attitudes concerning the existence of compact objects that must be heavier than the canonical maximum mass of a neutron star. And while there are useful indicators from X-ray observations as to the possible presence of black holes in X-ray binaries, the ultimate diagnostic has to be the dynamical study that has been described here. The next major advances will come from observing the many quiescent transients that are too faint for current 4m class telescopes and require access to the about to be completed VLT and Gemini telescopes.

Acknowledgments. I would like to thank Erik Kuulkers and Tariq Shahbaz for their help in preparing this manuscript.

References

ABRAMOWICZ, M.A., CHEN, X., KATO, S., LASOTA, J.-P. & REGEV, O. 1995, *ApJ*, **438**, L37.
BEEKMAN, G., SHAHBAZ, T., NAYLOR, T. & CHARLES, P.A. 1996, *MNRAS*, **281**, L1.
BEEKMAN, G., SHAHBAZ, T., NAYLOR, T., CHARLES, P.A., WAGNER, R.M. & MARTINI, P. 1997, *MNRAS*, **290**, 303.
BELLONI, T., ET AL. 1997, *ApJ*, **479**, L145.
BERRIMAN, G., SZKODY, P. & CAPPS, R.W. 1985, *MNRAS*, **217**, 327.
BRANDT, W.N., PODSIADLOWSKI, P.& SIGURDSSON, S. 1995, *MNRAS*, 277, L35.
CASARES, J., CHARLES, P.A. & NAYLOR, T. 1992, *Nature*, **355**, 614.
CASARES, J. & CHARLES, P.A. 1994, *MNRAS*,**271**, L5.
CASARES, J. ET AL. 1995, *MNRAS*, **274**, 565.
CASARES, J., MARTÍN, E.L., CHARLES, P.A., MOLARO, P. & REBOLO, R. 1997, *New Astron.*, 1, 299.
CHAKRABARTI, S.K. 1998, *Ind.J.Phys. (Reviews)* **72 B**, 183
CHAKRABARTI, S.K. & TITARCHUK, L.G. 1995, *ApJ*, **455**, 623.
CHEN, W., SHRADER, C. & LIVIO, M. 1997, *ApJ*, **491**, 312.
CHEN, X., ABRAMOWICZ, M.A., LASOTA, J.-P., NARAYAN, R. & YI, I. 1995, *ApJ*, **443**, L61.
EGGLETON, P.P. 1983, *ApJ*, **268**, 368.
ELVIS, M., PAGE, C.G., POUNDS, K.A., RICKETTS, M.J. & TURNER, M.J.L. 1975, *Nature*, **257**, 656.
FILIPPENKO, A.V., MATHESON, T. & BARTH, A.J. 1995a, *ApJ*, **455**, L139.
FILIPPENKO, A.V., MATHESON, T. & HO, L.C. 1995b, *ApJ*, **455**, 614.
FILIPPENKO, A.V., MATHESON, T., LEONARD, D.C., BARTH, A.J. & SCHUYLER, D.V. 1997, *PASP*, **109**, 461.
HARLAFTIS, E.T., HORNE, K. & FILIPPENKO, A.V. 1996, *PASP*, **108**, 762.
HARLAFTIS, E.T., STEEGHS, D., HORNE, K. & FILIPPENKO, A.V. 1997, *AJ*, **114**, 1170.
HARRIES, J.R., MCCRACKEN, K.G., FRANCEY, R.J. & FENTON, A.G. 1967, *Nature*, **215**, 38.
HERRERO, A., KUDRITZKI, R.P., GABLER, R., VILCHEZ, J.M. & GABLER, A. 1995, *A&A*, **297**, 556.
HJELLMING, R.M. & RUPEN, M.P. 1995, *Nature*, **375**, 464.
KAHABKA, P. & VAN DEN HEUVEL, E.P.J. 1997, *Ann.Rev.Astron.Ap.*, **35**, 69.
KING, A.R. 1988, *QJRAS*, **29**, 1.
KING, A.R. & RITTER, H. 1998, *MNRAS*, **293**, L42.
KOLB, U., ET AL. 1997, *ApJ*, **485**, L33.
KUULKERS, E. 1998, *New Astron.*, submitted.
MARSH, T.R., ROBINSON, E.L. & WOOD, J.H. 1994, *MNRAS*, **266**, 137.
MARTIN, A.C. ET AL 1995, *MNRAS*, **274**, L46.
MCCLINTOCK, J.E. 1991, *Ann.NY Acad.Sci.*, **647**, 495.
MCCLINTOCK, J.E. 1998, Proc. *8th Annual Astrophysics Conference in Maryland on*

"*Accretion Processes in Astrophysical Systems*".
MCCLINTOCK, J.E. & REMILLARD, R.A. 1986, *ApJ*, **308**, 110.
MCCLINTOCK, J.E. & REMILLARD, R.A. 1990, *ApJ*, **350**, 386.
MILLER, J.C., 1998, Proc. *12th Italian Conference on General Relativity and Gravitational Physics* World Scientific (in press).
MILLER, J.C., SHAHBAZ, T. & NOLAN, L.A. 1998, *MNRAS*, **294**, L25.
MIRABEL, I.F. ET AL. 1997, *ApJ*, **477**, L45.
MIRABEL, I.F. & RODRÍGUEZ, L.F. 1994, *Nature*, **371**, 46.
MORGAN, E. & REMILLARD, R. 1996, *ApJ*, **473**, L107.
NARAYAN, R., BARRET, D. & MCCLINTOCK, J.E. 1997, *ApJ*, **482**, 448.
OROSZ, J.A. ET AL. 1994, *ApJ*, **436**, 848.
OROSZ, J.A., BAILYN, C.D., MCCLINTOCK, J.E. & REMILLARD, R.A. 1996, *ApJ*, **468**, 380.
OROSZ, J.A. & BAILYN, C.D. 1997, *ApJ*, **477**, 876 (and *ApJ*, **482**, 1086).
OROSZ, J.A., JAIN, R.K., BAILYN, C.D., MCCLINTOCK, J.E. & REMILLARD, R.A. 1998, *ApJ*, in press.
PACZYNSKI, B. 1971, *Ann.Rev.Astron.Ap.*, **9**, 183.
REMILLARD, R.A., OROSZ, J.A., MCCLINTOCK, J.E. & BAILYN, C.D. 1996, *ApJ*, **459**, 226.
RHOADES, C.E. & RUFFINI, R. 1974, *Phys.Rev.Lett.*, **32**, 324.
SANWAL, D. ET AL. 1996, *ApJ*, **460**, 437.
SHAHBAZ, T., NAYLOR, T. & CHARLES, P.A. 1994a, *MNRAS*, **268**, 756.
SHAHBAZ, T., NAYLOR, T. & CHARLES, P.A. 1997, *MNRAS*, **285**, 607.
SHAHBAZ, T. ET AL. 1994b, *MNRAS*, **271**, L10.
SHAHBAZ, T., BANDYOPAHYAY, R. & CHARLES, P.A. 1998b, *MNRAS*, submitted.
SHAHBAZ, T., CHARLES, P.A. & KING, A.R. 1998a, *MNRAS*, submitted.
SHAHBAZ, T. & KUULKERS, E. 1998, *MNRAS*, **295**, L1.
SHAHBAZ, T., NAYLOR, T. & CHARLES, P.A. 1993, *MNRAS*, **265**, 655.
SHAHBAZ, T., ET AL. 1996, *MNRAS*, **282**, 977.
TANAKA, Y. & LEWIN, W.H.G. 1995, in *X-ray Binaries*, 126: CUP.
THORSETT, S.E. & CHAKRABARTY, D. 1998, *ApJ*, submitted.
THORSETT, S.E., ET AL. 1993, *ApJ*, **405**, L29.
VAN DER HOOFT, F. ET AL. 1997, *MNRAS*, **286**, L43.
VAN DER HOOFT, F. ET AL. 1998, *A&A*, **329**, 538.
VAN PARADIJS, J. & MCCLINTOCK, J.E. 1994, *A&A*, **290**, 133.
VAN PARADIJS, J. & MCCLINTOCK, J.E. 1995, in *X-Ray Binaries*, 58: CUP.
VERBUNT, F., BELLONI, T., JOHNSTON, H.M., VAN DER KLIS, M. & LEWIN, W.H.G. 1994, *A&A*, **285**, 903.
WADE, R.A. & HORNE, K. 1988, *ApJ*, **324**, 411.

OUTBURSTS IN BLACK HOLE X-RAY TRANSIENTS: CLUES FROM MULTIWAVELENGTH OBSERVATIONS

CAROLE A. HASWELL
Astronomy Centre, University of Sussex,
Brighton, BN1 9QJ, U.K.
chaswell@star.cpes.susx.ac.uk

Abstract. Observational work aiming to illuminate the mechanism(s) causing the dramatic luminosity variations of the BHXRTs is described. The relevant theoretical considerations, the disc limit cycle instability and the possibility of advective flows, are briefly introduced. Results on four individual systems, X-ray Nova Muscae 1991, GRO J1655-40, A0620-00 and GRO J0422+32, are described, focussing on the recent results from the 1996 outburst of GRO J1655-40 and 1994 observations of GRO J0422+32 near quiescence. The GRO J1655-40 outburst optical/UV continuum could be interpreted as thermal or possible *non-thermal* emission. If thermal, the optical continuum spectrum is suggestive of an irradiated disc, and correlations between the rapid variability in the X-ray and optical/UV light curves indicate reprocessing of X-ray flux in a thick disc. However the X-ray flux rose during the optical decline, making an irradiation interpretation problematical. Complications in the disc instability model arising from the large disc in GRO J1655-40 may explain the optical continuum within the context of thermal emission. There is strong evidence for self-absorbed synchrotron emission dominating the optical continuum of GRO J0422+32 near quiescence, as predicted by the the advective models. Many aspects of the outbursts in these systems remain incompletely understood.

1. Introduction

The Black Hole X-Ray Transient (BHXRT) class of interacting binary stars, and the dynamical evidence that these systems harbour black holes are described in the contribution to this volume by P.A. Charles. This paper will focus on the dramatic outbursts these systems exhibit, and attempt to use observational data to illuminate the physical mechanisms underlying these long term luminosity variations. The prototypical outburst light curves for the BHXRTs show a fast (a few days) rise and an exponential decay with time constant ∼ 30 days (Cannizzo, Chen, and Livio 1995).

This paper will begin with an outline of the main theoretical considerations proposed to explain the outbursts, and will proceed to discuss their application to observations of four individual systems. The discussion attempts to summarise the current overall understanding.

2. Models

2.1. DISC INSTABILITY MODEL

Standard accretion disc theory was largely developed in the context of cataclysmic variables (CVs), semi-detached interacting binaries in which a white dwarf accretor is fed via an accretion disc by a Roche lobe-filling mass donor. The structure of a BHXRT is similar to that of a CV, with the white dwarf replaced by a black hole.

Standard accretion disc theory makes the *assumption* that the heat generated by viscous dissipation in every annulus of the disc is promptly radiated away. The assumption allows the effective temperature distribution in the disc to be deduced once the mass transfer rate and the accretor mass are specified. The emitted spectrum of the disc can then be computed by simply summing the black body spectra emitted by each annulus. The resulting black body disc spectrum appears as a 'stretched' black body spectrum with the Rayleigh Jeans tail of the coolest (outermost) disc annulus and the Wien tail of the hottest (inner) disc annulus. At intermediate frequencies the temperature distribution of a steady optically thick accretion disc leads to a spectrum with slope $F_\nu \propto \nu^{1/3}$, see Frank, King, and Raine (1992) for details. An example of such a standard steady state accretion disc spectrum is shown in Fig. 2.

A subset of the CVs, the dwarf novae (DN), show outbursts which are qualitatively similar to those seen in the BHXRTs, and the Disk Instability Model (DIM) was developed from the standard black body disc theory to explain the DN outbursts. There has been much work published developing this model (see Cannizzo, 1993) and I will attempt here only a brief qualitative summary.

The DIM explains the outbursts in DN as the result of temperature-dependent viscosity. If largely ionised, a disc annulus is able to settle into a stable hot, high-viscosity, radiative state, where the mass transfer rate through the annulus and hence the viscous dissipation is large. On the other hand, if the disc material is too cool for hydrogen and helium to be ionised, the annulus can settle into into a stable cool, low-viscosity, convective state. At intermediate temperatures the material is partially ionised and the disc is both thermally and viscously unstable. When the externally imposed average mass transfer rate through any disc annulus leads to a temperature corresponding to partial ionisation, the disc cannot settle into a steady equilibrium configuration, and executes a limit-cycle instability. In the cool state the disc annulus accumulates mass and slowly heats up until the material ionises, at which instant the annulus executes a rapid transition to the hot state. Now the viscosity, mass transfer rate, and effective temperature are much increased, and the annulus empties out and appears much more luminous than it did in the cool state. As the annulus empties, the temperature gradually decreases until recombination occurs, triggering a rapid transition back to the cool state. Annuli respond to the changes in the mass transfer rates through their neighbours, so transitions lead to heating or cooling waves propagating radially through the disc.

The time spent in the cool state corresponds to the low luminosity, quiescent intervals in the light curve; during outburst the disc is in the hot state. For DN the time between outburst and the outburst durations have been matched well by the model. The longer recurrence timescales for BHXRTs and the shapes and durations of their outburst light curves, however, provide a challenge to the DIM (Lasota 1996).

The DIM makes definite quantitative predictions for the temperature distribution, and hence the expected broad band spectrum, throughout the outburst cycle (*e.g.* Cannizzo, Chen, & Livio 1995), hence can be tested by observing the spectral evolution through outburst and decline. Including the effects of irradiation complicates this, however, as illustrated in Fig. 2.

2.2. ADVECTIVE MODELS

The assumption that the viscously dissipated energy is promptly radiated away can be relaxed, to give models in which the flow 'advects' some of this energy, carrying it inwards as heat. Allowing advection introduces an additional unknown, namely what proportion of the viscously dissipated energy is advected. Hence an additional constraint or assumption must be introduced in order to generate solutions. Probably for this reason there was little work published on advective accretion flows until their possible relevance to the BHXRTs was realised in the mid 1990s. See Chakrabarti (1996) for a comprehensive discussion of theoretical considerations. Recent work (*e.g.* Hameury et al. 1997, Chakrabarti 1997) attempts to combine the DIM with an inner advective or sub-Keplerian flow to explain observations of BHXRTs. The inner accretion flow in these models is expected to produce non-thermal emission.

3. X-Ray Nova Muscae 1991

This object was monitored in the optical/UV through the decline from its 1991 outburst. The spectral evolution was analysed by Cheng et al. (1992) who compared it with the predictions of the DIM. The data appeared consistent with steady-state optically thick accretion disc spectra and the deduced mass transfer rate fell monotonicly during the decline. The DIM predicts, however, that the declining mass transfer rate is accompanied by a cooling wave propagating through the disc as successive hot, high viscosity, annuli make the transition to the cool, low viscosity, state. The consequent changing temperature distribution should have produced an observable cooling wave signature at the long wavelength end of the spectrum. The cooling wave was not observed, despite data of sufficient range and quality to detect it, suggesting problems with the straightforward application of the DIM to BHXRTs.

4. A0620-00

The prototype of the BHXRT class, A0620-00, was observed extensively in the optical in quiescence. However, it was not until 16 years after the 1975 outburst that the UV emission was probed with HST and a contemporaneous ROSAT soft X-ray observation allowed the broad band quiescent emission from the accretion flow to be analysed by McClintock, Horne, & Remillard (1995). After subtracting the contribution of the K5 V mass donor star, they found the 1100 − 4500Å accretion spectrum could be modeled as a 9000 K blackbody, with an area of only $\sim 1\%$ of the disc area. The low UV flux emitted by this accreting black hole was a surprise. By analogy with quiescent DN, a mass transfer rate into the outer disc of $\dot{M}_{disc} \sim 10^{-10}$ $M_\odot$ yr^{-1} was inferred. Meanwhile, the ROSAT soft X-ray flux implied a mass transfer rate through the inner disc of only $\dot{M}_{BH} < 5 \times 10^{-15}$ $M_\odot$ yr^{-1}. Qualitatively, therefore, these findings were in agreement with the DIM, suggesting the accumulation of material in the quiescent disc. The extremely low $\dot{M}_{BH}$ seemed improbable, however, and the authors pointed out that isolated black holes might well accrete more than this from the ISM.

A new explanation was advanced by Narayan, McClintock, & Yi (1996), who postulated that the standard disc model is only applicable to the outer flow, and that within ~ 3000 R_{Sch} the flow is advective: *i.e.* the viscously-generated thermal energy is carried with the flow rather than being promptly radiated away. For black hole accretors, this advected energy can be carried through the event horizon. With this hypothesis, therefore, the extremely low quiescent accretion fluxes do not necessarily demand the extremely low mass transfer rates inferred from the standard accretion disc

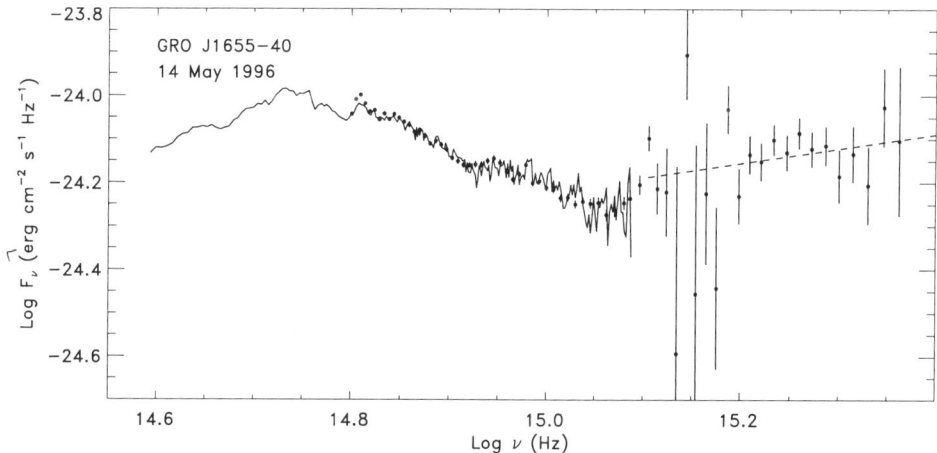

Figure 1. The dereddened spectrum of GRO J1655-40. The spectral slope changes dramatically at $\log \nu \sim 15.05$ ($\lambda \sim 2600$Å). The dashed line shows a $f_\nu \propto \nu^{1/3}$ fit to the UV data. Adapted from Hynes et al. 1998a.

model. While there has been much debate about the details of these models, the basic advective idea appears promising.

5. GRO J1655-40

GRO J1655-40 was discovered in outburst in 1994 July when it reached 1.1 Crab in the 20-200 keV band (Harmon et al. 1995); since then it has undergone repeated outbursts to a similar level and is apparently an atypical BHXRT. Superluminal radio jets were associated with the 1994 outburst (Hjellming and Rupen, 1995). Following the onset of X-ray activity in April 1996, a coordinated HST and RXTE campaign was mounted with several visits between 1996 May 14 and July 22. Fuller descriptions of these observations and their interpretation are given in Hynes et al. (1998a,b).

5.1. BROAD BAND SPECTRAL EVOLUTION

GRO J1655-40 is a highly reddened source, so an accurate correction for interstellar extinction is a prerequisite to any analysis of the spectrum. The 2175Å feature gives a sensitive measure of the extinction: E(B-V)=1.2±0.1, a value consistent with direct estimates of the visual extinction and with measurements of interstellar absorption lines (Hynes et al. 1998a).

Figure 1 is the 1996 May 14 dereddened UV-optical spectrum. Though the UV portion of the spectrum is consistent with the $\nu^{1/3}$ power-law predicted by the steady-state blackbody disc model, the optical ($\lambda > 2600$Å)

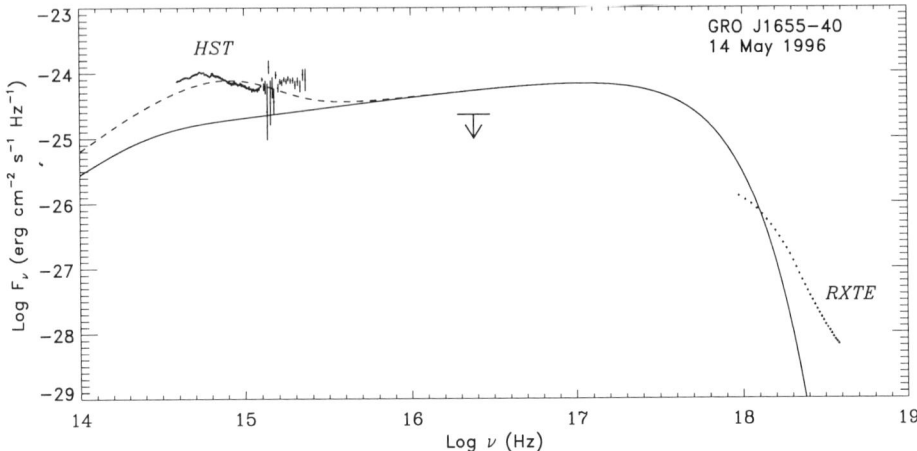

Figure 2. Composite spectrum for the May 14 observations. The solid lines shows a steady state accretion disc spectrum, whilst the dashed line shows an irradiated disc spectrum. These spectra are **not** fitted to the data, and are merely illustrative.

spectrum rises to longer wavelengths in contrast to the predictions of the model. Ignoring the $\lambda > 2600$Å data, a $\nu^{1/3}$ model can be fit to the UV data, implying a mass transfer rate of $1 \times 10^{-7} M_\odot \, \text{yr}^{-1} \leq \dot{M} \leq 7 \times 10^{-6} M_\odot \, \text{yr}^{-1}$, if the compact object mass is $7 \, M_\odot$ as the quiescent orbital light curve implies (Orosz and Bailyn 1997, hereafter OB97). The dominant source of uncertainty in $\dot{M}$ arises from interstellar extinction. Assuming an accretion efficiency of 10%, the Eddington rate is $\dot{M}_{\text{Edd}} = 1.6 \times 10^{-7} M_\odot \, \text{yr}^{-1}$, so near the peak of the outburst this interpretation of the UV spectrum implies $\dot{M} \approx \dot{M}_{\text{Edd}}$.

We need to invoke something other than a pure steady-state optically thick accretion disc in order to explain the optical light. The shape of the spectrum is qualitatively suggestive of an irradiated disc; irradiation can alter the temperature profile of the outer disc producing a rise in flux towards longer wavelengths as illustrated in Figure 2. The multiwavelength light curves for the outburst (Fig 3, from Hynes et al. 1998a) do not, however, appear to support a simple irradiation model: the optical and UV flux declines while the X-ray flux rises! This behaviour is contrary to the predictions of King and Ritter (1998) whose analysis of the effects of irradiation concluded that the UV and optical light curves should resemble the X-rays. Nontheless correlated X-ray and optical/UV variability was detected (see section 5.3 and Fig. 6) indicating at least some of the optical/UV flux is due to reprocessing.

In order to characterise the optical component of the spectra we fit

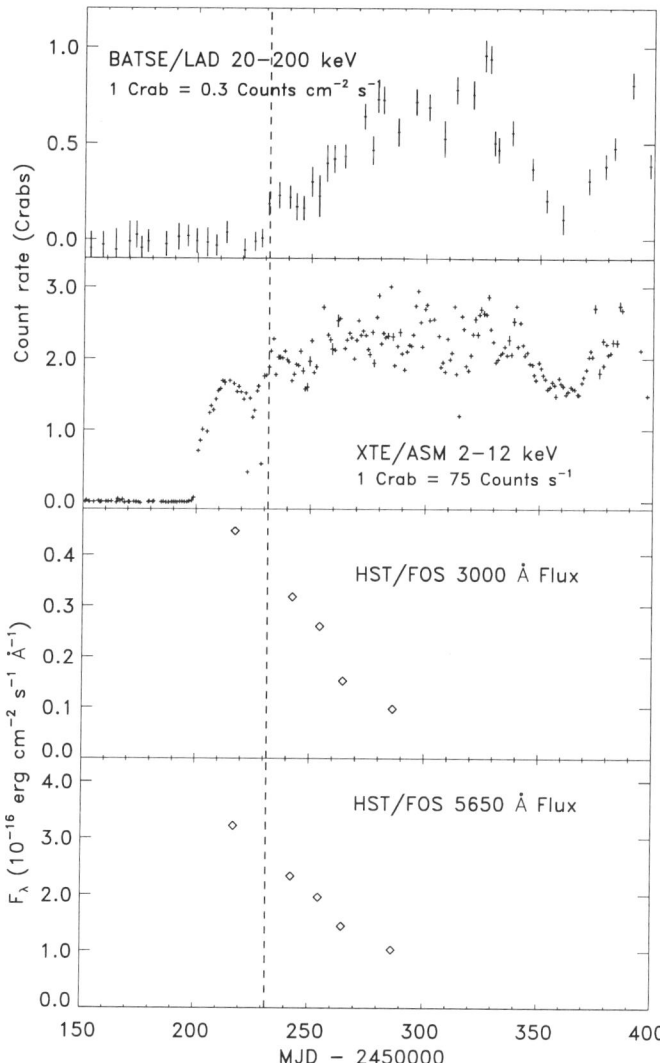

Figure 3. Long term light curves of the 1996 outburst of GRO J1655-40. Statistical errors in the UV and optical points are comparable to or smaller than the symbols used. The time-axis begins at 1996 March 8. The optical and UV light curves approximate the standard fast rise exponential decay behaviour, but the discrepant behaviour of the X-ray data is clear. The dotted line shows the first radio detection of this outburst. The closest non-detection was 9 days before this, making the exact beginning of radio flare difficult to pinpoint. From Hynes et al. 1998a.

blackbody spectra to the $\lambda > 2600$Å data for each visit (Figure 4). While the fluxes fell by about a factor of three between our first and last visit, the colour temperature remained almost constant, dropping from 9800 K to 8700 K; the emitting area dropped from 5.0×10^{23} cm^2 to 2.2×10^{23} cm^2.

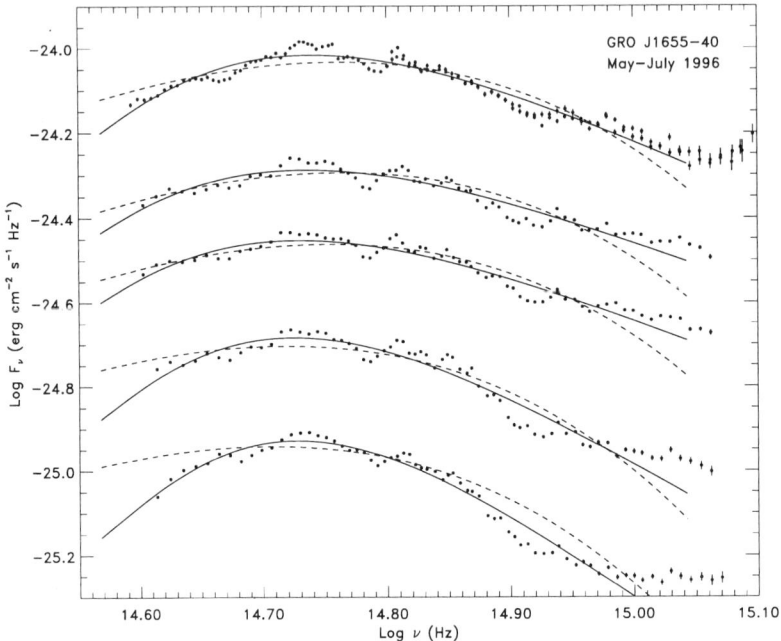

Figure 4. Blackbody (dashed line) and synchrotron (solid line) fits to the $\lambda > 2600$Å spectra of GRO J1655-40. The bumps in the observed spectra are attributable to spectral line features in the source and to diffuse interstellar bands. In order to separate the successive visits clearly, a downward shift of 0.1 has been introduced in each visit relative to the one above it *i.e.* the lowest visit has been shifted downwards by 0.4. Adapted from Hynes et al. 1998a.

The system parameters for GRO J1655-40 are well constrained (OB97, Hjellming and Rupen 1995) and the total available emitting area of the disc and secondary star is $\sim 5 \times 10^{23}$ cm^2, so it is possible to explain the optical emission at the peak of the outburst as thermal emission, but only if both the secondary star and the majority of the disc area have essentially the same isothermal temperature distribution. Attributing the $\nu^{1/3}$ UV component to a steady-state disc is not necessarily ruled out, as this requires only the hot inner regions of the disc.

Close inspection of Fig. 4 reveals a suggestion that our optical spectra are more strongly peaked than a single temperature blackbody, so we considered non-thermal mechanisms. Self-absorbed synchrotron emission from a cloud of relativistic electrons with a power-law energy distribution produced good fits to the optical component; the deduced electron energies are $\gamma \sim 60-90$, the magnetic field is $B \sim 40-60$ kG, and the size of the cloud is $50-100$ R_{Sch}. While the synchrotron models fit better, they have an extra free parameter, and there is no external check comparable to that provided by the emitting area required for the black body interpretation.

Attributing substantial optical flux to this compact nonthermal source relieves the requirement for a large isothermal emitter in the system. It is interesting to note that intrinsic VRI band linear polarisation ($> 3\%$) was detected in July 1996 (Scaltriti et al. 1997), consistent with the hypothesis of optical synchrotron emission.

On the other hand, the DIM may *require* a large isothermal outer disc for a long-period system like GRO J1655-40. Since the disc in such a system is large, the temperature for a steady state disc falls below the minimum temperature for the hot, high viscosity, state long before the outer disc is reached. Even for an Eddington mass transfer rate in GRO J1655-40 the temperature drops below that of the hot state at a radius less than a quarter of the Roche lobe radius. This means that there is no global steady-state solution for $\dot{M} \leq \dot{M}_{Edd}$. It is not clear what will happen to the temperature distribution in such a case, but it is possible that in outburst much of the outer disc could be maintained just in the hot state. Hence one might expect the outer disc to appear as an approximately constant temperature, shrinking area emitter as the decline proceeds. This hypothesis need to be tested with self-consistent numerical modeling, but until this is done our data cannot rule out a thermal interpretation.

5.2. A DISC WIND?

During our 1996 May 14 observation we found the C IV 1550Å, Si IV 1400Å, and Si III 1300Å UV resonance lines all show likely P-Cygni profiles (Figure 5). The peak to trough separation in all three cases is ~ 5000 km s^{-1}, slightly larger than seen in outbursting dwarf novae. Line profiles produced by biconical accretion disc winds were calculated by Shlosman and Vitello (1993) who found 'classical' P Cygni profiles only for inclinations around 60–70°, in striking agreement with the inclination determined for GRO J1655-40 (OB97). We conclude that there was likely a biconical accretion disc wind at the peak of the UV outburst, when $\dot{M} \sim \dot{M}_{Edd}$.

5.3. LIGHT ECHOES

Figure 6 shows the evidence for the correlated rapid variability between the optical/UV and X-ray emission which was detected in simultaneous RXTE and HST data from 1996 June 8. The correlations occurred with a mean delay of the optical/UV of up to 25 s (Hynes et al. 1998b). Hence, the correlations are likely due to reprocessing of the X-rays into optical and UV

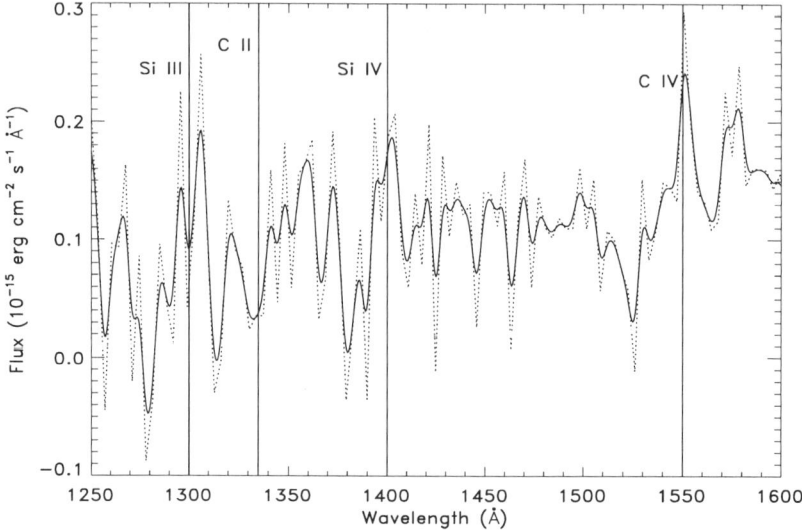

Figure 5. Likely P-Cygni profiles detected in the UV resonance lines in the 1996 May 14 UV spectrum of GRO J1655-40. The dotted line shows the raw data; the solid line is the same smoothed with a width equal to a resolution element. From Hynes et al. 1998a.

emission, with the delay being due to the finite light travel time between the X-ray source and the reprocessing regions. These light echoes can be used to constrain the morphology of the reprocessing regions. The lag of up to 25 s is consistent with that expected for reprocessing in the accretion disc. At the binary phase of the observations, lags of greater than 40 s are expected for light echoes from the mass donor star. This may imply that the X-ray absorbing material in the accretion flow is sufficiently vertically extensive to effectively shield the mass donor from direct X-ray illumination. For the Roche geometry deduced by OB97, this implies H/R is at least ~ 0.25 along the line of centres.

6. GRO J0422+32

HST observations of this target covering the vacuum UV and the entire optical region were obtained on 1994 August 25 and 26, approximately two years after the first observed outburst, and seven months after the last reported reflare (Callanan et al. 1995). A full discussion of these data is given in Hynes and Haswell (1998). Figure 7 shows the R band light curve including the point deduced from our spectrum. The vertical marks above the light curves indicate the times of mini-outbursts according to the suggestion (Chevalier and Ilovaisky 1995) that they recur on a 120 day

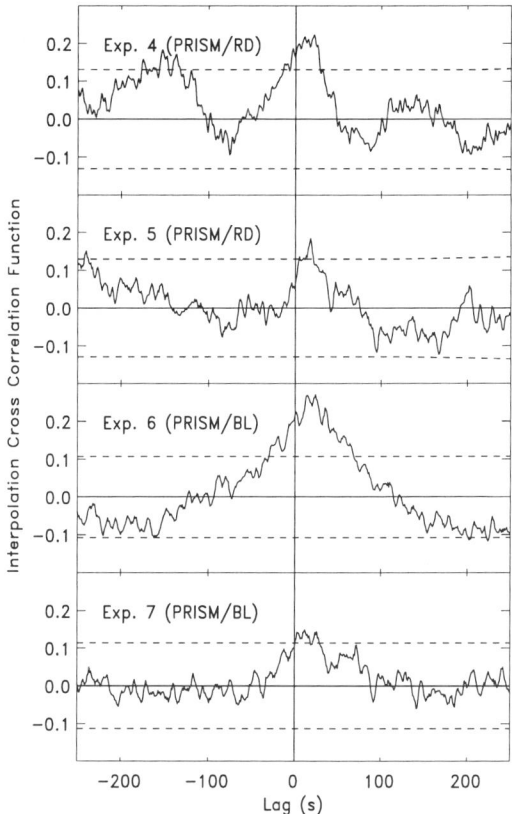

Figure 6. Interpolation cross correlation functions for 1996 June 8 GRO J1655-40 data. Dashed lines show 3σ confidence limits expected for uncorrelated variability. All four plots show features at $\sim 20\,\mathrm{s}$ which are significant at the 3σ level, although only Exposure 6 can be considered more than marginal at this level. From Hynes et al. 1998b.

period. Our observation clearly lies above the subsequent points, and it appears we saw the end of a previously unreported mini-outburst.

Our spectrum is shown in Figure 8 along with the estimated contribution from the mass donor (Casares et al. 1995, Filippenko et al. 1995) . Figure 9 (which has logarithmic axes for ease of comparison with Figs. 1, 2, and 4) shows the spectrum after dereddening and subtracting the mass donor star flux, hence represents the intrinsic accretion spectrum. The spectral shape in Fig. 9 is distinctive: there is a very pronounced peak in the optical at around ~ 4500Å and since the interstellar reddening towards this system is moderate, $E(B-V) = 0.3 \pm 0.1$, it is unlikely that this is an artifact of an improper reddening correction. The best fitting black

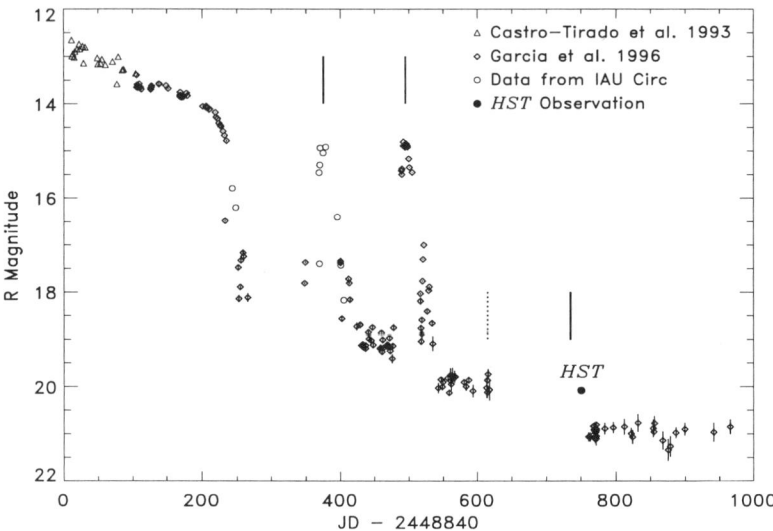

Figure 7. The R band light curve of the outburst, adapted from Fig. 1 of Garcia et al. (1996). Vertical marks indicate observed or extrapolated times of mini-outburst. Our measurement clearly lies significantly above the subsequent photometry, and is consistent with the expected time of a mini-outburst. From Hynes and Haswell 1998.

body shown in Figure 9 is clearly less sharply peaked than the observed spectrum from GRO J0422+32, while a simple self-absorbed synchrotron spectrum can successfully reproduce the continuum shape. The parameters of the best fitting synchrotron spectrum were $l \sim 40\,\mathrm{km}$, $n_e \sim 10^{20}\,\mathrm{cm}^{-3}$, $p \sim 10$, $B \sim 20\,\mathrm{kG}$; the electrons contributing to the observed spectrum have energies 40–80 MeV. We stress, however, that we do not claim great significance for these values, concluding simply that the continuum spectrum is highly suggestive of a self-absorbed synchrotron source, rather than black body emission.

7. Discussion and Future Work

The quiescent continuum spectrum of GRO J0422+32 is highly suggestive of self-absorbed synchrotron emission rather than thermal emission from an optically thick black body accretion disc. Since the advective models of BHXRTs in quiescence predict that the optical emission will be dominated by self-absorbed synchrotron emission from a hot, optically thin, inner accretion flow, the GRO J0422+32 spectrum must be considered supportive of the advective idea for these systems in quiescence.

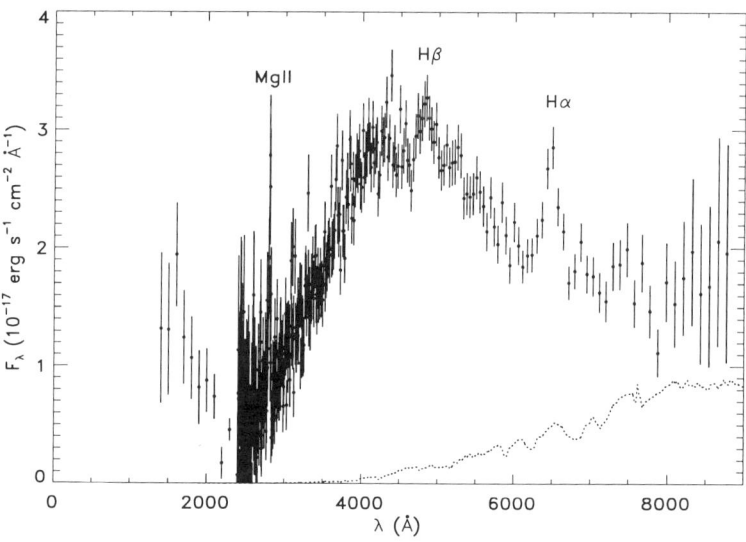

Figure 8. The observed UV-optical spectrum of GRO J0422+32 in early quiescence. Also shown dotted is our estimate of the spectrum of the companion star, reddened by E(B-V)=0.3. Much of the light at 8000Å is due to the mass donor star, but we are clearly detecting accretion flux at shorter wavelengths. Adapted from Hynes and Haswell 1998.

The spectra of GRO J1655-40 on the decline from the 1996 outburst could also be partially due to such non-thermal emission, though in this case the evidence is far less clear cut. In contrast, the optical/UV continuum emission from X-Ray Nova Muscae 1991 during the decline from outburst appeared consistent with the spectrum expected for a standard black body disc, though even in this case the detailed quantitative predictions of the DIM fail.

The observations described herein have provided some intriguing challenges to the theoretical models for transient outbursts. As we collect detailed multi-epoch data on more systems it is becoming clear that the BHXRTs exhibit complex and diverse behaviour, and it seems we need to consider a variety of mechanisms in order to properly understand the wealth of observational phenomena.

Acknowledgments: Support for this work was provided by the Nuffield Foundation and by NASA through grant numbers GO-6017.01-94A and GO-4377.02-92A from the Space Telescope Science Institute, which is operated by AURA under NASA contract NAS5-26555. I am grateful to my collaborators, particularly my D.Phil. student Rob Hynes, for their contributions to this work. Full discussions of the recent work described herein

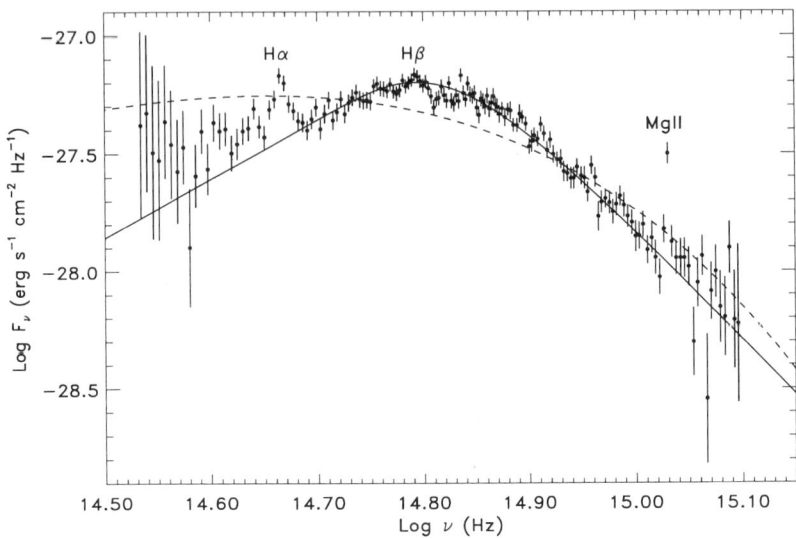

Figure 9. The near quiescent accretion spectrum from GRO J0422+32: dereddened data after subtraction of the companion contribution. The best fitting black body (dashed) and self-absorbed synchrotron (solid) spectra are overplotted. The data below 4000 Å ($\log \nu > 14.88$) has been averaged into 25 Å bins for clarity. The large error bars at low frequencies are systematic and arise from assuming a ±1 pixel miscentring uncertainty.

is in Hynes et al. (1998a,b) and Hynes and Haswell (1998).

References

Callanan, P.J. Garcia, M.R., McClintock, J.E., Zhao, P., Remillard, R.A., Bailyn, C.D. Orosz, J.A. Harmon, B. A., Paciesas, W.S. 1995, ApJ, 441, 786.
Cannizzo J. K., 1993, in Accretion Disks in Compact Stellar Systems, Wheeler J. C., World Scientific Publishing, Singapore, p. 318
Cannizzo J. K., Chen W., Livio M., 1995, ApJ, 454, 880
Casares J., Martin A. C., Charles P. A. Martín E. L., Rebolo R., Harlaftis E. T., Castro-Tirado A. J., 1995, MNRAS, 276, L35
Chakrabarti, S.K. 1996, Physics Reports, 266, 229
Chakrabarti, S.K. 1997, ApJ 484, 313.
Cheng,F.-H., Horne, K., Panagia, N., Shrader, C.R., Gilmozzi, R., Paresce, F., Lund, N. 1992, ApJ, 397, 664.
Chevalier, C., Ilovaisky, S.A. 1995, A&A 276, 103.
Frank, J., King, A., and Raine, D. 1992, 'Accretion Power in Astrophysics', CUP.
Filippenko A. V., Matheson T., Ho L. C., 1995, ApJ, 455, 614
Garcia, M.R., Callanan, P.J., McClintock, J.E., and Zhao, P., 1996, ApJ, 460, 932. 1995, ApJ, 455, 614.
Harmon B. A., et al., 1995, Nat, 374, 703
Hameury, J.-M., Lasota, J.-P., McClintock, J.E., and Narayan, R. 1997 ApJ, 489, 234.
Hjellming R. M., Rupen M. P., 1995, Nat, 375, 464

Hynes, R.I., Haswell, C.A., Shrader, C.R., Chen, W., Horne, K., Harlaftis, E.T., O'Brien, K., Hellier, C., and Fender, R.P. 1998a, MNRAS, in press.
Hynes, R.I., O'Brien, K., Horne, K., Chen, W., Haswell, C.A. 1998b, MNRAS, submitted.
Hynes, R.I., Haswell, C.A. 1998, MNRAS, submitted.
King, A.R., and Ritter, H. 1998, MNRAS, 293, 42.
Lasota, J.P. 1996, in Compact Stars in Binaries, J. van Paradijs et al., p. 43.
McClintock, J., Horne, K., and Remillard, R., 1995, ApJ, 442, 358.
Narayan, R., McClintock, J., and Yi, I. 1996, ApJ, 457, 821.
Orosz J. A., Bailyn C. D., 1997, ApJ, 477, 876 (OB97)
Scaltriti F., Bodo G., Ghisellini G., Gliozzi M., and Trussoni E. 1997, A&A, 325, 29.
Shlosman I., Vitello P., 1993, Ap.J, 409, 372

L. Titarchuk

K. Borozdin

THE HIGH-ENERGY SPECTRA OF ACCRETING BLACK HOLES: VIEWING THE MATTER AS IT DISAPPEARS DOWN THE SCHWARZSCHILD DRAIN

L.G. TITARCHUK
NASA Goddard Space Flight Center
NASA/GSFC, code 660, Greenbelt, MD 20771, USA
and George Mason University Center. Earth Obs. & Space Res.
GMU, CSI, Science & Tech. I Fairfax VA 22030-4444, USA

C.R. SHRADER
NASA Goddard Space Flight Center
NASA/GSFC, code 660, Greenbelt, MD 20771, USA
and Universities Space Research Association, Lanham MD, USA

AND

S. TRUDOLYUBOV, M. REVNIVTSEV AND K. BOROZDIN
Space Research Institute, Moscow, Russia

Abstract. Many attempts to demonstrate the existence of black holes have proven difficult, largely because the very defining property of black holes prevents any outward propagation of radiation. Dynamical studies of binaries have led to convincing stellar-mass s BHCs (Orosz & Bailyn, 1997), and kinematic studies from water masers in several active galactic nuclei (Miyoshi et al. 1995) are strongly suggestive, but it is desirable to identify a generic signature with a broader range of applicability. Recent efforts to interpret asymmetric line profiles in terms of gravitational redshift effects within a few Schwarzschild radii of the black hole have attracted considerable attention (Bromley, Miller & Pariev, 1998). These efforts however, neglect some important effects, which may invalidate the conclusions. The fundamental signature of accreting black holes should be direct observation of the matter "drain" (Titarchuk, Mastchiadis & Kylafis 1997, Titarchuk & Zannias, 1998) - that is to observe the matter as it flows inward with velocities approaching the speed of light near the event horizon. We have shown (Titarchuk & Zannias, 1998) that the behavior of the matter flowing down this "drain" is linked to the very nature of the black hole itself and that a distinct observational signature is produced. Here we present the observational data that demonstrate the existence of black holes in the five binary systems GRO J1655-40, GRS 1915+105, GRS 1739-278, 4U 1630-47 and XTE J1755-32.

1. Introduction

Proving the existence of black holes in the universe, and establishing the observable properties associated with their immediate environments, has been a long-standing problem in modern astrophysics. To explore this issue, we have interpreted the spectra of a sample of black hole candidates within the framework of the bulk-motion accretion model (Titarchuk, Mastichiadis & Kylafis, 1997; Titarchuk & Zannias, 1998).

The fundamentally new and unique constituents of our model derived from first principles, by using the general and special relativistic kinetic formalism, predict the spectral and timing properties of BH systems in their soft state. They are.: (i) The observed spectrum consists of a fraction of the thermal-disk radiation which is viewed directly, and a smaller fraction which is upscattered in the converging inflow before propagating outwards, towards the observer. (ii) The time variability of the extended power law tail associated with the black hole horizon site is a few times the light crossing time $(3 \times 10^{-5} M/M_\odot$ s). Our main objective in applying this model has been to test its predictive power regarding the observable spectral and temporal features of relativistic inflows in realistic astrophysical environments. Black holes are most commonly associated with stellar binaries (BHCs) and Active Galactic Nuclei (AGN), where there is widely believed to be a disk of matter surrounding the compact object. The spectrum emitted by such a disk is quite well understood both theoretically (Shakura & Sunyaev, 1973) and from observations. In the high (soft) state of BHCs, the disk is an effective generator of soft photons: of energy E$\sim$ 1 keV for solar mass BHCs, and E$\sim$ 10 − 100 eV for supermassive AGNs. The outgoing soft photon flux emanating from the inner accretion disk cools the surrounding medium (Chakrabarti & Titarchuk, 1995, Titarchuk, Lapidus, & Muslimov, 1998). The majority of these photons are detected in observations as a blackbody-like soft spectral component. Some photons, however, should unavoidably experience upscattering to higher energies through interaction with the bulk-motion inflow in the close neighborhood of a black hole. These upscattered photons form the power-law component extending up to energies of E$\lesssim$ 500 keV, the average electron energy in the bulk-motion inflow. The formation of the power-law spectrum due to bulk-motion upscattering has been recently proved by using the formalism of the general- and special-relativistic kinetic theory (Titarchuk & Zannias, 1998) Relativistic bulk inflow is likely to be occurring in a wide variety of black hole environments, where the strong gravitational field dominates the pressure forces, so we point to the existence of a characteristic high-energy spectroscopic signature: an extended power-law tail with the high energy cutoff at $\lesssim 500$ keV.

2. Bulk-Motion Spectral Model

As explained above, this signature originates from upscattering of low energy photons by fast moving electrons with velocities, v, approaching the speed of light, c. A soft photon of energy E, in the process of multiple scattering off the electrons, gets substantially blue-shifted to energy

$$E' = E \frac{1 - (v/c)\cos\theta}{1 - (v/c)\cos\theta'} \qquad (1)$$

due to Doppler effect provided at least one photon is scattered in the direction of electron motion (i.e. when $\cos\theta' \approx 1$). For example, in the first scattering event we assume the direction of incident photon, θ_1, is nearly normal to the electron velocity, and the direction of the scattered photon is nearly aligned with the electron velocity. In the process of the photon propagation through the ambient cloud, the angle between the photon and electron velocities increases. Thus, in the second event the cosine angle, $\cos\theta_2$, tends to approach zero. The angle of outgoing photon, θ_2', has to be large enough, in order for the Doppler boosted photon to reach an observer.

Any system having a disk structure around a compact object is expected to have a source of low-energy photons. The spectral index of the boosted photon distribution is determined only by the mass accretion rate and the plasma temperature of the bulk flow (. The presence of this high-energy power-law component is a generic feature of the model. Other extended power-law components, which may be related to the relativistic electron motion, e.g. in a jet, are not uniquely constrained to this energy band because they are not tied with the electron rest mass $m_e c^2$.

We thus see that the spectrum of an accreting black hole in its high state, when we have a sufficient amount of soft photons from the accretion disk, is formed as a sum of two components :(1) thermal disk component and (2) the upscattered component. The latter is formed by some fraction of the disk photons upscattered off high velocity electrons of the converging inflow, and it is seen as the extended power-law at energies much higher than the characteristic energy of the soft photons.

A review of the observational literature on high-energy studies of Galactic BHC, reveals that the basic characteristics described above are typical of high-state black hole spectra. In fact, as has been suggested by many authors (White, Kaluzienski, & Swank, 1984; Tanaka & Shibazaki, 1996) these characteristics is *a possible black hole signature*. We assert that this phenomenological evidence gains a strong theoretical support from the model we describe.

3. Application to Recent High Energy Observations

In Fig. 1, we present the spectra of five X-ray transients observed by Rossi X-ray Timing Explore. That have been obtained by the PCA and HEXTE instruments. Our calculations perfectly fit the data, with the detailed discussions of observations and data reduction procedure is presented separately (Borozdin, 1998; Revnivtsev, Gilfanov & Churazov, 1998). The fitting parameters of our model are listed in Table 1. Note that the spectra presented here are just some examples from a larger set of data readily approximated by this model. In fact, we have already analyzed most of available broad-band high-state X-ray spectra (see also Ebisawa, Titarchuk & Chakrabarti, 1996; Shrader & Titarchuk, 1997) and can conclude that *bulk-motion model successfully represents the sample as a whole*.

TABLE 1. Results of RXTE data fitting with the Bulk motion Comptonization model

Parameter	GRO J1655-40 28/05/97	GRS 1915+105 07/10/96*	GRS 1739-278 31/03/97	4U 1630-47 29/05/96
T_c, keV	1.002 ± 0.004	0.90 ± 0.01	0.92 ± 0.01	1.07 ± 0.01
α	1.27 ± 0.04	1.17 ± 0.02	1.58 ± 0.03	2.07 ± 0.10
f	0.040 ± 0.001	0.56 ± 0.01	0.14 ± 0.01	0.11 ± 0.02
$N_H, \times 10^{22} cm^{-2}$	0.5 (fixed)	5.0 (fixed)	4.0 (fixed)	4.0 (fixed)
χ^2 (d.o.f)	574.4(538)	311.5(291)	308.7(295)	335.1(285)

– because the source was highly variable during this observation, we have accumulated data corresponding to the one level of the total X-ray flux ($\sim 30 - 40\%$ of total observational time)

Having demonstrated the capability of our model in interpreting the high-state BHC spectra, we would like to discuss the low(hard)-state case. In this case the flux of soft photons from accretion disk is insufficient to cool the environment ambient to the disk. Soft photons are intensively Comptonized (upwards in energy) by electrons of the hot cloud surrounding the disk. The emerging spectrum has the well-known thermal-Comptonization shape (Sunyaev & Titarchuk, 1980; Sunyaev & Trumper, 1979). The close-in environment of the black hole is screened and cannot be observed directly. It is thus clear that overall spectral shape in this case does not depend on the nature of a central object, the fact which is bourn out by observations, e.g. of X-ray bursters in low state (White, Stella & Parmar, 1988; Revnivtsev et al. 1997) which resemble the low-state BHCs.

Using our three model parameters, the color temperature T_c, the spectral index α, and the parameter f characterizing the fraction of the soft flux illuminating the inflow site; along with the measurement of the total flux from the source, we can restore, *in principle*, the mass accretion rate, BH mass and source distance within the framework of standard accretion disk theory (Shakura & Sunyaev, 1973). We will discuss this point in much finer detail in separate paper (Borozdin *et al.* in preparation).

4. Discussion an Conclusion

Our basic conclusions are also consistent with observations of extragalactic black hole systems. For example, in the narrow-line Seyfert 1 Galactic nuclei (NLS1s) the X-ray power-law is significantly steeper and its normalization is more variable, with time scale of order 10^4 s, than in broad-line Seyfert 1 galaxies. This suggests that NLS1s may represent the high spectral state of extragalactic supermassive black holes (Chakrabarti & Titarchuk, 1995; Pounds, Done & Osborne, 1995; Brandt, Mathur & Elvis, 1997; Comastri et al. 1998).

One of the interesting consequences of our model is that QPOs supposed to be generated in the inner edge of the accretion disk should have a pronounced hard-X-ray variability signature (Morgan, Remillard & Greiner 1997). This is because the seed photons for the converging flow upscattering come from the same inner disk region. Soft-to-hard photon lag times (Cui, Zhang, Focke, & Swank, 1997) are expected to be on the order of the light-crossing time for the converging inflow region. The direct soft-photon flux detected from the source, ($\gtrsim 5\%$ of the Eddington) would necessarily be linked to the generic power-law component.

It seems worth noting that our fits do not imply as high values of disk color temperature as have been reported for the two superluminal jet sources GRO J1655-40 and GRS1915+105, where the observations have been interpreted in terms of the simple power law plus blackbody or multi-color disk model (Zhang et al., 1997a). The values we derived here are very typical for BHC and do not require any exotic interpretation in contrast to what had been suggested to explain the abovementioned results treated as evidence for fast speed spinning black holes (Zhang, Cui & Chen, 1997b). Instead we suggest that the disk temperature has been overestimated due to the failure to model the soft and hard spectral components in a unified and a self-consistent manner.

Arguments to demonstrate the existence of black-hole horizons based on X-ray nova flux histories (Narayan, Garcia, & McClintock, 1997), and the

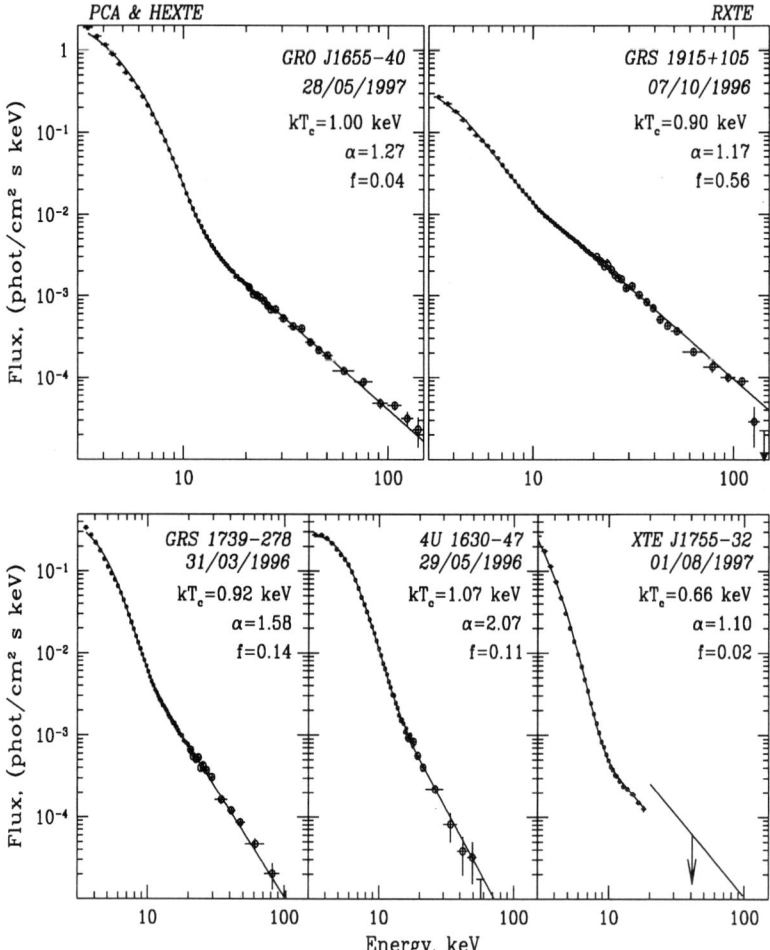

Figure 1. Broad-band high-energy spectra of Galactic black hole transient sources obtained by PCA and HEXTE experiments aboard the RXTE satellite. The axes are photon flux (ph/cm² /s/keV) versus energy (keV). PCA data are marked by *solid circles*, HEXTE data - by *hollow circles*. Solid lines represent bulk-motion model fits of the data. Source names, dates of observations and related best-fit model parameters are written on the panels. For XTE J1755-32 (right bottom panel) the upper limit for the flux detected with HEXTE in high energy band is shown by inclined line with arrow.

interpretation of asymmetrical line features as gravitational redshift effects in massive blackhole environments (Tanaka et al. 1995; Fabian et al. 1995; Bromley, Miller & Pariev, 1998), have recently been advanced. However, to interpret the asymmetrical line features as gravitational redshift effects the authors (Bromley, Miller & Pariev, 1998) have tried to avoid the effects of high velocities arguing that "the gravitational redshift is most important when the inclination angle is low; as the line-of-sight component is relatively small". We demonstrate here, that on the contrary, Doppler blueshift effects are similarly important in the continuum formation. A large number of restrictions must be met in order for the line profile method to be valid. (1) The contribution of the first scattering must be dominant, (2) the viewing angle must be very small and (3) the cold disk radiation must somehow be transmitted through the hot cloud, (4) the continuum and line emission must vary within the time period comparable to the light crossing time. There are additional uncertainties: models of the primary hard X-ray radiation, its reflection from the disk, K_α emission from species other iron and time independent surface emissivity. Also in terms of fitting observational data, numerous parameters are required and lead only to the interpretation of a residual.

It is worth noting that those line spectra exhibiting broadened line features (Bromley, Miller & Pariev, 1998) are very similar to the line spectra calculated in the context of downscattering model (Sunyaev & Titarchuk 1980, Figs. 4, 9). These downscattered (redshifted) spectral features are formed as a result of outward propagation of iron-line photons, produced for example inside the disk due to illumination of the inner disk edge by the hard-X-ray coming, for example, from the ambient Compton cloud. In this case the inner part of disk is not necessarily located within a few Schwarzschild radii.

We wish to point out that in neither case have arguments been invoked that are tied directly to the most basic accreting black hole property: relativistic inward flux of matter falling onto the event horizon. We further note that recent studies cast doubt on the result based on soft X-ray transients quiescent/peak luminosity ratio (Narayan, Garcia, & McClintock, 1997), as the effect tends to disappear for larger samples (Chen et al. 1998; Chen, Shrader & Livio, 1997).

In conclusion, we wish to emphasize once again that remarkable agreement of observational data with predictions of the relativistic theory leads us to believe that *we have identified a generic spectral signature of accreting black hole, and that in viewing spectra of accreting black holes in their high state we are seeing direct evidence of matter sinking down the black hole drain.*

ACKNOWLEDGMENTS

This research has made use of data obtained through the High Energy Astrophysics Science Archive Research Center Online Service, provided by the NASA/Goddard Space Flight Center. Portions of this work were supported by the Rossi X-Ray Timing Explorer Guest Investigator Programs and by RBRF grant 96-15-96343.

References

Borozdin, K.N. et al. (1998), X-ray nova GRS1739-278 close to Galactic Center. *Astronomy Letters* (submitted)

Brandt, W.N., Mathur, S. & Elvis, M. (1997), Spectral slopes of broad-and narrow-line Seyfert 1s. *Mon. Not. R. Astron. Soc.*, **285**, L25-L30

Bromley, B. C., Miller, W.A. & Pariev, V.I. (1998), The inner edge of the accretion disk around a supermassive black hole. *Nature* **391**, 54-56

Chakrabarti S.K. & Titarchuk, L. G. (1995), Spectral properties of accretion disks. *Astrophys J.* **455**, 623-639

Chen, W., Shrader, C.R., & Livio, M. (1997), The properties of x-ray and optical light curves of x-ray novae. *Astrophys J.* **491**, 312-338

Chen, W. et al. (1998), Evidence for event horizon in soft x-ray transients? in *Accretion Processes in Astrophysics*, (eds. Holt, S. & Kallman, T.) (in the press)

Comastri, A. et al. (1998) BeppoSAX observations of Narrow-Line Seyfert 1 galaxies:I. Ton S 180 *Astronomy & Astrophysics* (in the press).

Cui, W., Zhang, S.N., Focke, W., & Swank, J.H. (1997), Temporal properties of Cyg X-1 during the spectral transition. *Astrophys J.* **484**, 383-393

Ebisawa K., Titarchuk L. G., & Chakrabarti, S.K. (1996), On the spectral slopes of hard x-ray emission from black holes candidates. *Pub.Astron.Soc.Japan* **48**, 59-65

Fabian, A. C. et al. (1995), On broad iron K-alpha lines in Seyfert 1 galaxies. *Mon. Not. R. Astron. Soc.* **277**, L1-L15

Miyoshi, M. et al. (1995), Evidence for a black hole from high rotation velocities in sub-parsec region of NGC4258. *Nature* **373**, 127-129

Morgan, E. H., Remillard, R. A., & Greiner, J. (1997), RXTE observations of QPOs in the GRS 1915+105. *Astrophys J.* **482**, 993-1010

Narayan, R., Garcia, M.R., & McClintock, J.E. (1997), Advection-dominated accretion and black hole event horizons. *Astrophys J.* **478**, L79-L82

Orosz, J.O. & Bailyn, C.D. (1997), Optical observations of GRO J1655-40 in quiescence. 1 A precise mass for black hole primary. *Astrophys J.* **477**, 876-896

Pounds, K.A., Done, C., & Osborne, J. (1995), RE 1034+39: a high state Seyfert galaxy? *Mon. Not. R. Astron. Soc.* **277**, L5-L10

Revnivtsev, M. et al. (1997), Detection of the hard x-ray emission from Z-ray burster 4U1705-44 with GRANAT/SIGMA. in *Proceedings 2nd INTEGRAL Workshop "The transparent Universe"* (eds. Winkler, C., Courvoisier, T. & Durouchoux, Ph.) 277-278 (ESA SP-382)

Revnivtsev, M., Gilfanov, M. & Churazov, E. (1998), RXTE broad-band observations of X-ray Nova XTE J1755-324. *Astronomy & Astrophysics* (submitted)

Shakura, N.I., & Sunyaev, R.A. (1973), Black holes in binary systems. Observational Appearance. *Astronomy & Astrophysics* **24**, 337-355

Shrader C. & Titarchuk L. G. (1997), The high energy spectra of accreting black holes. *Bul.Amer.Astron.Soc.* **191**, 1223-1224

Sunyaev R.A.,& Trümper J. (1979), Hard x-ray spectrum of Cyg X-1. 1979, *Nature* **279**, 506-508

Sunyaev R.A. & Titarchuk, L.G. (1980), Comptonization of x-rays in plasma clouds: Typical radiation spectra. *Astronomy & Astrophysics* **86**, 121-138

Tanaka, Y. et al. (1995), Gravitationally redshifted emission implying an accretion disk and massive black hole in active galaxy MCG-6-30-15, *Nature* **375**, 659-661

Tanaka Y. & Shibazaki N. (1996), X-ray novae. *Annu Rev Astr* **34**, 607-644

Titarchuk, L.G., Mastchiadis, A. & Kylafis, N.D. (1997), X-ray spectral formation in a converging fluid flow: spherical accretion into black holes. *Astrophys J.* **487**, 834-846

Titarchuk, L.G., Lapidus, I.I. & Muslimov, A. (1998), Mechanisms for high-frequency QPOs in neutron star and black hole binaries. *Astrophys J.* **499**, (astro/ph-9712348)

Titarchuk, L.G., & Zannias, T. (1998), The extended power law as an intrinsic signature for a black hole. *Astrophys J.* **493**, 863-872

White, N.E., Kaluzienski, L.J., & Swank, J.H., (1984), The spectral of X-ray transients. in *High Energy Transients in Astrophysics* (ed S.Woosley) 31-48 (AIP:New York).

White, N., Stella, L., & Parmar A. (1988), The X-ray spectral properties of accretion disks in X-ray binaries. *Astrophys J.* **324**, 363-378

Zhang, S.N., et al, (1997a) Broadband high-energy observations of GRO J1655-40 during outburst. *Astrophys J.* **491**, 312-338

Zhang, S.N., Cui, W. & Chen, W. (1997b), Black hole spin in x-ray binaries: observational consequences. *Astrophys J.* **482**, L155-L158

B. Paul

M. Gilfanov at the Panel Discussion

X–RAY SPECTRAL VARIABILITY OF BLACK HOLE BINARIES

M.GILFANOV, E.CHURAZOV AND R.SUNYAEV
Max-Planck-Institut für Astrophysik, Karl-Schwarzschild-Str. 1, 85740 Garching bei Munchen, Germany
and
Space Research Institute, Profsouznaya 84/32, 117810 Moscow, Russia

1. Introduction

Observations of black hole binaries in the X–ray/low γ–ray energy domain revealed complicated pattern of the spectral variability. The most prominent spectral changes are associated with well known and intensively studied transition between soft/hard spectral states (e.g. Tanaka, 1989, Grebenev et al., 1993, Sunyaev et al., 1993). This transitions manifest themselves as a dramatic redistribution of the emitted energy over wavelength (e.g. Fig.1). Along with these less evident "subtle" variations of the spectral properties are often observed (e.g. Fig.2). In many cases spectral changes are accompanied with changes in the short term aperiodic variability properties (e.g. Miyamoto et al., 1995, van der Klis, 1995, Belloni et al., 1996). The spectral and short term variability characteristics of the emission are directly measurable quantities, provided sufficient spectral and temporal resolution, sensitivity and energy coverage.

In many theoretical models the geometry and conditions in the X–ray emission generation zone are defined primarily by the mass accretion rate. Opposite to spectral and temporal characteristics the mass accretion rate is not directly measurable quantity. In some cases it can be determined in the model dependent way. Related quantity is the bolometric luminosity which is not directly measurable either due to limited energy coverage or due to low energy interstellar absorption. The bolometric luminosity can be in principle estimated via extrapolation of the spectral model – best fit to the observed spectrum. That obviously requires certain assumptions to be made about spectral behavior beyond the energy range covered by the instrument, i.e. is spectral model dependent. The only directly measurable

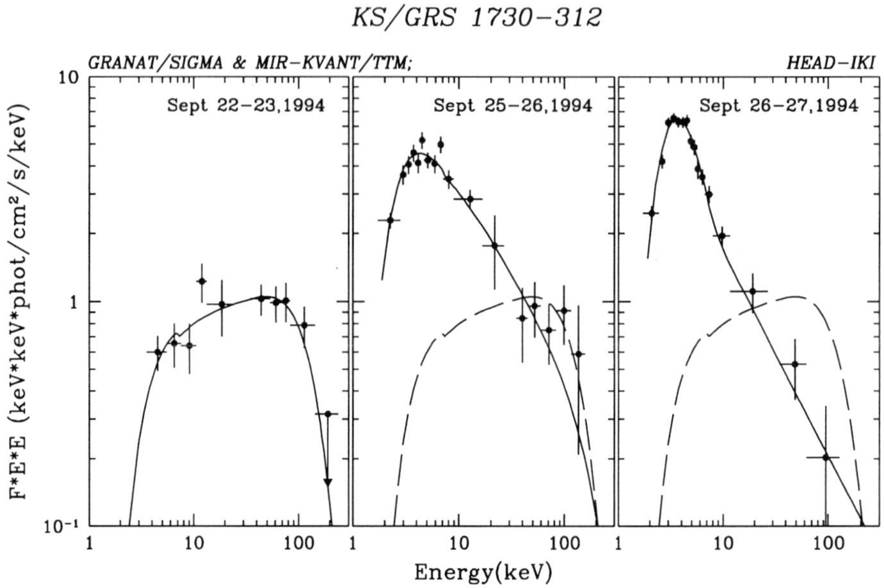

KS/GRS 1730-312

Figure 1. The spectra of KS1730-312 in different spectral states (TTM and SIGMA data, from Trudolyubov et al., 1996.

quantity is the luminosity in some restricted energy range defined by the bandpass of the instrument.

As is well known, the Comptonization theory and, in particular, the simplest approximation given by Sunyaev & Titarchuk, 1980 formula was quite successful in describing individual spectra of the black hole candidates in the low spectral state (e.g. Sunyaev & Truemper, 1979; Grebenev et al., 1993). The Compton up scattering of low frequency radiation in hot optically thin part of the accretion flow is very likely a mechanism of generation of the hard spectral component. However the structure of the accretion flow and in particular origin of the hot electrons and geometry of the Comptonization region are still unclear.

Below we discuss spectral variability of several black hole candidates basing on the MIR-KVANT/TTM, GRANAT/SIGMA and ASCA observations.

2. Spectral states

The fact of existence of different spectral states for black hole candidates was established more than two decades ago (e.g. Tananbaum et al., 1972, Holt et al., 1976, Ogawara et al., 1982). Nonetheless the luminosity change

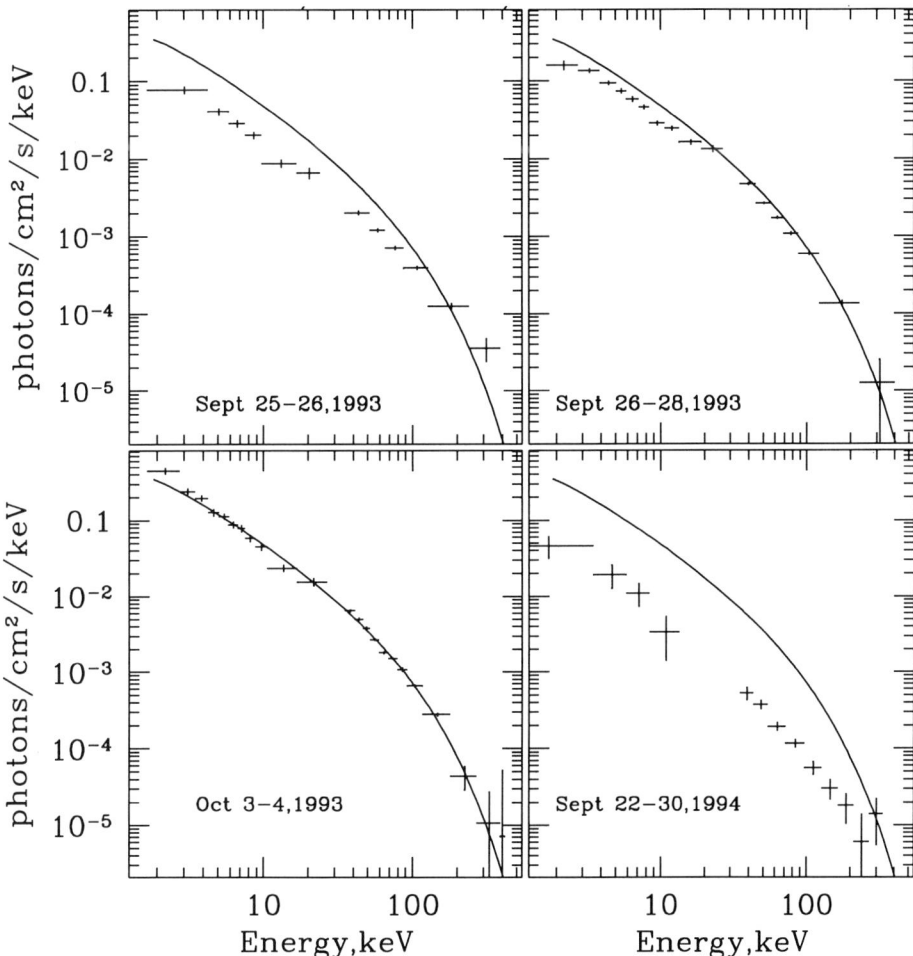

Figure 2. The spectra of GRS1716-249 (Nova Oph 1993) at different epoch (TTM and SIGMA data, from Revnivtsev et al., 1998a).

(and it's sign even) between spectral states was unclear. The primary purpose of this paragraph is to estimate typical luminosity levels corresponding to each spectral state for several sources. For completeness we review briefly the pattern of spectral states with emphasis on the spectral properties and compare with the spectra states observed for weakly magnetized neutron star binaries.

2.1. BLACK HOLE BINARIES

The spectral states are roughly classified according to relative importance of the soft and the hard spectral components. Three spectral states are usually distinguished:

1. *low* or *hard* – the emission is dominated by the hard spectral component which shape is adequately described by the Comptonization model.
2. *high* or *soft* – the primary contribution to the emergent emission is due to the soft spectral component, the hard component being absent or steep and weak.
3. *superhigh* – both the soft and the hard spectral components are present, with the major fraction of the X–ray luminosity being emitted in the soft spectral component. Properties of the hard spectral component are different from that in the low state.

The commonly accepted point of view is that the soft spectral component is due to emission from optically thick and geometrically thin (part of the) accretion disk of the type predicted by the standard model of Shakura & Sunyaev, 1973. The hard spectral component is believed to result from Compton upscattering of soft photons in optically thin Comptonization region. The temporal variability properties of these two spectral components are very different (e.g. van der Klis, 1995, Miyamoto et al., 1991). Besides that, spectral and short term variability properties of the hard spectral component observed in the low and superhigh spectral states are also quite different (Miyamoto et al., 1995; van der Klis, 1995). In particular, sufficiently steep photon index (≈ 2.5) and small fraction of the luminosity emitted in the hard power law component in the superhigh state could be easily accommodated in the disk–corona model without postulating that major part of the gravitational energy of accreting matter is dissipated in the rarefied corona (e.g. Gilfanov et al., 1991).

Since the bulk of the emitted energy might shift over the photon energy by $\sim$ two orders of magnitude (typical black body temperature of the soft component is $\sim 0.3 - 1.0$ keV, the $F_E \times E^2$ for the hard component peaks at ~ 100 keV) accurate luminosity estimate requires broad energy coverage – from $\lesssim 1$ keV to $\gtrsim$ few hundred keV. Besides that accurate estimate of luminosity of the soft spectral component ($kT_{bb} \sim 0.3 - 1$ keV) requires in many cases proper account for the interstellar absorption. The latter was especially important for the recently observed soft (or intermediate, Belloni et al., 1996, Zhang et al., 1997) spectral state of Cyg X–1.

The Table 1 presents luminosity estimates for several sources. Note that the numbers given in the table do not correspond to the whole range of the luminosities for a particular spectral state. They rather represent some luminosity levels at which particular source was picked up by different obser-

TABLE 1. Luminosity in different spectral states for several black hole candidates

Spectral state	Nova Mus	KS1730-312	Cyg X-1	XTE J1755-324
low	$\approx 1 \cdot 10^{37}$	$\approx 4.4 \cdot 10^{37}$	$\sim (2-4) \cdot 10^{37}$	$2 \cdot 10^{36} - 10^{37}$
high	$\approx 8 \cdot 10^{37}$	$\approx 1.2 \cdot 10^{38}$ (?)[1]	$\sim 8 \cdot 10^{37}$ (?)[1]	
superhigh	$\approx 5 \cdot 10^{38}$			$\sim 10^{38}$

The source distance was assumed 5 kpc for Nova Mus, 8.5 kpc for KS1730-312 and XTE J1755-324 and 2.4 kpc for Cyg X-1. The energy range is 1–300 keV for Nova Mus and KS1730-312 and 0.5–300 keV for Cyg X-1. No correction for interstellar absorption was applied to Nova Mus and KS1730-312 data.

[1] Probably intermediate state between low and high spectral states (see Belloni et al., 1996 for Cyg X-1).

vatories/instruments. The high state luminosity for Cyg X-1 was calculated using the best fit parameters for the soft and hard spectral components from (Dotani et al., 1996) assuming hydrogen column density of $6 \cdot 10^{21}$ cm^{-2}. The Table 1 demonstrates, that the luminosity increases from the low state to high and superhigh spectral states.

2.2. NEUTRON STAR BINARIES

The spectral study of X-ray bursters, (e.g. Langmeier et al., 1987, Mitsuda et al., 1989, Ford et al., 1996), demonstrate, that X-ray bursters (weakly magnetized neutron star binaries) show bimodal spectral behavior similar to that observed for black hole binaries (see also Barret & Vedrenne, 1994, Tanaka & Shibazaki, 1996, Revnivtsev et al., 1998b). At least two distinct spectral states of X-ray bursters may be identified – soft/high and hard/low (cf. spectral states of black hole candidates). The 4U1705-44 and 4U1608-56 are discussed below as two relatively well studied examples (Revnivtsev et al., 1998b).

The high/soft state spectrum corresponds to luminosity level of $L_X \sim (3-9) \times 10^{37}$ erg/s[1] or $L_X \sim (0.2-0.6) \times L_{Edd}$ and roughly[2] could be represented as a blackbody spectrum with temperature of $kT_{bb} \sim 1-2$ keV possibly with superimposed weak and rather steep power law tail which diminishes above $\sim 20-30$ keV. In the case of 4U1705-44 the upper limit on the luminosity in the higher energy domain during the soft state is

---

[1] The hard X-ray luminosities and upper limits for these two X-ray bursters obtained by GRANAT/SIGMA and cited in this subsection are from Revnivtsev et al., 1998b

[2] Detailed discussion of the spectral behavior of the X-ray bursters at high luminosity is beyond scope of these paper (see e.g. White, Nagase & Parmar, 1995 for review)

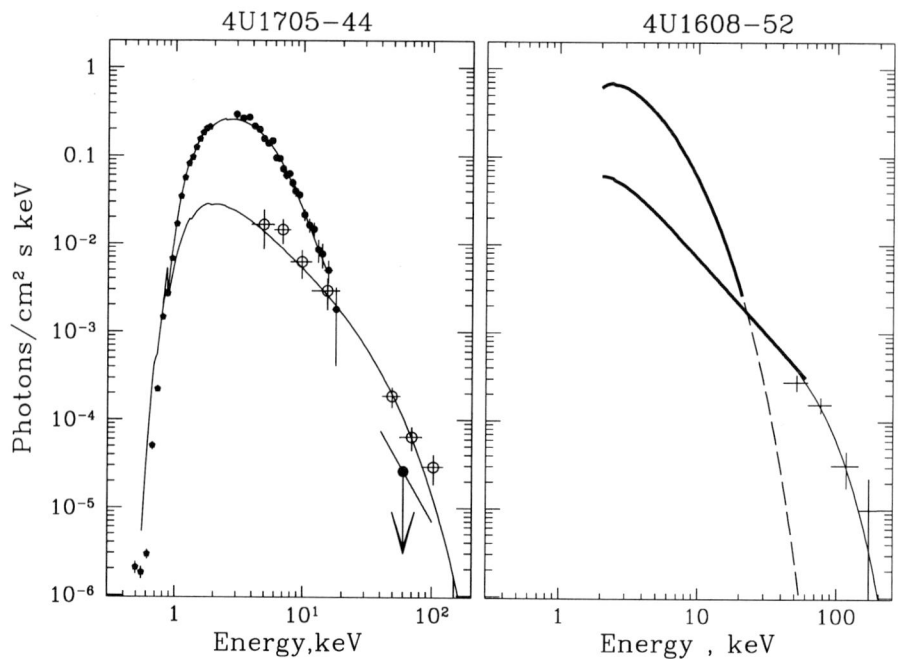

Figure 3. The spectra of 4U1705-44 and 4U1608-52 in soft and hard spectral states (from Revnivtsev et al., 1998b).

2×10^{36} erg/s (2σ, 35–100 keV, assuming the spectral shape similar to that observed in the low state).

At lower luminosity level ($L_X \sim (0.7-2) \times 10^{37}$ erg/s or $L_X \sim 0.1 \times L_{edd}$) the spectrum becomes considerably harder with photon index in the low energy limit of $\sim 1.7 - 2.0$ and exponential cut–off at $\sim 30 - 70$ keV. The low state spectrum could be generally described by the Comptonization model.

The overall spectral shape in the soft and the hard states is generally

TABLE 2. Luminosity in different spectral states for two X-ray bursters

Spectral state	4U1705-44	4U1608-52
low	$\sim 1 \cdot 10^{37}$	$\sim 0.7 \cdot 10^{37}$
high	$\sim 5 - 9 \cdot 10^{37}$	$\sim 3 \cdot 10^{37}$

The source distance was assumed 7.4 kpc for 4U1705-44 and 3.6 kpc for 4U1608-52. The energy range is 0.5–200 keV.

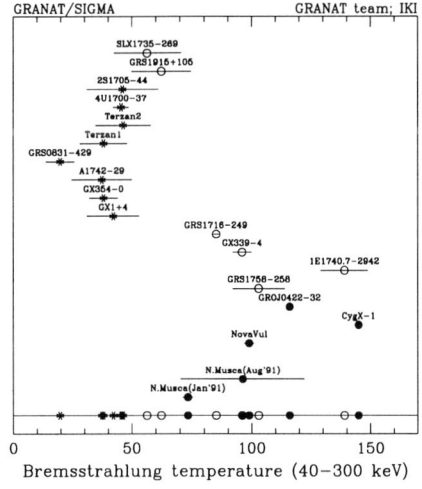 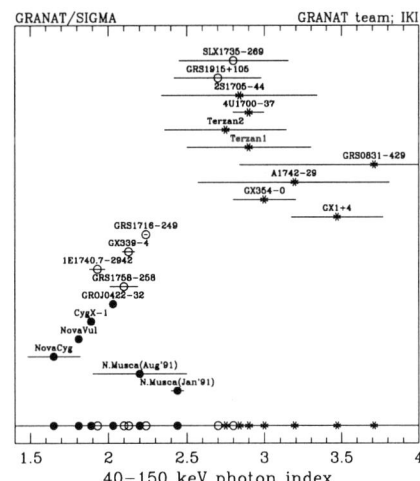

Figure 4. The hardness of the spectra of black hole binaries (in low or superhigh spectral state) and neutron star binaries (low spectral state) as observed in the hard X-ray domain expressed in terms of best-fit bremsstrahlung temperature (left) and 40-150 keV photon index (right). The binary systems with known nature of the compact object are marked by asterisk (a neutron star) or filled circle (a black hole). The GRANAT/SIGMA data; from Gilfanov et al., 1993.

similar to that of black hole candidates except for somewhat higher value of kT_{bb} in the soft state. Another important difference is that the X-ray bursters have ~ 2 − 3 times lower energy of the exponential cut-off of the spectrum in the hard state (Fig.4, Gilfanov et al., 1993). This difference is directly observable and opens a possibility to distinguish an accreting low magnetized neutron star from a black hole. On the other hand it's worth mentioning that in the hard state the spectral slope sufficiently below the cut-off energy is nearly the same as for the hard state spectra of black hole candidates.

The X-ray bursters undergo spectral transitions at approximately the same luminosity level as black hole binaries (Tables 1,2) although the mass of the compact object might differ by more than the order of magnitude. In terms of critical Eddington luminosity in the case of X-ray bursters transition occurs at L_X/L_{Edd} ~few×0.1 while for black hole binaries the threshold luminosity corresponds to $L_X/L_{Edd} \lesssim 0.1$.

3. Optically thick disk emission in the low spectral state

As was mentioned above the high/superhigh spectral states are characterized by presence of prominent soft spectral component which gives the dominant contribution to the $\lesssim 1$ keV to few hundred keV luminosity. In the low state the balk of X-ray luminosity is emitted in the hard Comptonized spectral component. There are reasons to believe (e.g. Tanaka, 1989, Ebisawa et al., 1994) that the soft spectral component originates from geometrically thin, optically thick part of the accretion disk. The spectra observed during the superhigh spectral state show that the optically thick disk might coexist with rarefied optically thin hot region where the hard spectral component originates from. In this context the search for the soft component – emission from the optically thick part of the accretion disk in the low spectral state is of certain interest for understanding the structure of the accretion disk.

3.1. CYGNUS X–1

The first indications of presence of the soft excess in the low state spectrum of Cyg X–1 were found about two decades ago (Priedhorsky et al., 1979, Balucinska–Church et al., 1995). Most apparently it could be noticed from comparison of the values of the hydrogen column density inferred by the low energy cut-off of the X-ray spectrum, $\sim 3 \cdot 10^{21}$ cm^{-2}, with that known from 21 cm observations and interstellar reddening of the optical companion, $\sim 6 \cdot 10^{21}$ cm^{-2}. The data of the ASCA observations fully support this conclusion. That excess may be ascribed to the presence of the soft emission with the best fit blackbody temperature $T_{bb} \sim 120$ eV and luminosity of the order of $\sim 20\%$ of the overall 0.5–300 keV luminosity of the source[3] in the nominal low spectral state.

The approximation of the soft excess (ASCA observation in November, 1994) by a multicolor disk model (Shakura & Sunyaev, 1973, Makishima et al., 1986) gives value of the temperature at the inner boundary of the accretion disk $kT_{in} \approx 136$ eV and the radius of the inner boundary $R_{in} \approx 440 \pm 60$ km. The best fit value of the inner disk radius corresponds to $\approx 15 R_g$ for the $10 M_\odot$ black hole i.e. exceeds considerably the $3 R_g$. That value might be interpreted as a radius at which the geometrically thin, optically thick disk approximation ceases to hold. As is well known (Shakura & Sunyaev, 1976), the inner part of the standard accretion disk (Shakura & Sunyaev, 1973) where the radiation pressure dominates is unstable. That inner part of the accretion flow could be responsible for generation of the hard spectral component observed in the low spectral state (Sunyaev et al., 1991; see recent discussions of the advection flows by Chakrabarti & Titarchuk, 1995, Narayan, 1996).

---

[3] It should be mentioned that the luminosity estimate is rather uncertain in this case because bulk of the soft component emission is absorbed by the line–of–sight gas.

Within that approximation the disk temperature depends upon the radius according to well known relation:

$$T(r) = \left(\frac{3GM\dot{M}}{8\pi\sigma r^3}\right)^{1/4} \left(1 - \left(\frac{r_0}{r}\right)^{1/2}\right)^{1/4}$$

(Shakura & Sunyaev, 1973). Therefore, knowing that $T(R = 440km) = 136$ eV we can estimate the disk mass accretion rate

$$\dot{M}_{disk} \sim (2-3) \cdot 10^{17} \, g/sec$$

(assuming 2.5 kpc distance and $10M_\odot$ black hole). On the other hand knowing that the absorption corrected energy flux from the source at that time was $F_X(E > 300 \text{ eV}) \sim (4-5) \cdot 10^{-8}$ erg/sec/cm^2 and assuming the accretion efficiency of $\eta \sim 0.1$ we can obtain *independent* estimate of the total mass accretion rate

$$\dot{M}_{total} \sim (3-4) \cdot 10^{17} \times \left(\frac{0.1}{\eta}\right) \, g/sec,$$

which is consistent with that derived above from analysis of the low energy part of the spectrum. Note, however, that these two values estimate two essentially different quantities and could suffer from different uncertainties. The $\dot{M}_{total}$, according to the way it was derived, accounts for contribution of the matter possibly accreting via rarefied optically thin halo and, on the other hand, depends on assumed value of the accretion efficiency, i.e. importance of the advection. The $\dot{M}_{disk}$ is an estimate of accretion rate in geometrically thin keplerian disk and is not affected by these two effects, but depends on the validity of simple multicolor disk approximation.

The above values of the accretion rate correspond to $\dot{M} \sim (0.01-0.03) \times \dot{M}_{crit}$.

3.2. GRS1716-249 (X-RAY NOVA OPH 1993)

The GRS1716-249 was observed by ASCA on Oct.5, 1993 near the maximum of it's X-ray luminosity. Opposite to Cyg X-1, the low frequency cut-off of the X-ray spectrum agrees quite well with the Galactic value of $N_H \approx 4.5 \cdot 10^{21}$ cm^{-2} (Della Valle et al., 1994). The upper limit on the luminosity of the soft component similar to the one observed in the spectrum of Cyg X-1 in 1994 during "standard" low spectral state (black body spectrum with $kT_{bb} \sim 100$ eV is $\sim 10^{36}$ erg/sec, assuming 2.5 kpc distance (Della Valle et al., 1994), which is less than $\sim 3\%$ of the total X-ray luminosity of the source.

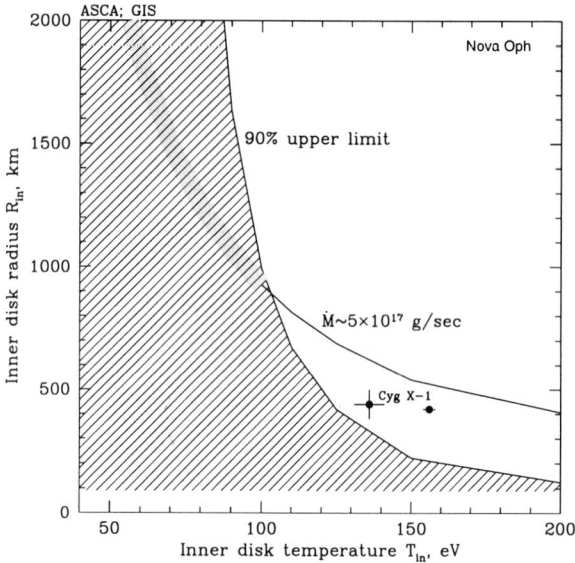

Figure 5. Constraints on the radius and temperature at the inner boundary of the optically thick part of the disk in GRS1716-249 (X-ray Nova Oph 1993). ASCA observation on Oct.5,1993 (GIS data). The hatched area under the curve marked "90% upper limit" is a range of parameters consistent with the ASCA non detection of the soft excess in the spectrum (90% confidence, assuming $N_H = 4.5 \cdot 10^{21}$ cm^{-2}). The line marked "$\dot{M} \sim 5 \times 10^{17}$ g/sec" shows relation between R_{in} and T_{in} for that value of the disk accretion rate. Two points marked "Cyg X-1" are values measured for Cyg X-1 in 1994 (left) and 1993 (right). The source distance was assumed 2.5 kpc, $10 M_\odot$ black hole, $cos(i) = 0.5$.

The range of values of the temperature and the radius of the inner boundary of the optically thick part of the accretion disk, consistent with the ASCA GIS spectrum of the source is shown in Fig.5 (hatched area below the curve marked "90% upper limit").

The multicolor disk parameters could be restricted further, assuming that the total luminosity of the source may be used as an estimator of the disk mass accretion rate. The 0.5–300 keV source luminosity at the date of ASCA observation was $\approx 4 \cdot 10^{37}$ erg/sec (the high energy luminosity was measured by GRANAT/SIGMA – Revnivtsev et al., 1998a) which corresponds to $\dot{M}_{total} \sim 5 \times 10^{17}$ g/sec. The relation between T_{in} and R_{in} corresponding to that value of $\dot{M}_{disk}$ is shown in Fig.5 by a curve marked "$\dot{M} \sim 5 \times 10^{17}$ g/sec". Only part of that curve within the hatched area is consistent with non detection of the soft excess in the spectrum of the source by ASCA. Under this assumption, the accretion disk parameters are constrained by $T_{in} \lesssim 100 - 110$ ev, $R_{in} \gtrsim 900$ km $\approx 30 R_g$ for $10 M_\odot$ black hole.

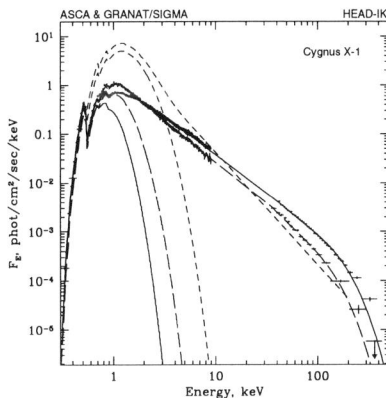 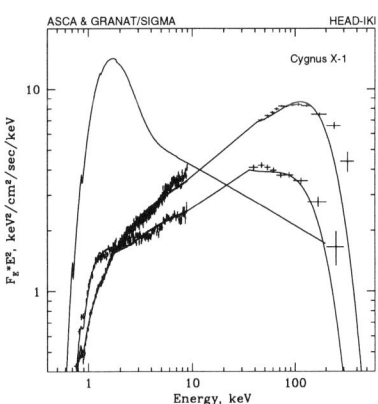

Figure 6. The broad band spectra of Cyg X–1 in the "nominal" hard state, during very low hard X–ray flux episode observed in 1993 and 1996. The data above 35 keV are from Kuznetsov et al., 1997.

On the other hand, assuming, that the inner disk radius is the same as for Cyg X–1, $\sim 15 R_g$, the upper limit on the inner disk temperature is $T_{in} \lesssim 120$ eV, which constraints the disk accretion rate: $\dot{M}_{disk} \lesssim 1 \cdot 10^{17}$ g/sec.

4. Broad band spectral variability of Cyg X–1.

The Cyg X–1 was observed with ASCA satellite in a number of occasions during 1993–1995. Some of this observations were performed during extended low hard X–ray flux episode (although not at the minimum of the hard X–ray flux; see the next paragraph). This gives a possibility to study variations of the broad band spectrum of the source corresponding to the spectral changes observed in the high energy domain (discussed in the next paragraph, Fig.8). We used for this purpose two ASCA observations performed in November 1993 (low hard X–ray flux episode) and one year later in November 1994 ("standard" hard spectral state of the source).

The broad band spectra of the source in 1993 and 1994 (ASCA and GRANAT/SIGMA data) are shown in Fig.6. The parameters of the spectral approximation of the ASCA data in the 0.5–10 keV energy range by the model consisting of the power law emission with reflected component and multicolor disk component (the hydrogen column density fixed at $6 \cdot 10^{21}$ cm^{-2}) are presented in Table 3.

The spectral changes observed by SIGMA in the high energy domain during low hard X–ray flux episode in 1993–1994 were accompanied with

TABLE 3. The best fit parameters for the ASCA data approximation with the model consisting of the power law with reflection and multicolor disk emission.

Parameters [1]	25-26/11/94 "nominal" hard state	11-12/11/93 ~beginning of the low hard X-ray flux 1993 episode	30-31/05/96 [5] ~middle of the low hard X-ray flux 1996 episode
photon index, α	1.64 ± 0.01	1.99 ± 0.01	~ 2.2
T_{in}, eV	136 ± 5	156 ± 2	~ 470
R_{in}, km [3]	440 ± 60	420 ± 15	~ 110
F_{pl} [2]	0.86 (1.4)	0.75 (1.6)	3.1 (3.3)
F_{disk} [2]	0.02 (0.7)	0.06 (1.3)	0.2 (5.7)
$\dot{M}_{disk}$, 10^{17} g/sec [4]	~ 1.7	~ 2.8	~ 5

[1] The hydrogen column density was fixed at $N_H L = 6 \cdot 10^{21}$ cm^{-2}.
[2] The 0.3–9 keV energy flux of the power law and disk emission components, 10^{-8} erg/sec/cm^2 (the absorption corrected value is given in parenthesis)
[3] Assuming the source distance of 2.5 kpc, binary system inclination angle of 70°
[4] Estimated from the parameters of the soft excess using the multicolor disk approximation
[5] The multicolor disk approximation parameters were *roughly* estimated using the spectral parameters reported by Dotani et al., 1996

corresponding changes in the standard X–ray band (Fig.6 and second and third columns of Table 3):

1. The slope of the Comptonized radiation changes from ≈ 1.6 for standard hard state spectrum to ≈ 2.0 (low hard X–ray flux episode). As well known the steepening of the spectrum may be due to decrease of either the electron temperature or the Thompson optical depth. The high energy data indicate that the position of the high energy cut–off of the spectrum shifts towards lower energy implying that the electron temperature in the Comptonization region is decreasing.[4]

2. Parameters of the soft spectral component – supposedly emission from optically thick, geometrically thin outer part of the accretion disk – change as well. Qualitatively, these changes correspond to increase of the luminosity and the mean photon energy of the soft component. The relative contribution of the soft component to total X–ray luminosity increases as well. In terms of the multicolor disk approximation these changes may be attributed to increase of the disk temperature.

---

[4] The ASCA and SIGMA data are not contemporaneous restricting therefore the possibility of reliable Comptonization model parameters estimate from the broad band spectral fitting

Behaviour of the radius of the inner boundary of the optically thick part of the accretion disk is unclear due to limited accuracy of inner radius determination. Within this model these changes are caused by increase of the mass accretion rate by a factor of $\sim 1.5 - 2$.

The sign of change of the overall X-ray luminosity is rather uncertain mainly due to quite large value of interstellar absorption and the fact that the soft spectral component might contribute non negligible fraction to the total X-ray luminosity[5]. The observed (uncorrected for the low energy absorption) 0.5–300 keV luminosity slightly decreased from $\approx 4.1 \cdot 10^{-8}$ erg/sec/cm^2 ($\approx 3.1 \cdot 10^{37}$ erg/sec) to $\approx 2.6 \cdot 10^{-8}$ erg/sec/cm^2 ($\approx 2.0 \cdot 10^{37}$ erg/sec) with decrease of the hard X-ray flux.

Another episode of very low hard X-ray flux from Cyg X-1 occurred in May–June 1996 (Zhang et al., 1997). The source was intensively observed by GRO/BATSE in hard X-ray energy domain and by XTE and ASCA in the standard X-ray band. The accretion disk and Comptonized emission parameters roughly estimated using the spectral parameters reported by Dotani et al. (1996) are given in the fourth column of Table 3. The ASCA observation in 1993 occurred two month before the source reached the lowest value of the hard X-ray flux, whereas in 1996 Cyg X-1 was observed with ASCA nearly at the minimum of the hard X-ray flux. Correspondingly, 1996 observation found steeper power law, higher disk temperature, smaller disk inner radius and higher value of the mass accretion rate as estimated from the disk parameters (Table 3). It seems very likely, that both 1993 and 1996 events correspond to the same phenomena – transition of the source to the soft spectral state caused by increase of the mass accretion rate by a factor of few.

5. Spectral variability of the hard spectral component.

In this paragraph we will consider variability of the hard spectral component at the energies $\gtrsim 40$ keV basing on observations of the GRANAT/SIGMA. Since often the SIGMA observations aren't complemented with observations in the standard X-ray band it is sometimes unclear if observed spectral variations are intrinsic to the low/hard spectral state or correspond to transition from the low to high spectral state (e.g. extended episodes of very low hard X-ray flux observed for Cyg X-1 and 1E1740.7–2942).

In order to quantify shape of the hard spectral component the best fit optically thin bremsstrahlung temperature is used. This approach provides simple though rather crude single parameter representation of the hardness of spectrum. The same approximation is used to estimate the energy flux

---

[5]Note that the for the hydrogen column density of $6 \cdot 10^{21}$ cm^2 more than $\approx 95\%$ of the disk emission with parameters given in the Table 3 is absorbed

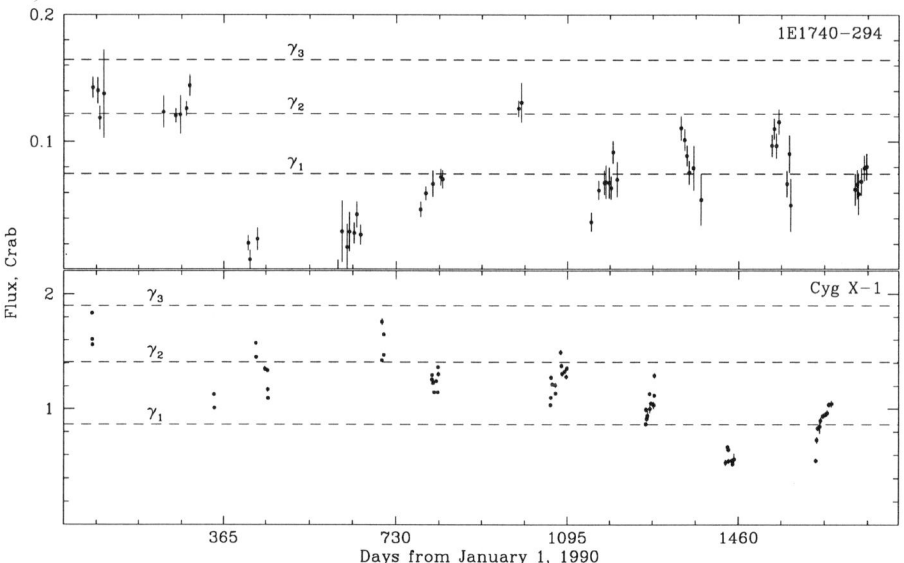

Figure 7. The 40–150 keV light curves of Cyg X-1 (top) and 1E1740.7–2942 (bottom). Each data point represents the average over $\approx 2 - 8$ hours (Cyg X-1) and ~ 50 hours (1E1740.7–2942). The date Jan.1,1990 corresponds to MJD 47892. Approximate flux levels corresponding to the three γ-states of Cyg X-1 are shown on both panels. From Kuznetsov et al., 1997.

or luminosity. The short term variability is characterized by relative rms in the $0.01 - 0.1$ Hz frequency range.

5.1. CYGNUS X–1 AND 1E1740.7-2942

The hard X–ray light curves of both Cyg X–1 and 1E1740.7–2942 have a complex structure with short term (time scales of days to weeks) variations superimposed on long term (time scale of years) intensity changes of generally larger relative amplitude (Kuznetsov et al., 1997).

The long-term light curve of Cygnus X-1 (Fig.7) recorded by SIGMA (Salotti et al., 1992; Vikhlinin et al., 1994; Ballet et al., 1996) shows variations of the 40-150 keV flux by a factor of $\sim$4. During 1990 – mid 1993 (Vikhlinin et al., 1994) the source was typically detected near the γ_2 intensity level. In Dec 1993 and in the first observations in June 1994 the minimal flux was detected, ~ 0.5 Crab. According to BATSE data (Crary et al., 1996) having much better time coverage these two observational sets occurred during an extended low hard X-ray flux episode. The lowest flux from Cyg X-1 during this period was detected in Feb.1994 ($0.2\gamma_1$ – Phlips et al., 1996). During almost all SIGMA observations considerable variability

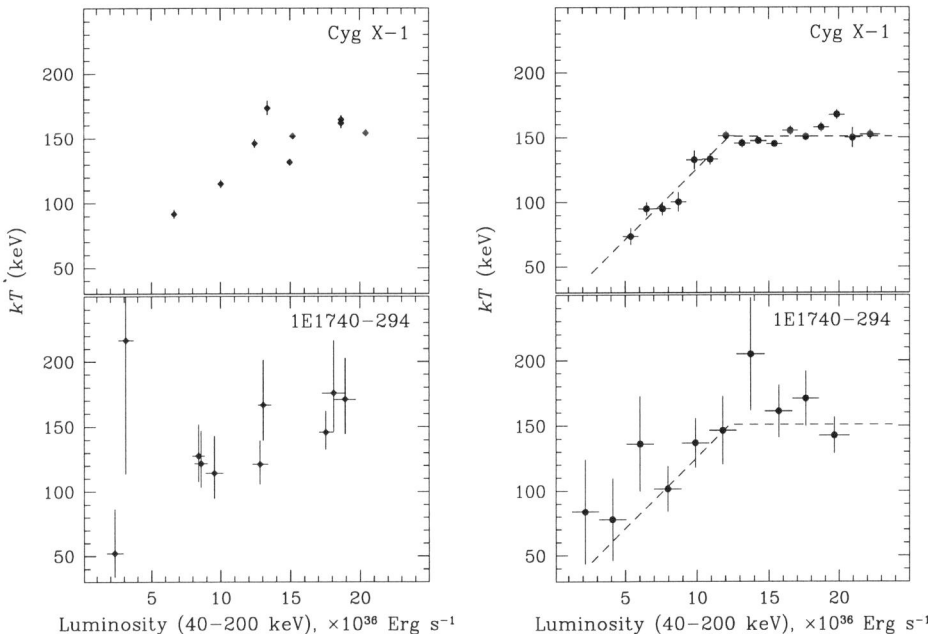

Figure 8. The best-fit bremsstrahlung temperature plotted against the hard X-ray luminosity (40-200 keV) for Cyg X-1 (upper panels) and 1E 1740.7–2942 (lower panels). The data were averaged over 1 to 20 days of consecutive observations (left panels) and according to the source intensity (right panels). From Kuznetsov et al., 1997, see also Ballet et al., 1996.

on the hours–days time scale was detected – by a factor of ~ 1.5.

The light curve of 1E 1740 recorded by SIGMA (Fig7) shows a qualitatively similar pattern (Cordier et al., 1993; Churazov et al., 1993). The 40-150 keV flux from 1E 1740 changed by a factor of ~ 10 on the timescale of $\sim 1/2$ year with the minimal flux corresponding to the extended minimum observed during 1991 (Churazov et al., 1993). Variability by a factor of 1.5 on a days-weeks time scale was detected during most of the observational sets.

The SIGMA observations provided on one hand rather sparse time coverage – especially for Cyg X-1, and, on the other, a limited time resolution restricted by the instrument time resolution (several hours for spectral information) and, especially for 1E 1740, by the accuracy of the spectral and variability parameters estimation. The latter leads to the necessity of further grouping of the data. In order to verify possible effects of the data averaging two grouping methods were applied to the data and the results are shown in Fig.8 and 9 (see Ballet et al., 1996, Kuznetsov et al., 1997 for the details).

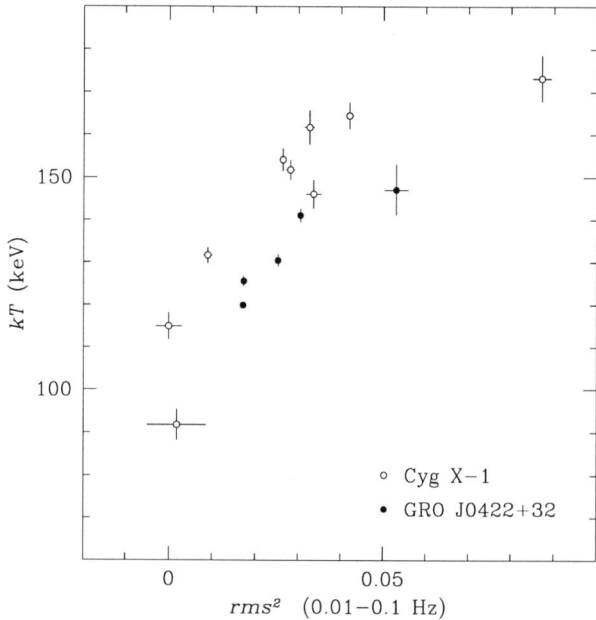

Figure 9. The best-fit bremsstrahlung temperature plotted against the rms^2 of the short-term flux variations in the 0.01–0.1 Hz frequency range. Open circles correspond to Cygnus X-1, filled circles – to X-ray Nova Persei 1992 (GRO J0422+32). From Kuznetsov et al., 1997.

Although no strict point-to-point correlations were detected certain general tendencies are evident. For both sources an approximate correlation between kT and L_X exists. At low hard X-ray luminosity – below $\sim 10^{37}$ erg/sec – kT increases with L_X. At higher luminosity the spectral hardness depends weaker or does not depend at all on the hard X-ray luminosity. On the other hand for Cyg X–1 the spectral hardness is in general positively correlated with the relative amplitude of short-term variability. The low luminosity end of these approximate correlations (low kT and low rms) corresponds to extended episodes of very low hard X-ray flux which occurred during SIGMA observations.

5.2. THE BLACK HOLE X–RAY NOVAE

The X-ray Novae form distinct and being intensively studied class of objects (see Tanaka & Shibazaki, 1996 for review). Typically they are X-ray binary systems composed of a low mass normal star and most likely a black hole. The presence of a black hole is dynamically proven in many cases – the X-ray Novae form most numerous so far class of objects known to har-

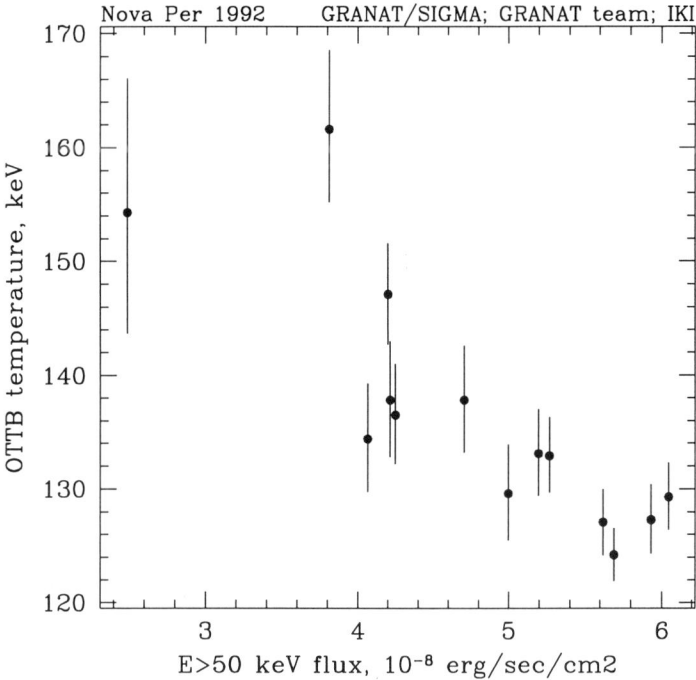

Figure 10. The relation between spectral hardness and high energy flux for Nova Per 1992 (GRO J0422+32) (SIGMA data, from Vikhlinin et al., 1995).

bor stellar mass black holes (Cowley, 1992, Tanaka & Shibazaki, 1996 and references therein).

Phenomenologically, from the viewpoint of the spectral evolution the X-ray Novae might be divided in to two subgroups according to the spectral state (low vs. high/superhigh) at the maximum of the X-ray light curve. The likely reason for that difference is the peak value of the dimensionless mass accretion rate $\dot{m} = \dot{M}/\dot{M}_{Crit}$. The X-ray Novae supposedly having higher $\dot{m}$ (e.g. Nova Vul 1989, Nova Mus 1991) pass through the entire range of the spectral states during their evolution, undergoing the spectral transitions similar to that observed for Cyg X-1 and GX339-4. The less luminous (referred below as "low $\dot{m}$") X-ray Novae (e.g. Nova Per 1992, Nova Oph 1993) never show soft spectral component and apparently are in the low/hard spectral state during entire outburst. The spectral variability of the latter is considered below.

An example of spectral evolution (X-ray Nova Per 1992) is shown in Fig.10 – the decrease of hard X-ray flux was accompanied with hardening of the spectrum (Vikhlinin et al., 1995). Similar relation between spectral hardness and luminosity was observed for X-ray Nova Oph 1993 (e.g. Fig.2, see Revnivtsev et al., 1998a for the details). Such behavior is apparently opposite to that observed for Cyg X-1 and 1E1740.7—2942 - cf Fig.8 and 10.

5.3. SPECTRAL VARIABILITY OF THE HARD SPECTRAL COMPONENT.

Study of the broad band spectra using the data of MIR-KVANT/TTM and ASCA observations close in time to the GRANAT/SIGMA observations indicates that change of the hard X-ray flux $\gtrsim 40$ keV traces change of the overall luminosity of the hard Comptonized spectral component (but not necessarily of the total X-ray luminosity). Therefore (Fig.8 and 10), the hardness of the Comptonized radiation and it's luminosity are *positively correlated* in the case of Cyg X-1 and *anti correlated* in the case of (at least some of) low $\dot{m}$ X-ray Novae.

It is very likely that in the case of the X-ray Novae the mass accretion rate is decreasing with time after the maximum of the light curve, i.e. change of the hard spectral component luminosity traces change of the mass accretion rate. For Cyg X-1 and 1E1740.7—2942 the change of the mass accretion rate is unclear and can't be directly determined. The study of the emission from the optically thick part of the disk (see below) indicates in the model dependent way that in the case of Cyg X-1 the low hard X-ray flux episode corresponded to increase of the mass accretion rate.

Thus, we may *tentatively* conclude that in the low spectral state of both low $\dot{m}$ X-ray Novae and Cyg X-1 the *increase* of the mass accretion rate leads to the *softening* of the spectrum of the Comptonized radiation. On the other hand relation between the mass accretion rate and the luminosity of Comptonized radiation is less unambiguous: the increase of the mass accretion rate might result in either increase of the Comptonized radiation luminosity (e.g. X-ray Nova Per) or it's decrease (Cyg X-1).

Regardless of the luminosity and the mass accretion rate change, the relation between the spectral hardness and the level of the short term aperiodic variability are qualitatively the same for Cyg X-1 and X-ray Nova Per (Fig.9). Similar behavior was recently found for GX339-4 (Trudolyubov et al., 1997).

6. Summary

Possible dependence of spectral characteristics of the soft and hard component upon the mass accretion rate is schematically shown in Fig.11. The range of $\dot{M}$ considered in Fig.11 covers low through high spectral states and doesn't include superhigh state supposedly corresponding to higher values of $\dot{M}$. These curves include a large degree of interpolation of the experi-

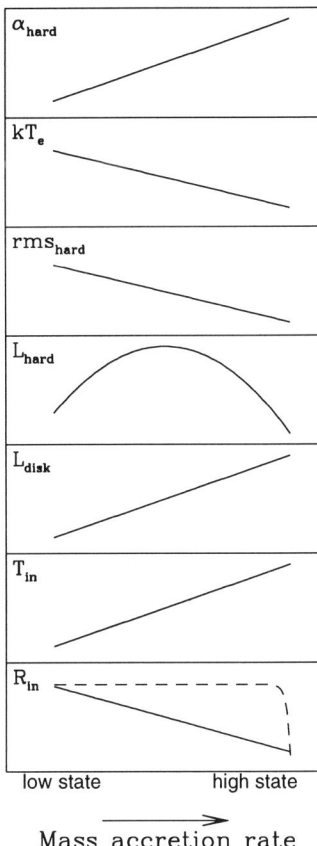

Figure 11. Possible dependence of parameters of the hard and soft spectral components on the mass accretion rate $\dot{m}$ (low through high spectral states; the very high spectral state corresponding to higher $\dot{m}$ is beyond considered range of the mass accretion rates).

mental data and therefore should be treated with certain caution. A fair amount of new broad band observations is needed to prove and detailize these dependences.

One of unclear points is dependence of radius of the inner boundary of the optically thick part of the accretion disk on the mass accretion rate. One possibility is that increase of the $\dot{M}$ results in smooth continuous decrease of R_{in}. The transition from the low state to the high state is consequently smooth in this case. On the other hand sudden increase of the soft X-ray luminosity observed by XTE/ASM during transition of Cyg X–1 to the soft state (Zhang et al., 1997) might indicate that two distinct accretion regimes exist: the lower $\dot{M}$ regime – the inner boundary is located sufficiently far from the compact object, at $\gtrsim (10-20) R_g$, and higher $\dot{M}$ regime, with the inner boundary being close to the compact object, $\sim 3R_g$. It should be mentioned though, that in any case change of the parameters of the hard spectral component (L_{hard}, α_{hard}, kT_e) is smooth (Zhang et al., 1997).

The opposite dependence of the hardness of the hard spectral component upon it's luminosity, observed for several low $\dot{m}$ X-ray Novae and, on the other hand, for Cyg X-1 and 1E1740.7-2942 could be understood assuming that these objects differ in value of $\dot{m}$. In both cases softening of the Comptonized radiation spectrum and decrease of the level of short term variability are caused by increase of the mass accretion rate, but sign of change of the Comptonized radiation luminosity is opposite at lower and higher values of the mass accretion rate (Fig.11). At low $\dot{m}$ fraction of the gravitational energy dissipated in the optically thick part of the accretion flow is small and change of the hard spectral component luminosity traces change of $\dot{M}$. Opposite to that, at higher $\dot{m}$ increase of $\dot{M}$ leads to increase of the fraction of gravitational energy dissipated in the optically thick part of the accretion flow and to decrease of the hard spectral component luminosity. This kind of non monotonic dependence of the hard component luminosity upon the mass accretion rate was probably observed for X-ray Nova Muscae after the secondary maximum of it's light curve (Paciesas et al., 1993; Ebisawa et al., 1994) and for GX339-4 (Harmon et al., 1994; Trudolyubov et al., 1997).

The consequence of non monotonic dependence of the hard X-ray luminosity upon the mass accretion rate is that at sufficiently small $\dot{M}$ Cyg X-1 and 1E1740.7-2942 should show hardness upon luminosity dependence similar to that observed for low $\dot{m}$ X-ray Novae. Such behavior was possibly observed for Cyg X-1 - the $kT - L_X$ dependence splits into two branches at $L_X \lesssim 10^{37}$ erg/sec (Ballet et al., 1996; Kuznetsov et al., 1996).

Acknowledgment: This research has made use of data obtained through the High Energy Astrophysics Science Archive Research Center Online Service, provided by the NASA/Goddard Space Flight Center.

References

Ballet J. et al., 1996, Makino F., Mitsuda K eds., Frontiers Science Ser. 19, Proc. 2nd ASCA Symp., X-Ray imaging and spectroscopy of cosmic hot Plasma. Universal Acad.Press, Tokyo, p.453
Balucinska-Church M. et al., 1995 Astron. Astrophys., 302, 5
Barret D. & Vedrenne G., 1994, ApJS, 92, 505
Belloni T. et al. 1996, ApJ, 472, 107
Chakrabarti S. & Titarchuk L., 1995, ApJ, 455, 623
Churazov E. et al., 1993, ApJ, 407, 752
Cordier B. et al., 1993, ApJ, 272, 277
Cowley A., 1992, Annu.Rev.Astron.Astrophys., 30, 287
Crary D.J. et al., 1996, ApJ, 462, L71
Della Valle M. et al., 1994, Astron. Astrophys., 290, 803

Dotani T. et al., 1996, IAUC 6415
Ebisawa K. et al., 1994, PASJ, 46, 375
Ford E. et al., 1996, ApJ 469, L37
Gilfanov M. et al., 1991, Pis'ma v Astron.Zhurn. 17, 1059
Gilfanov M. et al., 1993, in: Alpar, M., Kiziloglu, U., Van Paradijs, J., eds., AIP Conf. Proc., The lives of the neutron stars, Kluwer, Dordrecht, NATO ASI Ser., 308, p. 712
Harmon B. A. et al., 1994, ApJ, 425, L17
Holt S. et al., 1976, ApJ, 203, L63
Grebenev S. et al, 1993, A&A Suppl.Ser., 97, 281
Kuznetsov S. et al., 1996, Proc. of Röentgenstrahlung from the Universe. Garching, MPE Report 263, 157
Kuznetsov S. et al., 1997, MNRAS, 292, 651
Langmeier A. et al., 1987, ApJ, 323, 288
Makishima K. et al., 1986, ApJ, 308, 635
Mitsuda K. et al., 1989, PASJ, 41,97
Miyamoto S. et al. 1991, ApJ, 383, 784
Miyamoto S. et al., 1995, ApJ, 442, 13
Narayan R., 1996, ApJ, 462, 136
Ogawara Y. et al., 1982, Nature, 295, 675
Paciesas W. et al., 1993, Alabama Univ. Preprint
Phlips B. et al., ApJ, 1996, 465, 907
Priedhorsky W. et al., ApJ, 1979, 233, 350
Revnivtsev M. et al., 1998a, A&A, 331, 557
Revnivtsev M. et al., 1998b, to be submitted to Astron.Astrophys
Salotti L. et al., 1992, A&A, 253, 245
Shakura N. & Sunyaev R., 1973 Astron.Astrophys, 24, 337
Shakura N. & Sunyaev R., 1976 M.N.R.A.S., 175, 613
Sunyaev R. & Truemper J., 1979, Nature, 279, 506
Sunyaev R. & Titarchuk L., 1980, A&A, 86, 121
Sunyaev R. et al., 1991, A&A, 247, L29
Sunyaev R. et al., 1993, A&A, 280, L1
Tanaka Y., 1989, *in Proceedings of* "23rd ESLAB Symposium", Bologna, Italy, ESA SP-296, editors: J.Hunt & B.Battrick, v.1, p.3
Tanaka Y. & Shibazaki N., 1996, Annu.Rev.Astron.Astrophys., 34, 607
Tananbaum H. et al. 1972, ApJ, 177, L5
Trudolyubov S. et al., 1996, Pisma v Astron. Zhurnal, v.22, p.740
Trudolyubov S. et al., 1998, A&A, 334, 895
van der Klis M., 1995, in "X-ray binaries", Eds: Lewin W., van Paradijs J. & van den Heuvel E., Cambridge Univ. Press, p.252
Vikhlinin A. et al., 1994, ApJ, 424, 395
Vikhlinin A. et al., 1995, ApJ, 441, 779
White N., Nagase F, & Parmar A., 1995, in "X-ray binaries", Eds: Lewin W., van Paradijs J. & van den Heuvel E., Cambridge Univ. Press, p.1
Zhang S. et al., 1997, ApJ, 477, L95

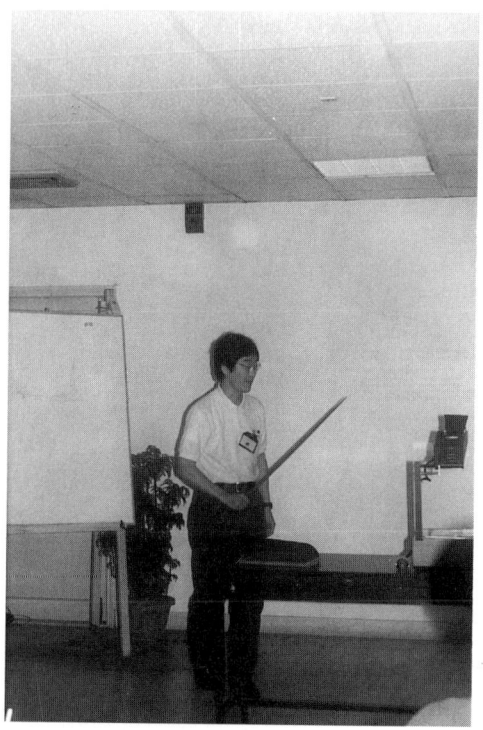

S. Kitamoto

Special Moments: D. Molteni (center) at the
Director's dinner

QUASI-PERIODIC OSCILLATIONS IN THE X-RAY FLUX FROM THE BLACK HOLE CANDIDATES

TADAYASU DOTANI
Institute of Space and Astronautical Science
3-1-1 Yoshino-dai, Sagamihara, Kanagawa 229-8510, Japan

Abstract. Quasi-periodic oscillations (QPOs) in the X-ray flux from the black hole candidates (BHCs) are reviewed. It is found that BHCs take basically two states: one in which the optically thick accretion disk extends close to the black hole, and the other in which the hot corona develops around the black hole to produce hard emission via the Compton up scattering. QPOs tend to become prominent near the transition between these two states.

1. Introduction

X-ray binary sources including black holes are important target in X-ray astronomy. Mass accretion onto a black hole is believed to be the energy source of various high energy phenomena in the universe. However, little is known about how the gravitational potential energy of the accreting matter is converted to X-ray and γ-ray radiation. One of the key characteristics to understand is that the emission mechanism varies with time. Rapid X-ray variabilities (≤ 1 sec) are considered to originate in the vicinity of the black hole, and convey useful information, yet difficult to obtain. In fact, X-ray emission from the black holes is notably variable. Its fractional root mean square (rms) amplitude often reaches 30–40 %. Among the time variations, quasi-periodic oscillations (QPOs), unlike the $1/f$-noise universal to natural phenomena, may afford quite important clues to understanding the emission mechanism of the black hole, because QPOs may be intrinsic to mass accretion.

The term quasi-periodic oscillations is used to indicate a well-defined and resolved peak in the power spectrum (van der Klis, 1989). They are different from coherent oscillations in the sense that they emerge as a broad peak in the power spectra. Relative width of the peak is a measure of the coherence. And a Q-value, defined by the ratio of the centroid frequency to the width, is used to indicate the coherence of the oscillation. Even if the power spectrum has a local maximum, it may be called as band-limited noise (BLN), if the Q-value is small (e.g. ≤ 1).

In this paper, we review the observational results of the QPOs from the black hole candidates (BHCs), *i.e.*, close binary system suspected to contain a black hole (for a review, see Tanaka & Lewin 1995). In §2, we explain the classification of the BHCs and spectral states. Characteristics of the QPOs are summarized for each class of BHCs in §3–5. Section 6 is dedicated to discussion on the nature of the QPOs from the BHCs.

2. Classification of the BHCs and the Spectral States

BHCs may be classified into several classes based on their X-ray characteristics (Tanaka & Lewin, 1995). This classification is important because the time variations, including the QPOs, seem to be different among these classes. Other than the conventional persistent and transient sources, we introduce here another class of BHCs, jet sources. As explained later, jet sources are also transient, but are unique in various aspects and deserve an independent class. In each class of BHCs, the X-ray source moves between several different spectral states, which is defined by the correlated spectral and timing behaviors of the source. The spectral states are analogous to the branches of the Z/atoll sources (van der Klis, 1989). It is also important to distinguish spectral states when we discuss the nature of the QPOs.

Persistent sources are, as indicated by the name, persistently bright BHCs. Only three sources are known to date as persistent BHCs, Cyg X-1, LMC X-1 and LMC X-3. All these sources are massive binaries, *i.e.*, the mass donating star is an early-type star (typically, O–B). Cyg X-1 is considered to be a prototype of the BHCs (Liang & Nolan, 1984). Its characteristics, such as high/low transition, ultra-soft spectrum, flickering, etc., are considered to be diagnostics of BHCs. Two spectral states, soft and hard states, are known in the persistent sources[1] . These two states have been referred to as high and low states in the literature according to the soft X-ray (≤ 10 keV) flux. However, wide-band observations covering from soft X-ray through γ-ray reveal that the bolometric luminosity does not change very much between these two states (Zhang *et al.*, 1997). Thus, the term of soft and hard states is more appropriate. The hard state, in which Cyg X-1 is mostly found, is characterized by a hard, power-law energy spectrum ($\Gamma \sim 1.7$) with rapid and large time variations, refereed to as flickering. The

---

[1]Some of the soft state in Cyg X-1 might be an intermediate state (Belloni *et al.*, 1997).

power spectrum is dominated by the band-limited noise (BLN), which is flat below ∼0.1 Hz and $1/f^\alpha$ ($\alpha \sim 1-2$) in higher frequencies. The soft state is characterized by the dominant soft X-ray emission from the optically thick accretion disk and a steep power law component ($\Gamma \sim 2.5$). Time variation is rather small, and the power spectrum shows a power-law shape. LMC X-1 and LMC X-3 are always found in the soft state.

BHCs in transient are sometimes called X-ray novae (Tanaka & Shibazaki, 1996). They are low-mass systems, *i.e.*, the mass donating star is mostly K or M dwarf star. Typically, they show an outburst once per several decades (some has shorter recurrence period) with a duration of a few months. Besides the hard and soft states, transient BHCs show a very high state (Miyamoto *et al.*, 1991). The very high state is characterized by a large luminosity close to the Eddington limit, simultaneous presence of the soft disk emission and a steep power law component in the energy spectrum, and large and rapid time variabilities. However, it is not clear whether or not the large luminosity is intrinsic to the very high state. GS2023+33 showed a hard, single power law spectrum characteristic to the hard state, even at the peak of the outburst (Tanaka, 1992). If the source luminosity is important to determine the spectral state, we may need to introduce an intermediate state (Méndez & van der Klis, 1997), in which the spectral and timing properties are similar to the very high state, but the source luminosity is lower by more than an order of magnitude. Transient sources spend most of the time in the quiescent state ($L_X \leq 10^{32}$ erg/sec; Tanaka & Shibazaki 1996). Characteristics of the quiescent state are not known due to the limited observational data.

Jet sources include two BHCs, GRS 1915+105 (Nova Aql 1992; Mirabel & Rodriguez 1994) and GRO J1655–40 (Nova Sco 1994; Tingay *et al.* 1995; Hjellming & Rupen 1995). Both sources show radio jets revealing the superluminal motion, which indicate the presence of the relativistic jets close to the speed of light. These sources are also transient sources, but their recurrence period seems to be significantly shorter than the X-ray novae. No consensus has been achieved on the spectral states of the jet sources.

3. Persistent Sources

Typical power spectra of Cyg X-1 in hard and soft states obtained with the ASCA GIS are shown in figure 1. The hard state power spectrum is characterized by the flat-top BLN (brake frequency, ∼0.1 Hz, is variable; Belloni & Hasinger 1990), whereas soft state power spectrum shows a power law shape. As shown later, these are two fundamental shapes of the power spectra in BHCs. QPOs with centroid frequency of 4–12 Hz have been observed from Cyg X-1 during the transition between the hard and the

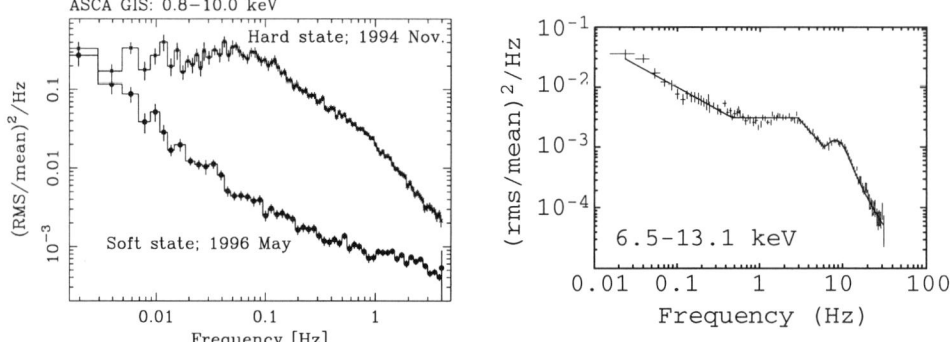

Figure 1. (Left) Representative power spectra of Cyg X-1 in hard and soft states obtained with the ASCA GIS. Power due to the Poisson fluctuation is subtracted. (Right) Example of the power spectra by RXTE PCA showing the QPO at 8 Hz and the accompanied BLN. Adapted from Cui *et al.* (1997).

soft states (Cui *et al.*, 1997). The power spectrum of the QPO obtained by RXTE PCA is also shown in figure 1. As seen in the figure, the QPO accompanies a BLN dominant around a few Hz. The break frequency of the noise and the QPO centroid frequency show a clear positive correlation. Thus the QPO and the BLN may have common origin. The QPO may be connected to the power law component in the energy spectrum, because it becomes more prominent at higher energies.

Low-frequency QPOs have been sometimes observed from persistent sources. QPO in 1–3 Hz was detected from Cyg X-1 in the hard state with Ginga when the spectrum was relatively soft (Rutledge *et al.*, 1998). Fractional rms amplitude of the QPO was about 10 %. Also QPO in ∼0.04–0.07 Hz was observed from Cyg X-1 by the SIGMA telescope in 40–150 keV in the hard state (Vikhlinin *et al.*, 1994). The QPO had a broad peak ($Q \sim 1$) with a rms fractional variation of 10–15 %. QPO of similar frequency (0.075 Hz) was detected from LMC X-1 with Ginga (Ebisawa *et al.*, 1989). The peak was relatively sharp ($Q \sim 8$) and had a significant second harmonic; no BLN was accompanied.

4. Transient Sources

Very prominent QPOs have been observed from the two transient BHCs, GS1124–68 and GX339–4[2] , in the very high state (Dotani, 1989; Miyamoto

---

[2] GX339–4 might be a jet source as indicated by the presence of a jet-like morphology in the radio band (Fender *et al.*, 1997) and a short recurrence time (Rubin *et al.*, 1998).

TABLE 1. Two types of QPOs in the very high state of the transient BHCs

	BLN associated QPO		PL associated QPO	
	GS1124−68	GX339−4	GS1124−68	GX339−4
Centroid Frequency[1]	3.0–7.6 Hz	6.3–7.6 Hz	4.7–6.6 Hz	5.4–6.1 Hz
Fractional rms Amplitude[1]	∼2–4 %	∼2–4 %	∼1–2 %	∼4–5 %
Presence of Harmonics	Sub and 2nd harmonics		2nd harmonics	
Intensity Dependence	$\propto \log(I)$	—[2]	$\propto I^{0.27}$	$(\propto I^{1.5})$[3]
Fractional rms Amp. of BLN	5–12 %	9–13 %	≤1 %	≤2 %

[1]Parameters for the fundamental oscillation in 1–14 keV.
[2]Intensity variation is not large enough to distinguish various models.
[3]Linear model can also fit the data.

et al., 1991; Takizawa et al., 1997). Fractional rms amplitude of the QPO reached 10 % at high energies (>10 keV). The spectral and time variabilities, including the QPOs, were very similar between GS1124–68 and GX339–4. Two patterns of correlated spectral and timing behaviors were identified in both sources. These two branches may be common and unique to BHCs in the very high state, like the spectral branches in the Z-sources.

Characteristics of the two types of QPO observed from GS1124–68 and GX339–4 are summarized in table 1. Figure 2 shows the power spectra calculated from the Ginga data of GS1124–68 (Takizawa et al., 1997). A clear contrast is seen between these two types of QPOs, especially in the amplitude of the BLN, intensity dependence, and the presence of sub-harmonics. We refer these two types of QPOs, respectively, as BLN associated QPO and PL (power law) associated QPO. In spite of these differences, energy dependence of the QPO amplitude and time lags are rather similar between the BLN and PL associated QPOs (Dotani, 1992). Especially, co-existence of soft and hard lag behind the ∼3 keV photons is unique to the QPOs in the very high state.

It was found that the transition between these two branches can be very short (<1 sec) and frequent (a few times per minute). Thus the transition is called flip-flop transition. Flip-flop transition was observed in both GS1124–68 (Takizawa et al., 1997) and GX339–4 (Miyamoto et al., 1991). An example of the light curve and the dynamic power spectrum of the flip-flop transition is shown in figure 2. Correlation between the count rate and the power spectrum (QPO and BLN) is clearly seen in the figure. However, change of the energy spectrum with this flip-flop transition is rather small; a slight softening is seen in the lower intensity level.

Several low-frequency QPOs were also observed from GX339–4. A QPO

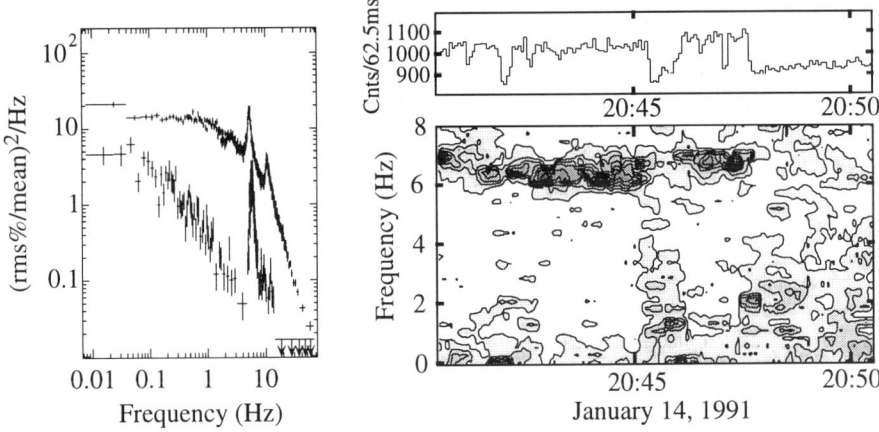

Figure 2. (Left) Two typical types of the power spectra of GS1124–68 in the very high state; both show clear QPOs. (Right) Dynamic power spectrum of GS1124–68 during the flip-flop transitions. Clear contrast of the QPO and BLN is seen in the power spectrum between high and low count rate levels.

of 0.8 Hz was detected by GRANAT when the source was in the hard state (low state) (Grebenev et al., 1991). The QPO had a fractional rms amplitude of 7 %, and accompanied a prominent BLN. Although the observational data are limited, this QPO may be similar to the 1–3 Hz QPO from Cyg X-1 detected by Ginga. Optical QPOs of time scales between 7–190 sec have also been detected from GX339–4 (Motch et al., 1983; Motch et al., 1985; Imamura et al., 1990; Steiman-Cameron et al., 1990).

5. Jet sources

Violent time variations have been detected from GRS 1915+105 on various time scales during the outburst (Greiner et al., 1996). QPO phenomena were also found to be rich in varieties. Chen et al. (1997) investigated the correlation between the hardness-intensity diagram and the power spectra, and identified two spectral branches, hard and soft branches, in GRS 1915+105. In the hard branch, the spectral hardness was anti-correlated with the X-ray intensity, and the power spectra showed QPOs of 0.5–6 Hz accompanied by the BLN (Morgan et al., 1997); second harmonics of the QPO were also prominent. Centroid frequency of the QPOs increased with X-ray intensity, but saturated around ~6 Hz. QPOs observed by the Indian X-ray Astronomy Experiment at 0.62–0.82 Hz may also be the same type (Paul et al., 1997; Paul et al., 1998b). On the other hand, in the soft branch the spectral hardness was positively correlated with the X-ray in-

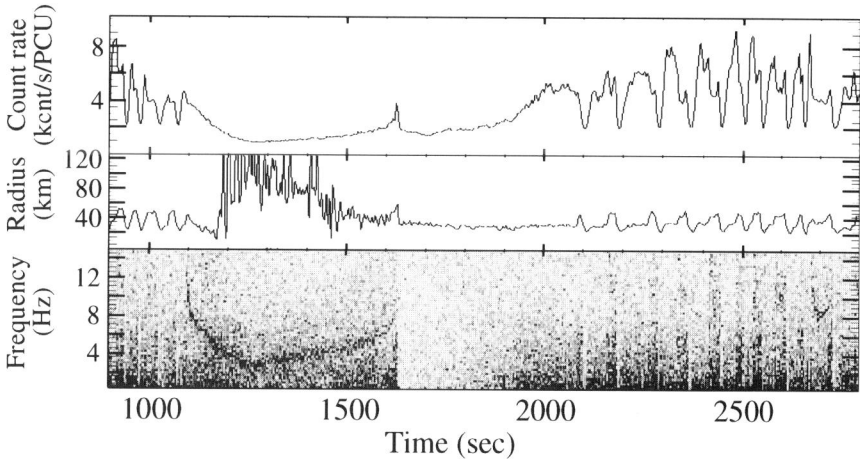

Figure 3. An example of the dynamic power spectrum of GRS 1915+106 during the "sputters". Note the changes of the BLN, QPOs and the disk radius between the high and low flux levels. Adapted from Swank *et al.* (1998).

tensity. The power spectra took approximately a power law shape, but a broad feature was sometimes seen around 1–6 Hz.

GRS 1915+105 stayed in between these two branches when the source showed repeating dip-like variations, called "sputters" (Greiner *et al.*, 1996). Spectroscopic analysis showed that the optically thick inner accretion disk disappeared and a power law spectral component became prominent during the low flux level of "sputters," while the optically thick disk extended inward during the high flux level (Belloni *et al.*, 1997; Taam *et al.*, 1997; Paul *et al.*, 1998a; Swank *et al.*, 1998). Time-resolved energy and power spectra analysis indicate that, when the inner disk disappears and the power law component become dominant, large hardness, intensity-dependent QPOs are observed in 2–14 Hz accompanied by the BLN (Swank *et al.*, 1998). On the other hand, when the optically thick disk extends inward and the power law component becomes weak and steep, corresponding to small hardness, neither the QPO nor the BLN is detected. An example of the dynamic power spectrum is shown in figure 3. These characteristics are in accordance with those of the hard and soft branches. Therefore, sputters are in fact rapid and repeated transitions between the hard and the soft branches.

Diverse and complex QPOs between 1 mHz and 10 Hz with various coherence ($1 < Q < 50$) have been observed in the bright state (Morgan *et al.*, 1997). This bright state, according to their classification, is considered to be in fact the soft branch by Chen *et al.* (1997). Continuum of the power spectra showed moderate strength of the BLN. Waveform of the

QPO was investigated in the light curve for those with a high Q-value (0.067 Hz QPO), and the mean "QPO-folded" profile was calculated. The profile contains a dip-like feature, whose repetition produced the QPO. The dip depth, $\sim$20–40 %, increases with the energy, thus the energy spectrum becomes soft at the bottom of the dip. A narrow QPO ($Q \sim 20$) centered at 67 Hz was also detected in several occasions in the bright state (Morgan et al., 1997). The centroid frequency was very stable; its variation was only 1 Hz. The fractional rms amplitude increased with energy from 1.5 % ($<$5 keV) to 6 % ($>$13 keV).

Rapid time variabilities of GRO J1655–40 have been investigated by Méndez et al. (1998). Around the peak of the outburst, the energy spectrum was hard, and the power spectrum took a power law shape with relatively small variabilities. During the decay of the outburst, the power spectrum changed to show prominent BLN with a break at $<$1 Hz, and a QPO feature moving from $\sim$6.5 Hz to $\sim$0.8 Hz appeared. The energy spectrum in this period was soft. These changes of the spectral and timing behaviors are interpreted as the transition from the soft state to the hard state.

A high frequency QPO centered at 298 Hz was detected from GRO J1655–40 (Remillard et al., 1997). The QPO feature was broad (width of 120 Hz) and had an amplitude of 0.8 %.

6. Discussion

6.1. CORONA AND DISK STATES

When we compare the correlation patterns of the timing (including the QPOs) and spectral behaviors of all the three classes of BHCs, we notice the presence of two basic states. In one state, power spectrum is dominated by the BLN with intensity dependent QPOs, and the energy spectrum has a relatively large contribution from the power law component resulting from the Compton up-scattering of soft photons by the hot corona. In the other state, the power spectrum has a power law shape with QPOs whose intensity dependence is small, and the energy spectrum is dominated by the emission from the optically thick accretion disk. From the analogy of GRS 1915+105 results (Belloni et al., 1997; Swank et al., 1998), we call the former a corona state and the latter a disk state. The key parameter which defines the corona/disk state seems to be the presence of the optically thick inner disk.

We reorganize the various spectral states and branches under the light of the corona/disk states in figure 4. The very high state (and the intermediate state) of the transient sources is considered to fall in between the corona and the disk states (Rutledge et al., 1998). Flip-flop transitions in GS1124–68 and GX339–4, "sputters" and the low-frequency QPO ($\sim$0.1 Hz) in GRS

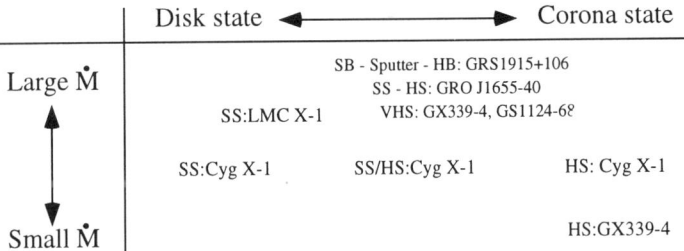

Figure 4. Spectral states of BHCs which showed QPOs are indicated in the scatter diagram between the corona/disk states and the mass accretion rate. Abbreviation in the figures are HS (hard state), SS (soft state), VHS (very high state), HB (hard branch) and SB (soft branch).

1915+105 may in fact be rapid and repeated transitions between the two states. As indicated in the figure, corona/disk states can not be determined only by the mass accretion rate. Some other parameters, such as abundance of the accreting matter, angular momentum of the black hole, etc. may affect the determination of the state. QPOs seem to be most prominent in between the corona and the disk states (Rutledge *et al.*, 1998). This means that QPOs are more easily excited when the inner disk and/or the hot corona becomes unstable ready for the transition.

6.2. COMPARISON WITH Z-SOURCES

There may be similarities between the rapid time variabilities of the BHCs and the Z-sources, bright neutron star binaries with low magnetic field (van der Klis, 1995). From the resemblance of BLN and QPO properties, it is suspected that the corona state of BHCs would correspond to the horizontal branch of Z-sources, and the disk state to the normal and flaring branches, respectively. However, the change in the energy spectrum of Z-sources along the branches is very subtle (Asai *et al.*, 1994; Hasinger *et al.*, 1990), and the drastic change like the one associated to the hard/soft transition of Cyg X-1 has never been observed. Copious X-ray photons from the neutron star surface may work to suppress instabilities of the accretion disk and the development of the hot corona, which may lead to the small change in the energy spectrum. This may also explain relatively small-time variabilities of Z-sources compared to those of the BHCs.

Acknowledgment

The author wish to thank Dr. K. Ebisawa for providing figure 1, and Mr. P. Hilton for the careful reading of the manuscript and useful comments.

References

Asai, K., Dotani, T., Mitsuda, K., et al. (1994), *Publ. Astron. Soc. Japan*, **46**, 479.
Belloni, T., & Hasinger, G. (1990), *Astron. Astrophys.* **227**, L33.
Belloni, T., Mendez, M., King, A. R., et al. (1997), *Astrophys. J.*, **488**, L109.
Chen, X., Swank, J. H., & Taam, R. E. (1997), *Astrophys. J.* **477**, L41.
Cui, W., Zhang, S. N., Force, et al. (1997), *Astrophys. J.* **484**, 383.
Dotani, T. (1989), *PhD Thesis*, University of Tokyo.
Dotani, T. (1992), in *Frontiers of X-ray Astronomy*, ed. Y. Tanaka, (University Academy Press: Tokyo), p.151.
Ebisawa, K., Mitsuda, K., & Inoue, H. (1989), *Publ. Astron. Soc. Japan* **41**, 519.
Fender, R. P., Spencer R. E., Newell, S. J., et al. (1997), *Mon. Not. R. Astron. Soc.* **286**, L29.
Grebenev, S. A., Syunyaev, R. A., Pavlinskii, M. N., et al. (1991), *Sov. Astron. Lett.* **17**, 413.
Greiner, J., Morgan, E. H., & Remillard, R. A. (1996), *Astrophys. J.* **473**, L107.
Hasinger, G., van der Klis, M., Ebisawa, K., et al. (1990), *Astron. Astrophys.* **235**, 131.
Hjellming, R. M., Rupen, M. P. (1995), *Nature*, **375**, 464.
Imamura, J. N., Kristian, J., Middleditch, J., et al. (1990), *Astrophys. J.* **365**, 312.
Liang, E. P., and Nolan, P. L. (1984), *Space Sci. Rev.* **38**, 353.
Méndez, M., & van der Klis, M., (1997), *Astrophys. J.* **479**, 926.
Méndez, M., Belloni, T., & van der Klis, M. (1998), *Astrophys. J. Lett.* in press.
Mirabel, I. F., & Rodriguez, L. F. (1994), *Nature* **371**, 46.
Miyamoto, S., Kimura, K., Kitamoto, S., et al. (1991), *Astrophys. J.* **383**, 784
Morgan, E. H., Remillard, R. A., & Greiner, J. (1997), *Astrophys. J.* **482**, 993.
Motch, C., Ricketts, M. J., Page, C. G., et al. (1983), *Astron. Astrophys.* **119**, 171.
Motch, C., Ilovaisky, S. A., Chevalier, C., et al. (1985), *Space Sci. Rev.* **40**, 219.
Paul, B., Agrawal, P. C., Rao, A. R., et al. (1997), *Astron. Astrophys*, **320**, L37.
Paul, B., Agrawal, P. C., Rao, A. R., et al. (1998a), *Astrophys. J.* **492**, L63.
Paul, B., Agrawal, P. C., Rao, A. R., et al. (1998b), *Astron. Astrophys. Suppl.*, **128**, 145.
Remillard, R. A., Morgan, E. H., McClintock, et al (1997), in Proc. 18th Texas Symp. on Relativistic Astrophyscs, ed. A. Olinto, J. Frieman, & D. Schramm (Chicago: Univ. Chicago), in press.
Rubin, B. C., Harmon, B. A., Paciesas, W. S., et al. (1998), *Astrophys. J.* **492**, L67.
Rutledge, R. E., Lewin, W. H. G., van der Klis, M., et al. (1998), to be submitted to *Astrophys. J.*
Steiman-Cameron, T., Imamura, J., Middleditch, J., et al. (1990), *Astrophys. J.* **359** 197.
Swank, J., Chen, X., Markward, C., et al. (1998), To be published in the proceedings of the conference "Accretion processes in the Astrophysics: Some Like it Hot", eds. S. Holt and T. Kallman, (astro-ph/9801220).
Taam, R. E., Chen, X., Swank, J. (1997), *Astrophys. J.* **485**, L83.
Takizawa, M., Dotani, T., Mitsuda, K., et al. (1997), *Astrophys. J.* **489**, 272.
Tanaka, Y. (1992), in *Ginga memorial symposium*, (ISAS: Tokyo), p.19.
Tanaka, Y., & Lewin, W. H. G. (1995) in *X-ray Binaries*, eds. W. H. G. Lewin, J. van Paradijs, & E. van den Heuvel, (Cambridge Univ. Press), p. 126.
Tanaka, Y., Shibazaki, N. (1996), *Annu. Rev. Astron. Astrophys.* **34**, 607.
Tingay, S. J., Jauncey, D. L., Preston, R. A., et al. (1995), *Nature* **374**, 141.
Van der Klis, M. (1989), *Annu. Rev. Astron. Astrophys.* **27**, 517.
Van der Klis, M. (1995) in *X-ray Binaries*, eds. W. H. G. Lewin, J. van Paradijs, & E. van den Heuvel, (Cambridge Univ. Press), p. 252.
Vikhlini, A., Churazov, E., Gilfanov, M., et al. (1994), *Astrophys. J.* **424**, 395.
Zhang, S. N., Cui, W., Harmon, B. A., et al. (1997), *Astrophys. J.* **477**, L95.

X-RAY PROPERTIES OF GRS 1915+105

B. PAUL

Tata Institute of Fundamental Research
Homi Bhabha Road, Mumbai 400005, India

Abstract.

Observational evidence for the presence of black holes in the universe has remarkably improved in the recent past. Presence of large concentrated mass at the centre of many galaxies and in some of the X-ray binaries has been deduced from dynamical evidences. Observations in the X-ray band also provide strong evidence for the presence of black holes in some X-ray binaries. It is firmly believed that advection effects are very important in the innermost part of the black hole accretion disks and in some of the binary black hole candidates, called soft X-ray transients, evidence for the presence of an event horizon is very strong. The extended power-law type energy spectrum in the hard X-ray band, observed in the black hole candidate sources are thought to be due to upscattering by relativistic bulk motion of material approaching the event horizon and provides additional support to the black hole hypothesis. More direct evidence for the presence of stellar mass black hole has been obtained very recently, from the detection of a unique type of X-ray burst from the Galactic transient X-ray source GRS 1915+105. This object is also called a *micro-quasar* because of the superluminally moving radio jets in this source and the short time scale phenomena observed in this source give us the opportunity to understand the analogous phenomena in quasars at a much larger time scale. We discuss the interesting results obtained from observation and study of the X-ray properties of GRS 1915+105, with various X-ray astronomy instruments leading to very strong evidences for the presence of a stellar mass black hole in this system.

1. Introduction

It is a well accepted postulate that the compact objects in X-ray binary systems whose mass exceed the limit for them to be neutron stars are black hole candidates. There are two types of such objects, one with a companion star of high mass and these are persistently bright sources (e.g. Cygnus X-1, LMC X-3), the other type is the soft X-ray transient sources with a low mass companion (e.g. A 0600-00, GRO J1655-40). About 15 such objects have been detected in our Galaxy and in the neighbouring galaxy LMC (Tanaka and Lewin 1995). Even very conservative calculations indicate the presence of more than 10^7 stellar mass black holes in the galaxy (van den Heuvel 1998). However, the black hole objects are doubly handicapped compared to the neutron stars as far as their detectability is concerned. Unlike the garden variety of radio pulsars (newly born neutron stars), isolated black hole are not observable. And unlike the recycled neutron stars in binary systems (millisecond pulsars) even the black hole sources which are in close binaries are undetectable when the companion star has evolved to form another compact object. Only those which are in close binary systems and are X-ray bright, fuelled by accretion of gas from the companion, are capable of revealing their existence. Though the most compelling evidence for the existence of a black hole in Galactic X-ray binaries normally comes from the measured mass function, in the absence of measured binary parameters phenomenological arguments are normally used, which, though compelling for a class of objects, are not conclusive enough for individual cases. This is mainly due to the fact that the accretion disk around a black hole has properties quite similar to that around a low magnetic field neutron star.

Evidence for the presence of supermassive compact objects is very strong at the center of many galaxies (see the article by L. C. Ho in this book for a review). Spatial distribution and velocity of stars in the centre, or rotation of a disk of gas around the centre in some nearby galaxies indicate the presence of a central dark object of mass in excess of 10^7 $M_\odot$. A very compelling evidence for a supermassive black hole in the centre of NGC 4258 has been obtained from the very accurate measurement of gas motion by tracing the water maser lines (Miyoshi et al. 1995). Strong evidence for the presence of a large concentrated mass in the centre of the Milky Way has been obtained from accurate measurement of proper motion of stars near the centre (Eckart and Genzel 1996). Other alternative explanations for these observations like a stable and long lived star cluster with a huge mass is very unlikely to be the case.

X-ray observation of AGNs however reveals processes going on near the very core of these objects, and the broad asymmetric emission line observed with ASCA indicate gravitational redshift and the presence of a relativistic disk (Tanaka et al. 1995). In stellar black hole candidates, precession of the accretion disk can produce quasi-periodic oscillations, the frequency of which will depend on the mass, angular momentum and sense of rotation of the disk. This interpretation of QPOs in black hole sources indicate the presence of a relativistic disk in these sources (Cui et al. 1998). In spite of all these circumstantial evidences of black holes, both stellar mass and supermassive, definite proof that black holes exist require confirmation of the existence of the event horizon.

The defining characteristic of a black hole is that it possesses an event horizon through which matter and energy can be advected but nothing can come out. Recent progress in the understanding of accretion onto black holes, has indicated a possible way of uniquely separating the properties of accretion onto black holes and neutron stars. It is found that the black hole accretion disks are cooled by advection in their innermost parts and it has been realized that advection is one of the most fundamental features of the black hole accretion (Chakrabarti 1996; Narayan et al. 1997). We discuss here a possible evidence for the direct detection of advection in a transient X-ray source GRS 1915+105.

2. Discovery and early X-ray observations of GRS 1915+105

The X-ray transient source GRS 1915+105 was discovered in 1992 using the WATCH (Wide Angle Telescope for Cosmic Hard X-rays) experiment onboard the GRANAT satellite (Castro-Tirado et al. 1994). Hard X-ray studies in 20-100 keV band have shown erratic intensity variations on time scales of a few hours to months (Sazonov et al. 1994; Foster et al. 1996). In two years of hard X-ray observations by WATCH, two powerful bursts were discovered during which the source luminosity reached as high as 10^{39} erg s^{-1}. The hard X-ray spectrum was found to fit well with a power-law of photon index $\alpha = -2.5$. Superluminal motions of two symmetric radio emitting jets of GRS 1915+105 were discovered by Mirabel & Rodriguez (1994).

A probable optical counterpart of the source has been observed only in the I band at 23.4 magnitude (Boeer et al. 1996). The optical faintness is probably due to very high extinction at a distance of 12.5 kpc. Emission lines of H and He in the infrared band were found to be narrow and no Doppler shift was observed (Castro-Tirado et al. 1996). High resolution spectral observations in the near-infrared have revealed similarity with the spectra of very high mass stars of Oe or Be spectral type. The relative strength of the emission lines observed in K band spectrum of this source, and slope of the continuum are found to be identical to those found in the spectra of late Oe or early Be type stars (Mirabel et al. 1997).

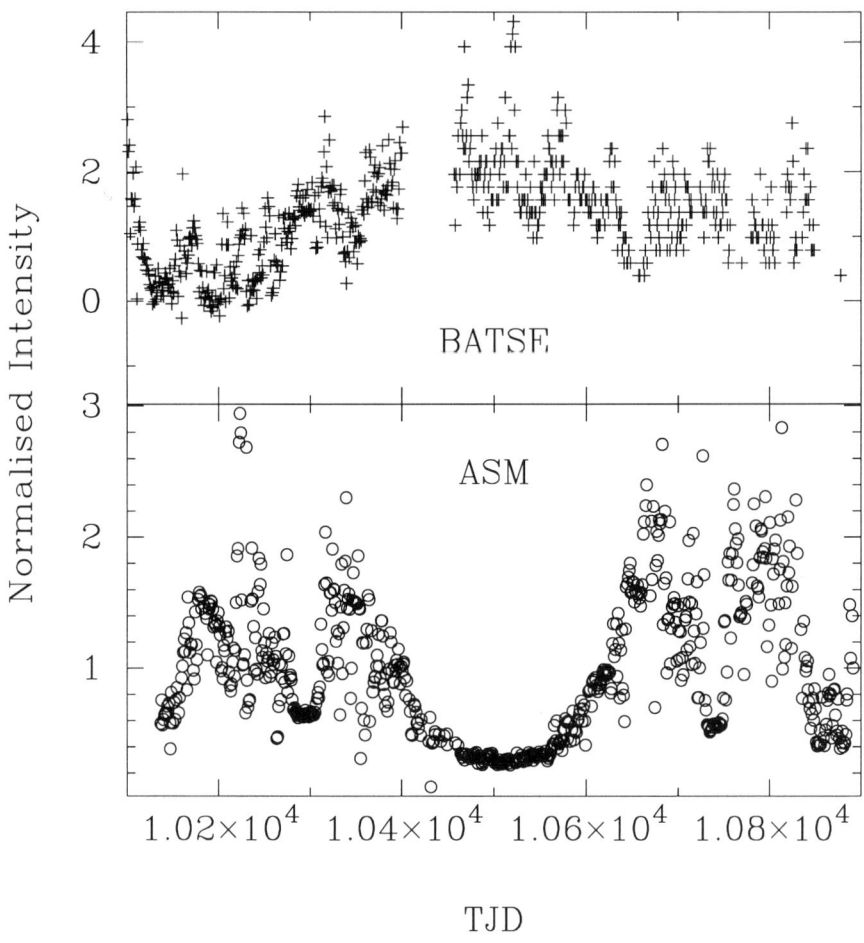

Figure 1. The normalised BATSE and ASM light curves of GRS 1915+105 for 800 days.

3. Long time variability

The X-ray intensity of GRS 1915+105 was found to vary on a variety of time scales and the spectrum also changed during the brightness variations which are believed to be due to major disk instabilities. The All Sky Monitor (ASM) observations in the 2–10 keV energy band and the Burst and Transient Source Experiment (BATSE) occultation observation in the 20–100 keV band for a long time as shown in Figure 1, clearly identifies many different intensity states of this source. Sometimes the intensity varies erratically around a mean intensity level. The erratic intensity variations are also found to have varying nature. During the high intensity states, flares have been observed during which the X-ray intensity is found to increase by a factor as large as 6 with time scale of a few days. The source also has shown a few relatively quiet state of duration between 10 to 150 days during which the spectrum is harder. The two light curves in Figure 1, made from the archival data of the ASM and the BATSE detectors show no clear correlation or anti-correlation between the intensity in the soft and hard X-ray bands.

4. Recent X-ray observations

GRS 1915+105 was observed on several occasions by the All Sky Monitor (ASM) and the Proportional Counter Array (PCA) of the Rossi X-ray Timing Explorer (RXTE). PCA observations in its bright and flaring state in 1996 showed drastic variations in its intensity on a variety of time scales ranging from sub-second to days and were suggested to be due to major disk instabilities (Greiner et al. 1996). We mention here some of the important results obtained from the observation of this source with the BATSE, ASM and the PCA detectors.

4.1. CORRELATED HARD X-RAY AND RADIO EMISSIONS

Correlated enhanced radio and hard X-ray emissions were discovered from the source with a near simultaneous monitoring over a long period (Foster et al. 1996). In a previous outburst during 1993 December to 1994 April decreases or dips in the hard X-ray flux were observed during the radio flares. This observation suggested an interaction between the hard X-ray emission and the jet production. Redirection of the accreted material onto jets can cause the observed X-ray and radio intensity fluctuations (Harmon et al. 1997).

4.2. QUASI-PERIODIC OSCILLATIONS

Intensity dependent narrow QPOs were detected with the PCA. Strong harmonics were also seen at the lower frequencies which vanish with increase in the intensity and QPO frequency. When the source is in the hard branch of the colour-intensity diagram, the QPO features are narrow and the low frequency part of the PDS is flat. In the soft branch, the QPO feature is broad or absent and the low frequency part of the PDS has a power law form with index in the range -1.25 to -1.05 (Chen et al. 1997). QPO frequency is found to vary from time to time, sometimes two different frequency QPOs are superposed. The lowest frequency observed is 0.0016 Hz and the highest is 67 Hz. Multiple harmonics of the narrow QPO features are common. QPO folded profiles are seen to be nearly sinusoidal. At higher energy the QPO amplitude is large and shows a lag of about 0.03 in phase. Also the profiles are more complicated and deviates more from a sinusoidal at higher energy. The arrival time of the QPO pulses show a random variation rather than any systematic trend. The pulse profile analysis of the QPOs produced similar results over a large range of frequency, 0.067 to 1.8 Hz (Morgan et al. 1997).

4.3. THE 67 HZ QPO AND MASS OF THE BLACK HOLE

In some of the observations with the PCA, QPOs at a frequency of 67 Hz are observed. The feature is very sharp and the variation in the frequency from observation to observation is only 1 Hz. The rms amplitude of the oscillations is 0.5–1.6% and is more at higher photon energy. Associating these stable QPOs of 67 Hz to the Keplerian period of the innermost stable orbit around a black hole, the mass for a non-rotating black hole has been estimated to be 33 $M_\odot$ (Morgan et al. 1997). Another possible explanation is the frequency of the lowest radial g-mode oscillation of the disk, and this interpretation leads to a mass of 10 $M_\odot$ for a non-rotating black hole at the center (Nowak et al. 1997).

4.4. AN UNSTABLE CENTRAL DISK MODEL FOR THE BURSTS

Belloni et al. (1997a) have explained the complicated X-ray light-curve of GRS 1915+105 as due to rapid instabilities in the inner part of the accretion disk. Thermal-viscous instabilities in the disk can cause sudden removal of matter which is replenished by mass accretion from the companion. They have described the rise time of a burst as the propagation time of heat waves generated by sudden in-fall of matter (2 sec), the decay time is the matter in-fall time (0.5 sec) and the recurrence time is viscous time scale on which matter is replenished. Spectral changes observed in the source are also in agreement with the above explanation for the intensity sputterings. In this model only a small fraction of the accretion energy is converted into X-rays, the rest of it being either advected onto the black hole or used in driving the ejection of matter. Recently, Belloni et al. (1997b) have shown that all spectral changes during the different types of bursts in GRS 1915+105 can be attributed to rapid disappearing of the inner part of the accretion disk followed by refilling from the outer disk. Duration of events and inner radius of the emptied disk are found to match well with radius dependence of the viscous time scale.

The multiple component structure of the bursts in GRS 1915+105 are studied by Taam et al. (1997). In their observation, the spectrum is found to become softer during the primary component of the burst and it becomes harder during the successive components of the bursts. They have explained the primary component of the complicated burst structure and spectral evolution of the burst in terms of thermal-viscous instability in the disk. They suggested that an inward shift in the inner edge of the disk causes the observed bursting components following the primary one. However, the accretion disk models which solve the inner boundary conditions and invoke the advection effects explicitly (Chakrabarti 1996; Narayan et al. 1997), find that thermal-viscous instabilities are removed completely by the addition of advection effects.

In the following we describe the observation of rapid intensity variations, QPOs and quasi-regular bursts in GRS 1915+105 with the Indian X-ray Astronomy Experiment (IXAE). The latter were explained in terms of advection effects in an accretion disk around the black hole (Paul et al. 1998a).

5. IXAE observations of GRS 1915+105

The observations were carried out using the 3 PPCs of the Indian X-ray Astronomy Experiment (IXAE) onboard the Indian satellite IRS-P3 launched on 1996 March 21 from India. For details of the PPCs and the observation methodology see Agrawal et al. (1997). Observations with the PPCs are usually made in about 5 orbits of the satellite every day and each observation has a duration of about 20 minutes. Useful source exposure of 8,850 s was obtained during the observation period of 1996 July 20 to July 29, 39,300 s was obtained during 1997 June 12 to June 29 and 19,500 s was obtained during 1997 August 8 to August 12. Data were recorded with a time resolution of 1 s for approximately half of these and observations and with 0.1 s resolution for the rest.

5.1. RESULTS

We describe here briefly the rapid intensity variations and quasi-periodic oscillations discovered in the 1996 July observations (Agrawal et al. 1996) of this source and strong quasi regular bursts of four different types discovered in the 1997 June and August observations (Paul et al. 1998a).

5.1.1. *Rapid intensity variations*
A search was made to find intensity variations in the source largely exceeding the photon counting statistics over shorter time scales of seconds to subseconds in the 1996 data. Each individual time bin was inspected with respect to a running average in the light curve around that bin and intensity variations in time bins exceeding 4σ above the average were identified. Consistency was checked with similar variations in other detectors at the same instant. A few typical light curves showing sub-second intensity vari-

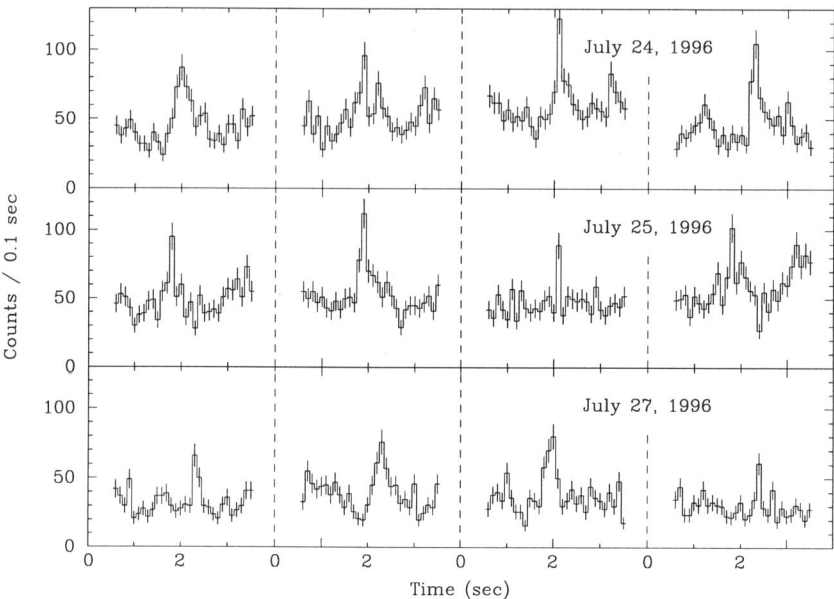

Figure 2. The subsecond flux variations seen in GRS1915+105 with the PPC during July 1996. Each panel shows a few of the flares seen on the days mentioned in the figure. Data of 3 seconds are plotted with a bin size of 100 msec around the flares. A factor of 2 or more increase in the X-ray flux is seen for a duration of about 100 to 400 msec.

ations are shown in Figure 2. These variations were detected independently in each detector with similar count rate profiles. It will be noticed from the light curves that GRS 1915+105 shows frequent flaring activity on time scales of less than a second and occasionally over 0.1 sec. During the flares the intensity varies by a factor of upto 3 in less than a second (Paul et al. 1997). However, no intensity variation is observed at a longer time scale of a minute or more (Paul et al. 1998b).

5.1.2. *Quasi-periodic oscillations*

During the 1996 July observations the source was in a low intensity state. We discovered quasi-periodic oscillations in GRS 1915+105 with the peak frequency varying between 0.62 to 0.82 Hz in the 1996 July observations. The periodogram obtained from the PPC observations is shown in Figure 3. The QPO feature is very narrow (Q = $\frac{\nu}{\delta\nu} \approx 5$) and strong with a rms variability of $\sim 10\%$. The QPOs are clearly detected independently at the same frequency in the data of each PPC as well as in the summed data. The QPO frequency varies in an erratic manner from day to day.

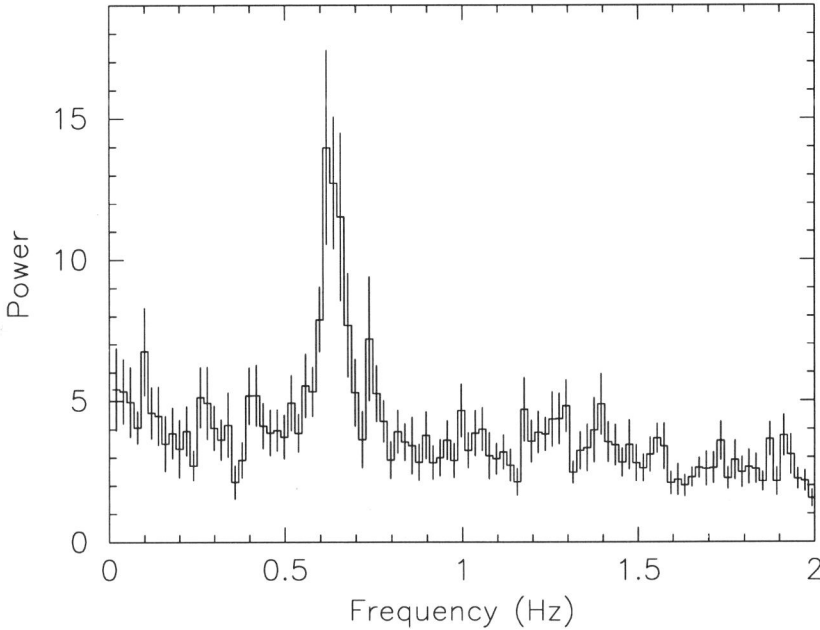

Figure 3. The periodogram of GRS 1915+105 showing QPOs at 0.7 Hz in 1996 July observations.

5.1.3. *Quasi-regular X-ray bursts*

During the 1997 June and August observations the source was in a bright state compared to that in 1996 July. A unique characteristic noted in the light curve of 1997 observations was the presence of strong quasi regular X-ray bursts. The bursts are unambiguously detected in all the PPC observations over the entire period of 1997 June 12−29 and August 8−12. Four types of bursts are observed during the PPC observations - (a) regular bursts, having a slow rise and sharp decay lasting for $\sim$ 10 s and recurring every 45 s, (b) irregular bursts of variable duration, slow rise, flat top and sharp decay, (c) long bursts, with duration of a few tens to a few hundred seconds, followed by sharp decay and (d) short bursts, with about 20 s duration and a double peaked structure. Regular bursts were detected during June 12−17 and again during June 22−26, the irregular bursts during June 18−21, long bursts during June 27−29 and short bursts were detected during August 8−12.

Representative light curves of 500 s duration obtained on different days are shown in Figure 1. A secondary peak near the end of the bursts is a common feature of all the bursts. A total of 635 regular bursts (in 28,200 s of observation), 78 irregular bursts (in 6,200 s), 40 long bursts (in 4,900

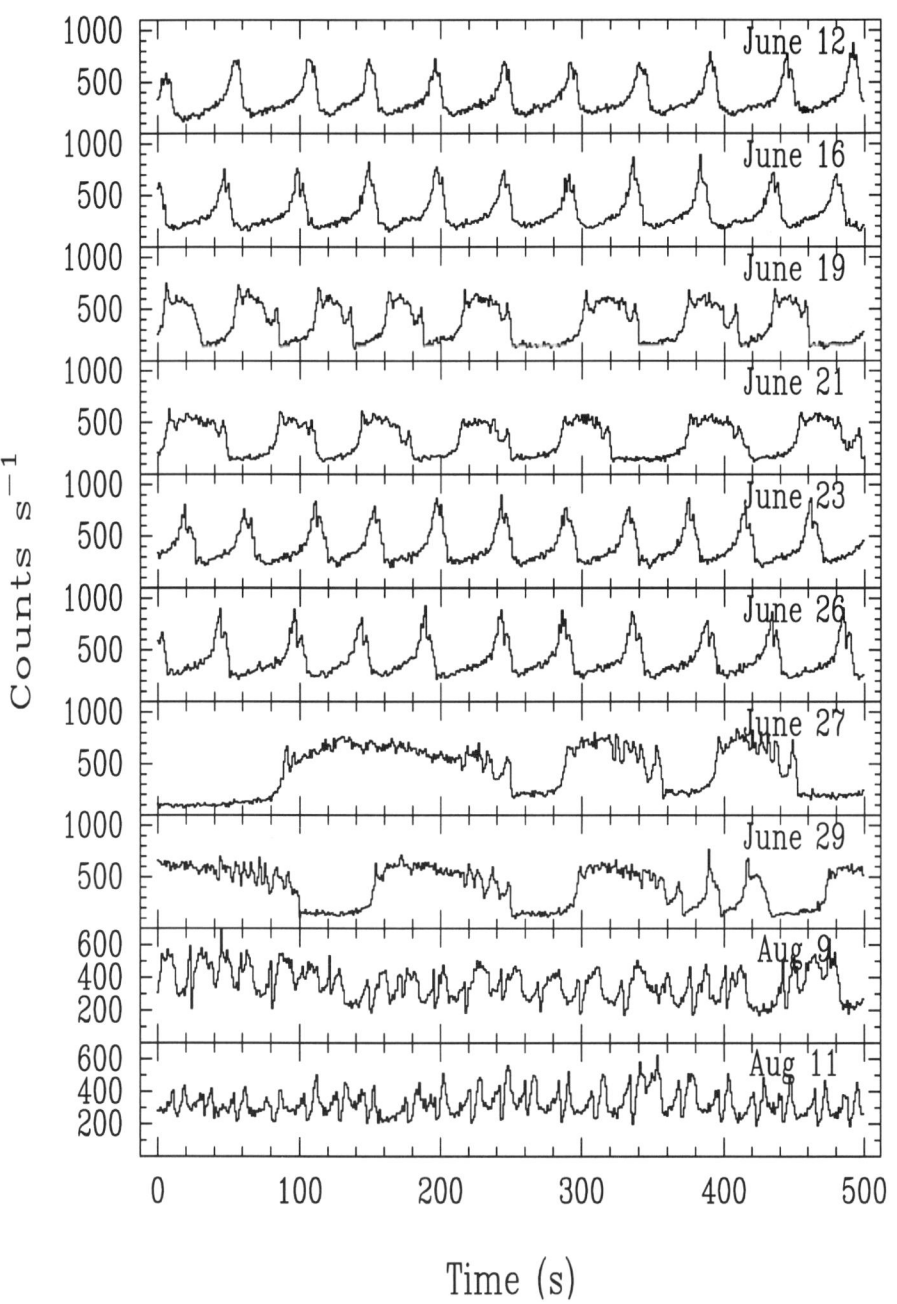

Figure 4. The regular (first, second, fifth and sixth panel from the top), irregular (third and fourth), long (seventh and eighth) and short (ninth and tenth panel) bursts observed in GRS 1915+105 with one of the PPCs. Date of each observation is given in the respective panels.

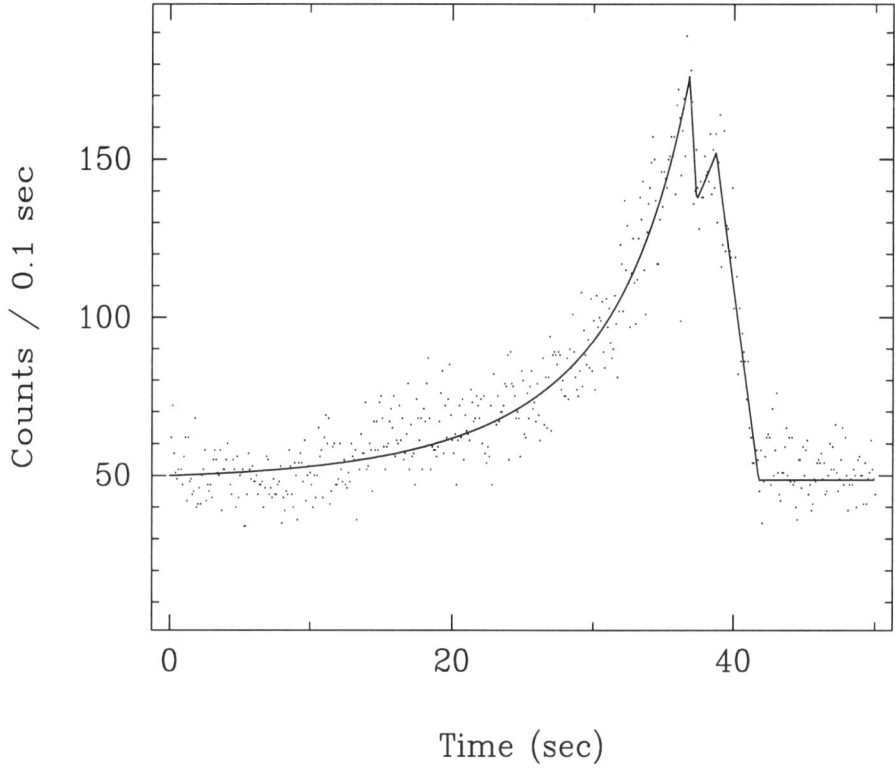

Figure 5. Profile of a regular burst is shown along with the best fit model (continuous line) consisting of two burst with exponential rise and linear decay, and a constant intensity.

s) and about 900 short bursts (in 19,500 s) have been detected. The long bursts also show strong quasi-periodic intensity variations near the end of each burst, which is absent in the beginning. It is also noticed that in case of long bursts there is a correlation between the burst duration and the duration of the quiescent state following the previous burst.

5.1.4. *Regular bursts*
Profile of the regular bursts : The regular bursts are detected in all the observations during June 12–17 and again during June 22–26. The peak intensity of the bursts is about 3–5 times the quiescent emission. In all the bursts, a dip is observed, just before the decay of the burst. But the most remarkable feature of our observations is the persistence of the regular bursts for several days with similar shape, structure and period. The separation between the successive bursts shows a random walk in time instead

of any regular pattern. Time separation between the bursts averaged over one day is found to be in the range of 40 s to 52 s, with a large scatter. The distribution of burst interval for each day fits well with a Gaussian, with a tail on the higher side, having a mean in the range of 40 s to 50 s and $\sigma \sim$ 3 s.

All the regular bursts have slow exponential rise with a time scale of 7–10 s to a primary peak, a dip followed by a secondary peak and a very fast linear decay within 2–3 seconds to the persistence level. A typical burst profile as shown in Figure 5 is well fitted to a sum of two bursts with exponential rise and very fast decay along with a constant emission. The number of photons detected in the regular bursts is about 30–50% of the number of photons detected in the quiescent period.

Spectral changes during the regular bursts : To find the spectral changes during the bursts with improved statistical accuracy, we added a large number of regular bursts by matching the peak of the fitted profiles. The co-added burst profiles in two different energy ranges (2–6 keV and 6–18 keV) are shown in the top two panels while the hardness ratio is shown in the third panel of Figure 6. The sharp features are smeared due to addition of bursts of different duration. Intensity changes are more prominent at higher energy and the energy spectrum becomes harder as the burst progresses. The burst is hardest near the end of its decay. This is a unique feature of these bursts which distinguishes them from the bursts seen in LMXBs which become softer in the decaying phase (Lewin et al. 1995).

We have calculated the possible temperature and radius distribution assuming a multi-temperature disk black-body model for the burst emission with the temperature (kT_{in}) and radius (R_{in}) of the innermost region of the accretion disk as free parameters. For this purpose, the burst count rate and the hardness ratio are obtained after subtracting the quiescent value from the observed count rates. From the response matrix of the PPC detectors, the observed total count rate and hardness ratio profiles are converted to the R_{in} and kT_{in} of the innermost disk. The hardness ratio profile of only the burst component, temperature of the inner part of the disk (kT_{in}) and the inner radius of the disk (R_{in}) are shown in the three panels at the bottom of Figure 6. The data after 30 s are taken as the quiescent level and hence they are not used for the hardness ratio calculation and plotting. It can be seen from the figure that the temperature increases sharply during the burst decay phase. The radius R_{in} remains constant during most of the rising phase of the burst, but decreases sharply during the burst decay.

The hardness-intensity plot of this source shows a hysteresis with the progress of the burst. In Figure 7 the hardness ratio of the bursting component is plotted against the burst intensity. The burst development is

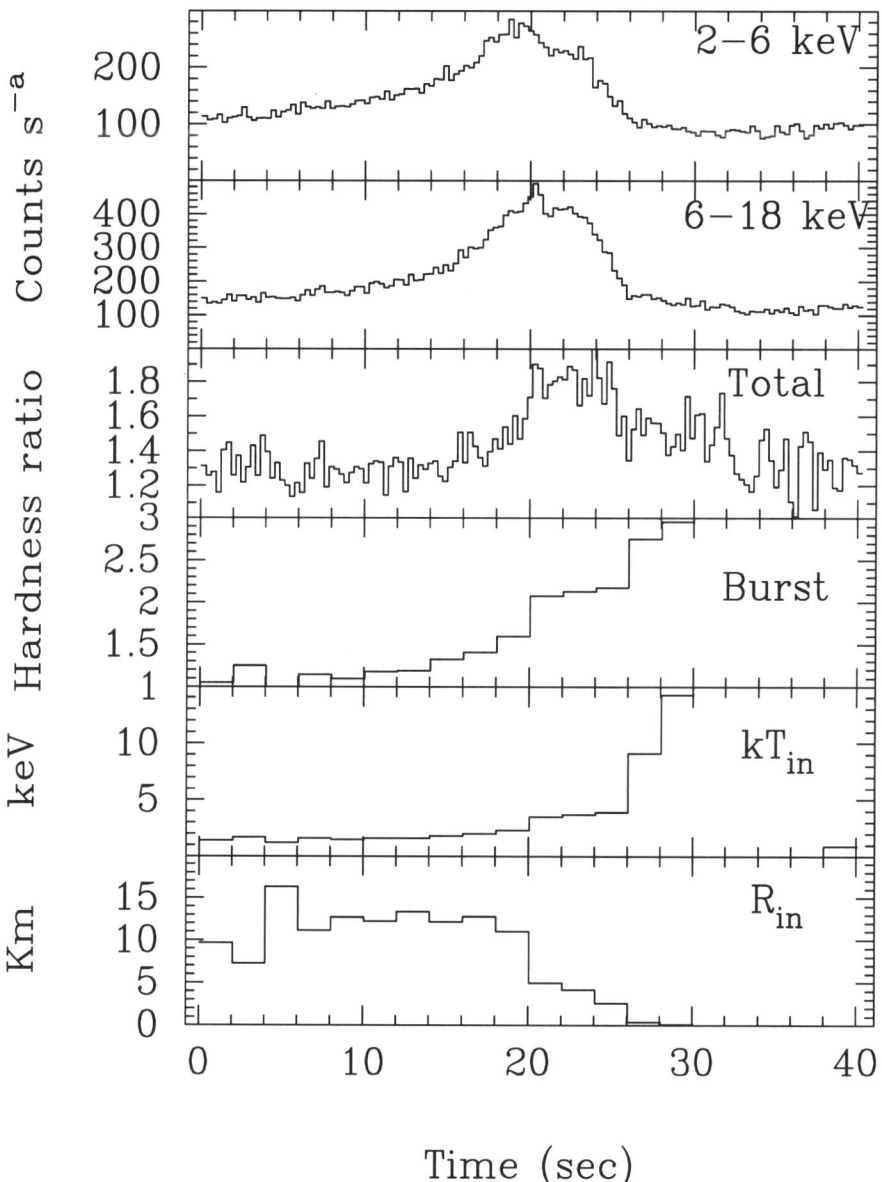

Figure 6. The burst profile in two different energy ranges are shown in the top two panels. The middle two panels show the hardness ratio of the total light-curve and hardness ratio of only the burst. The quiescent intensity (30–40 s time range in the figure) was subtracted from the light curves and the resultant profiles were taken to generate the hardness ratio of only the bursting component. This is why the fourth panel extends only upto 30 s. The temperature (kT_{in}) and radius (R_{in}) of the innermost disk of an assumed multi-temperature disk emission, are plotted in the bottom two panels.

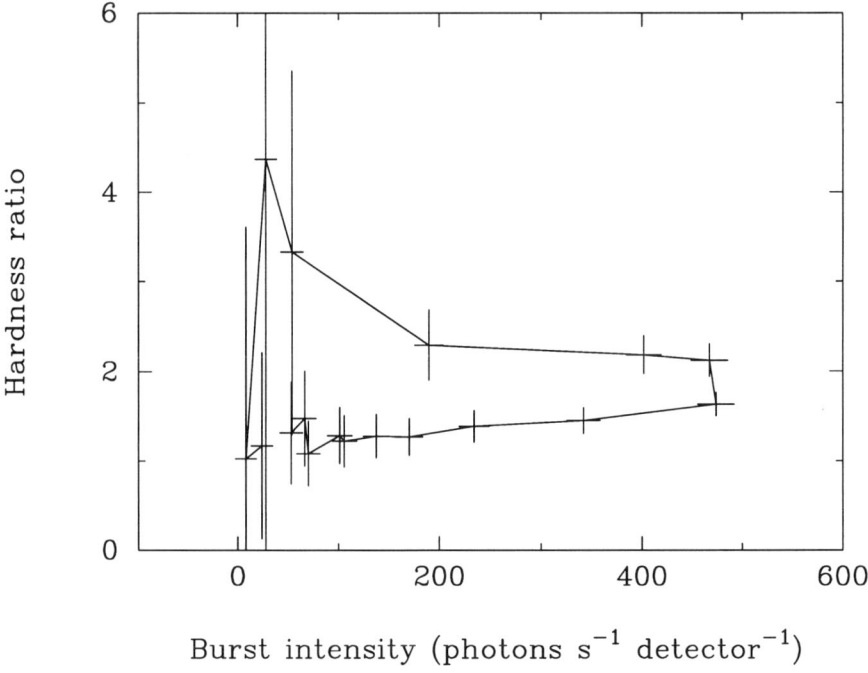

Figure 7. Hysteresis in the hardness ratio and intensity is shown during the regular bursts. The quiescent flux has been subtracted to obtain the hardness ratio of only the burst. The burst propagation is anti-clockwise in the figure.

anti-clockwise in this figure. There is a gradual hardening of the spectrum with the development of the burst and near the peak of the burst the spectrum suddenly becomes very hard. The spectrum remains hard during the burst decay and softens dramatically at the end of the burst.

5.1.5. *The power density spectra in the bursting phase*

Using the light curves of GRS 1915+105 with 1 sec and 0.1 sec bin size, we have generated the power density spectra. The two spectra are superimposed to obtain power in the broad frequency band of 0.001 to 5 Hz. The power spectra were normalized with respect to the photon counting rate and then averaged. The resultant spectra for the four different types of bursts are shown in Figure 8.

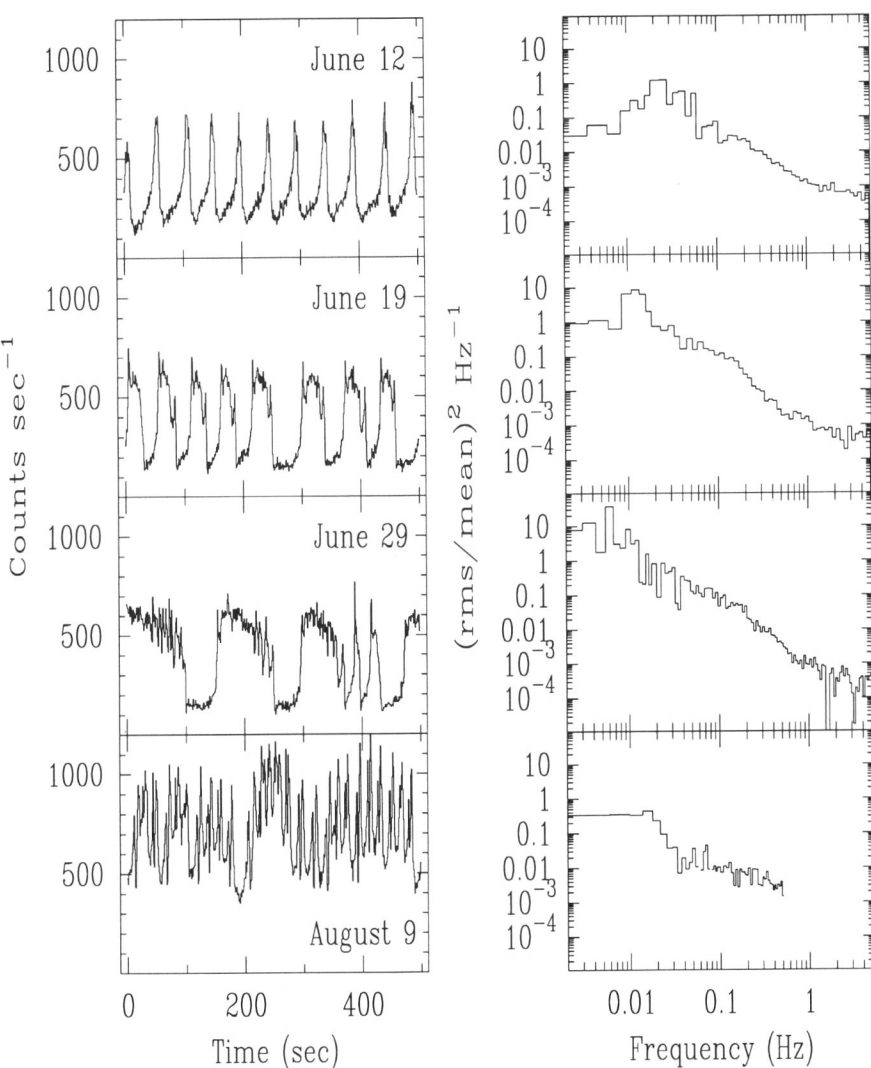

Figure 8. The light curves showing four different types of bursts observed in GRS 1915+105 are plotted in the left column along with the day of observation in 1997. The corresponding power density spectra are shown in the right column.

6. Discussion

The fast intensity variations, low frequency QPOs, hard energy spectrum, super-Eddington luminosity, superluminal radio jets all indicate that the compact object in GRS 1915+105 is a black hole. The bursts observed in

GRS 1915+105 with the IXAE are very different from the type I X-ray bursts seen in about 40 LMXBs and type II X-ray bursts in the Rapid Burster (MXB 1730−335) because of their unique feature of slow rise and fast decay. All the bursts in the LMXBs have fast rise time of less than a second to 10 seconds and slow decay of 10 seconds to a few minutes (Lewin et al. 1995). The two other peculiarities of the bursts in GRS 1915+105 are the regularity with which they occur over time scale of several days, and the presence of a secondary peak in all the bursts. In the classical bursts, the spectrum is initially hard and becomes softer as the burst decays (Lewin et al. 1995). In sharp contrast, the bursts in GRS 1915+105 remain hard till the end and it is, in fact, the hardest near the end of the burst.

The time-averaged luminosity of the regular bursts detected from GRS 1915+105 is 0.3 to 0.5 times the luminosity of the quiescent emission and is much higher than the same ratio in type I bursts (where the thermonuclear process has much smaller efficiency compared to the gravitational process) and less than the same in type II bursts (where the burst emission is due to gravitational energy release). The emission process involved in producing the bursts here is not likely to be thermo-nuclear because of the energetics involved. If the energy generation process is gravitational (like in type II bursts), the difference in efficiency might indicate the absence of hard surface in the compact object. A process in which the energy produced is due to gravitational potential but not all the energy is emitted as radiation, part of it being advected into the event horizon as kinetic energy of the matter, is appropriate for this source.

We have suggested that the regular bursts observed in GRS 1915+105 can be due to periodic infall of matter onto a black hole from an oscillating shock front (Paul et al. 1998a). In the black hole accretion disk model of Chakrabarti and Titarchuk (1995), the disk has two components, an equatorial Keplerian disk and a sub-Keplerian component just above and below the Keplerian disk. The sub-Keplerian component of the disk experiences a shock due to centrifugal barrier and if the cooling time scale of the post-shock halo matches with the material infall time scale, oscillations can set in (Molteni et al. 1996). If the matter accreted from the companion has high angular momentum and low viscosity, the shock front can be far away from the black hole and oscillation period can be comparable to what is observed in GRS 1915+105.

One possibility is that at some particular phase of this oscillation the piled up matter behind the shock falls catastrophically onto the black hole and a burst is produced. As the matter goes in, temperature increases producing large X-ray intensity. The burst is suddenly terminated as the matter goes behind the event horizon of the black hole. In this scenario, as the burst progresses, the temperature of the infalling matter increases, giving rise to the observed spectral hardening.

Acknowledgements

I thank S. K. Chakrabarti and other organisers of the workshop on "Observational Evidence of Black Holes in the Universe" held at the S. N. Bose National Centre for Basic Scinences, Calcutta for their hospitality. My sincere thanks are due to P. C. Agrawal and A. R. Rao for the discussions which helped to improve this review. I acknowledge the effort put in by the members of the IXAE team at TIFR, Mumbai and ISAC, Bangalore for making the observations successful. I thank the ASM team and BATSE Earth occultation team for providing the light curves.

References

Agrawal, P. C., Paul, B., Rao, A. R., et al. 1996, IAU Circ. 6488
Agrawal, P. C., Paul, B., Rao, A. R., et al. 1997, Journal of the Korean Astronomical Society, 29, S429
Belloni, T., Mendez, M., King A. R., van der Klis, M. and van Paradijs, J., 1997a, ApJ 479, L145
Belloni, T., Mendez, M., King A. R., van der Klis, M. and van Paradijs, J., 1997b, ApJ 488, L109
Boeer, M., Greiner, J. and Motch, C., 1996, A & A 305, 835
Castro-Tirado, A. J., Alberto, J., Brandt, S., et al. 1994, ApJS 92, 469
Castro-Tirado, A. J., Geballe, T. R. and Lund, N. 1994, ApJ 461, L99
Chakrabarti, S. K. 1996, ApJ, 464, 664
Chakrabarti, S. K., & Titarchuk, L. G. 1995 ApJ, 455, 623
Chen, X., Swank, J. H. and Taam, R., APJ 477, L41
Cui, W., Zhang, S. N. and Chen W., 1998, ApJ 492, L53
Eckart, A. and Ganzel, R., 1996, Nature 383, 415
Foster, R. S., Waltman E. B., Tavani, M., et al. 1996, ApJ, 467, L81
Greiner, J., Morgan, E. H., & Remillard, R. A. 1996, ApJ, 473, L107
Harmon, B. A., Deal, K. J., Paciesas, W. S., et al., 1997 ApJ 477, L85
Lewin, W. H. G., Jan Van Paradijs, & Taam, R. E. in X-ray Binaries, eds. Lewin., W. H. G., Jan Van Paradijs, & van den Heuvel, Cambridge: Cambridge University Press, p. 175-232, 1995
Mirabel, I. F., & Rodriguez, L. F. 1994, Nature, 371, 46
Mirabel, I. F., Bandyopadhyay, R., Charles, P. A., Shahbaz, T. and Rodriguez. L. F. 1997, ApJ 477, L45
Miyoshi, M., Moran, J., Herrnstein, J., et al. 1995, Nature 373, 127
Molteni, D., Sponholz, H., & Chakrabarti, S. K. 1996, ApJ, 457, 805
Morgan, E. H., Remillard, R. A., & Greiner, J. 1997, ApJ, 482, 993
Narayan, R., Garcia, M. R. and Mcclintock, J. E., 1997, ApJ 478, L79
Nowak, M. A., Wagoner, R. V., Begelman, M. C. and Lehr, D. E., 1997, ApJ 477, L91
Paul, B., Agrawal, P. C, Rao, A. R., et. al. 1997, A & A, 320, L37
Paul, B., Agrawal, P. C., Rao, A. R., et al. 1998a, ApJ 492, L63
Paul, B., Agrawal, P. C., Rao, A. R., et al. 1998b, A & A 128, 145
Sazonov, S. Y., Syunyaev, R. A., Lapshov, I. Yu., et al., 1994, Astron Letters. 20(6), 787
Taam, R. E., Chen, X., & Swank, J. H. 1997, ApJ, 485, L83
Tanaka, Y., & Lewin, W. H. G. in X-ray Binaries, eds. Lewin., W. H. G., Jan Van Paradijs, & van den Heuvel, Cambridge: Cambridge University Press, p. 166, 1995

Tanaka, Y., Nandra, K., Fabian, A. C., et al. 1995, Nature 375, 659
van den Heuvel., E. P. J., in High Energy Astronomy and Astrophysics, eds. Agrawal, P. C. & Vishwanath, P. R., Universities Press, Hyderabad, India, 1998

X-RAY OBSERVATION OF BLACK HOLE NOVAE

S. KITAMOTO
Department of Earth and Space Science,
Graduate School of Science, Osaka University
1-1, Machikaneyama-cho. Toyonaka, Osaka, 560-0043, Japan
CREST: Japan Science and Technology Corporation (JST)
4-1-8 Honmachi, Kawaguchi, Saitama, 332, Japan

Abstract

X-ray novae are the best objects to study their compact stars. Some of the X-ray novae have been confirmed that they were containing a black hole based on dynamical mass estimations. Besides the dynamical mass estimation, some X-ray novae showed possible black hole natures. In this work, I attempt to explain the black hole natures found by the observation of X-ray novae.

1. Introduction; Why are the X-Ray Novae Important

The word "X-Ray Nova" means the X-ray star which suddenly appears (during several days) and gradually declines and eventually fades away. Some of them show a hard energy spectrum and pulsation in their X-ray intensity. They are considered to contain a magnetized neutron star. Some show X-ray bursts and they are considered to include a non- (or weak-) magnetized neutron star. There is another category in the X-ray novae, which shows neither pulsation or X-ray bursts. Sometimes, they show "ultra soft" energy spectrum during their bright phase. This character of the "ultra soft" energy spectrum is typical in some of the black hole candidates such as Cyg X-1 in its high state and LMC X-3. As well as its dramatic brightening, this suggestion of the black hole nature of the X-ray novae makes them exciting objects.

Actually, in order to study stellar black holes, the X-ray novae are especially important because of the following points:

One important fact is that X-ray novae provide us a good opportunity to determine their dynamical mass after fading the X-ray activity. Majority of the black hole novae are low mass X-ray binaries. Therefore, during the X-ray is bright, optical light from the accretion disk dominates that from a companion. Thus it is difficult to get an information on the spectral type of the companion. However, in the case of novae, X-ray luminosity becomes faint. After fading away, it becomes possible to get the information of the spectral type of the companion and information of the mass of the companion. This makes it possible to estimate the mass of the X-ray star from the mass function. The first demonstration had been achieved for A0620-00 by McClintock and Reminard (1986).

The second important point is that X-ray novae have a large dynamic range of the accretion rate in one particular source. The observed fluxes of novae have a dynamic range of more than 3 order of magnitude. Figure 1 shows light curves of five bright X-ray novae. Therefore the novae are good laboratories for study of the accretion flow.

Relating this topics, Narayan, Garcia and McClintock (1997) pointed out one possible difference between a neutron star binary and a black hole binary. Their key point is that the dependence of the luminosity on the mass accretion rate changes depending on the existence of the event horizon. If there is a hard core at the center of the accretion flow, at least half of the gravitational energy is released. However, if there is an event horizon at the center of the accretion flow, energy is released only in the accretion disk. Especially, in the case of an advection dominant disk, the released energy is substantially smaller than the total gravitational energy. This difference leads statistical differences of the value of the L_{min}/L_{max} of the neutron star novae and of the black hole novae.

The most important point is a fact that 70% of the stellar black hole candidate are X-ray novae. Actually, 10 black hole binaries have been confirmed by the dynamical mass estimation and among them, 7 are X-ray novae (Tanaka and Shibazaki, 1996; Barret, McClintock and Grindlay,1996).

Consequently, novae are the best laboratory for the study of the black hole nature!

In this work, I will introduce the properties of temporal and spectral variation of X-ray Novae, and discuss the black hole nature of the X-ray novae.

For more detailed review on the X-ray time variation of the black hole candidates, see Miyamoto (1995) and van der Klis (1995). For the hard X-ray properties, see Gilfanov et al. (1995), For the reviews on the X-ray novae, see Tanaka and Shibazaki (1996) and Tanaka and Lewin (1995).

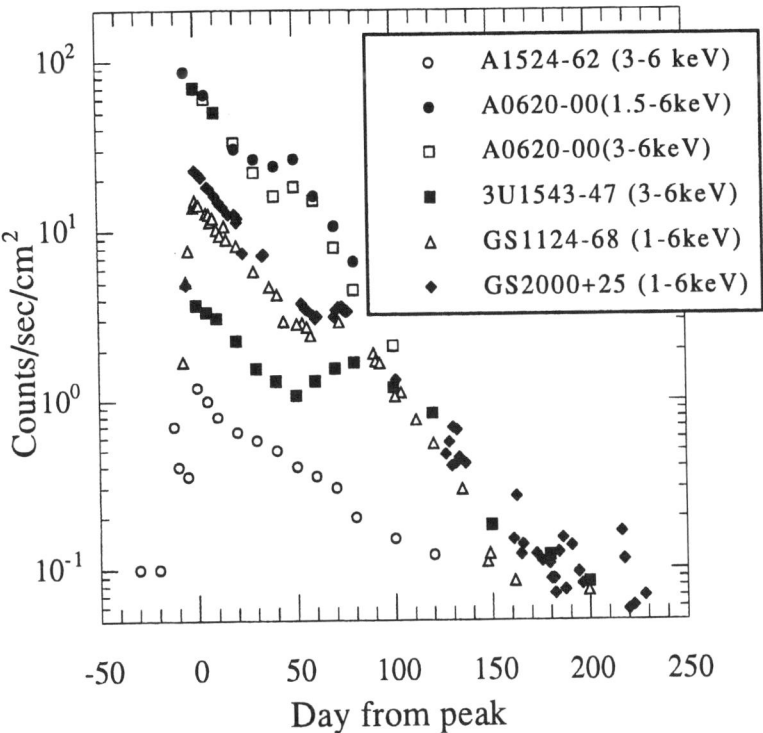

Figure 1. X-ray light curves of five bright X-ray Novae. (3U1543; Li, Sprott and Clark (1976), A1524-62; Kaluzienski et al. (1975), A0620-00; Matilsky et al. (1976), Kaluziensky et al. (1977), GS1124-683 and GS2000+25; Kitamoto (1992)).

2. Long Term Variation of Black Hole Novae

2.1. EXPONENTIAL-DECAY

The most notable fact is that many of the black hole novae had shown beautiful exponential-decays. Recent report of GRO J0422+32 also showed beautiful decay (van der Hooft, 1998). The decaying time scale is roughly 30 days (Kitamoto, 1992).

If we see the light curve more carefully, we can find that there are another peaks; second and third peaks. At least four sources showed the second peak on about 50 days after the main peak. However the reason of the second peak is not clear now. The third peak was observed in A0620-00, GS1124-683, GS2000+25 and GRO J0422+32 on about 150 days after the main peak. The third peak is accompanied with a hardening of the

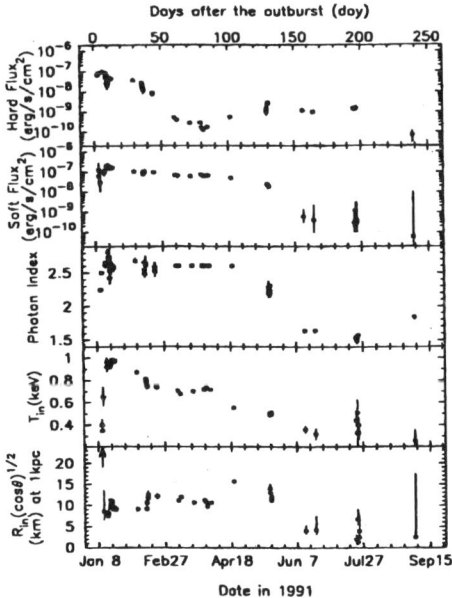

Figure 2. Spectral analysis results of GS1124-683 after Ebisawa et al. (1994). A two-component spectral model, which is composed of a disk blackbody (soft component) plus a power law (hard component), was applied to fit the data obtained with the LAC onboard *Ginga*.

energy spectrum. The good quality data of GS1124-683 clearly showed the energy spectral change at the third peak. Ebisawa et al.(1994) analyzed the *Ginga* data of GS1124-683. They applied a particular two component model; a soft component (disk-black body; Mitsuda et al. 1984)) + a hard component (power law), to fit the spectral data. Figure 2 shows their result. At the third peak (~130 days after the main peak) one can see prominent decreasing of the soft X-ray flux and changing the power law index of the hard component.

In figure 3, light curves and hardness rations of GS2000+25 and of GS1124-683 are plotted together. They seems to be twin brothers. One can see GS2000+25 also showed all properties such as beautiful decays, second and third peaks and change of the hardness ratio. These sudden change of the spectral properties at the third peak is implying the transition of the accretion disk structure.

As shown in figure 2, Ebisawa et al.(1994) derived the parameters of the disk black body model. The intensity decay is explained by the temperature decreases while the inner disk radius is constant until the third peak.

Here, we got an important result. During the decay phase, simple cooling model can explain the observation. This fact strongly supports that the

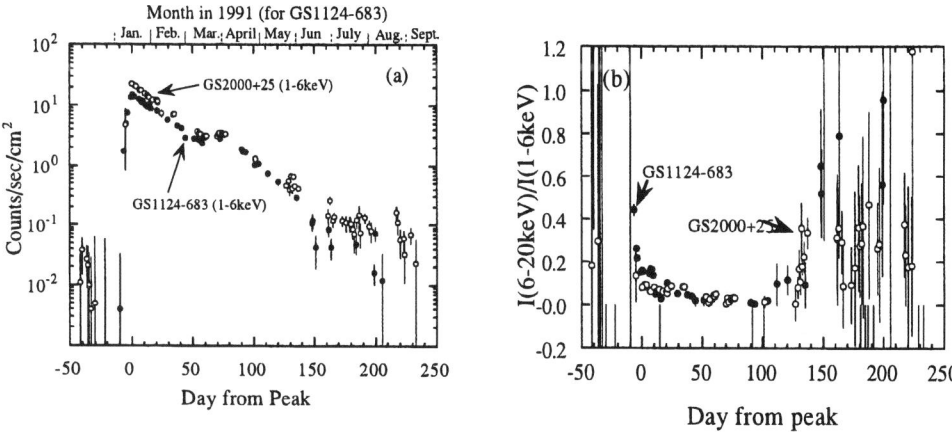

Figure 3. Light curves (a) and hardness ratio histories (b) of the GS1124-683 and of GS2000+25 obtained by the All Sky Monitor onboard *Ginga* after Kitamoto (1992).

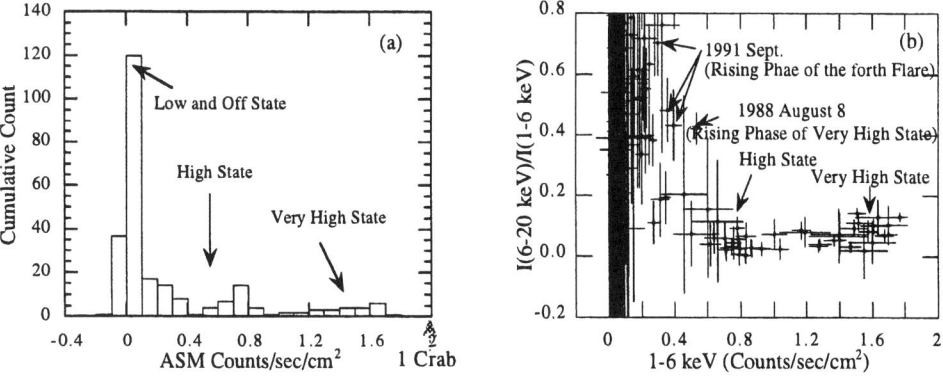

Figure 4. Cumulative count distribution (a) and a intensity-hardness map (b) of 4.5yr history of GX339-4 observed with the All Sky Monitor onboard *Ginga*, after Kitamoto et al. (1991).

disk blackbody model is correct and there is no hard core at the center of the accretion flow. Also if there is a neutron star, the difference of the cooling time scale between the neutron star and the accretion disk makes the intensity decay more complex.

2.2. CHANGE OF THE STATES

As shown in the previous subsection, there are some (at least two) states in the disk structure. More detailed study indicates that there are three states in the black hole X-ray novae. In the conventional two states; the hard (or low) and the soft (or high) states, the soft state can be divided into two different states, called very high state and high (quiet) state. The characteristics of these states are summarized in Table 1. In the very high state, the intensity is very bright and the spectrum is represented by a combination of a soft disk component and a power low component with a photon index of ~ -2.5. The temporal variation is large. The high (quiet) state is also bright state, but its power law component becomes small and its temporal variation is quiet. In the hard (or low) state, the intensity is faint and the spectrum can be represented by a power law with a photon index of ~ -1.5. The temporal variation is very variable.

TABLE 1. Three States in the Black Hole Binaries

State	Intensity	Spectrum	Variation	Photon Index
Very High	Very Bright	Soft +PL	Variable	-2.5
High (quiet)	Bright	Soft	Quiet	...
Low	Faint	PL	Variable	-1.5

Miyamoto et al.(1991) first pointed out the existence of the three states from the observation of GX339-4. In the figure 4, the cumulative counts distribution sorted by the intensity and intensity-hardness-map are plotted from the data obtained by the 4.5 yr observation by the ASM onboard *Ginga*. The cumulative counts distribution shows three distinct concentration and in the intensity hardness map we can see three concentration, indicating the existence of the three states (Kitamoto et al., 1991). Figure 5 shows the another light curve of the GS1124-68. The X-ray fluxes were decomposed into the power law component and the disk black body component. It showed a very high state, high quiet state. After the transition, it became a low state (Miyamoto et al., 1993). I like to note that GS2023+338 also showed a possible soft sate. The spectrum at the brightest phase was represented by a power law model with a photon index of about -2.5 (Kitamoto et al., 1989).

Consequently three stable states can be seen in the X-ray novae.

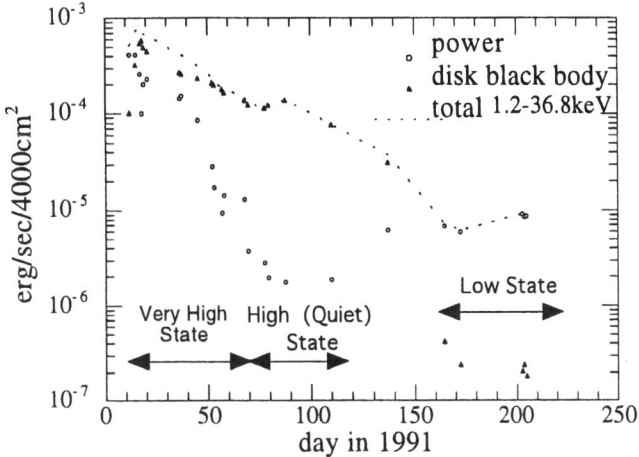

Figure 5. Light curve of the GS1124-683, after Miyamoto et al. (1993). A two-component model was applied to fit the data obtained wit the LAC onboard *Ginga*.

2.3. HYSTERETIC BEHAVIOR

This behavior was first pointed out by Miyamoto et al. (1995). If we look at the spectral evolution of GS1124-683 shown in figure 5 more carefully, we can find that the hard state continues till around the peak of the flare. Also a possible hard spectra were observed in the rising phase of GS2000+25 (see figure 3) and of GX339-4 (see figure 4(b)). However the hard state appear after decreasing down to 1/100. This behavior can be stated that there is a hysteretic behavior on the transition between the state.

3. Short Term Variation of Black Hole Novae

3.1. CANONICAL TIME VARIATION ON THE LOW STATES

In the low state, Miyamoto et al.(1992) pointed out that there is a canonical time variation. Figure 6 shows the light curves, power spectral densities (PSDs) and phase lag of the hard component, of the three black hole candidates; Cygnus X-1, GX339-4, and GS2023+338. Note that the PSDs are normalized by the mean intensities. In the Cygnus X-1 panel, two different data obtained from two different observations are shown. Both are the data in the low state. In the GX339-4 panel, both the data in the low state and in the high state are plotted together with the Cygnus X-1 data. The PSD in the high state is completely different from the low state data. However, if we look at the low state data above 0.2 Hz, the PSDs show very similar shapes and values. Also similar phase lags were obtained in the low states.

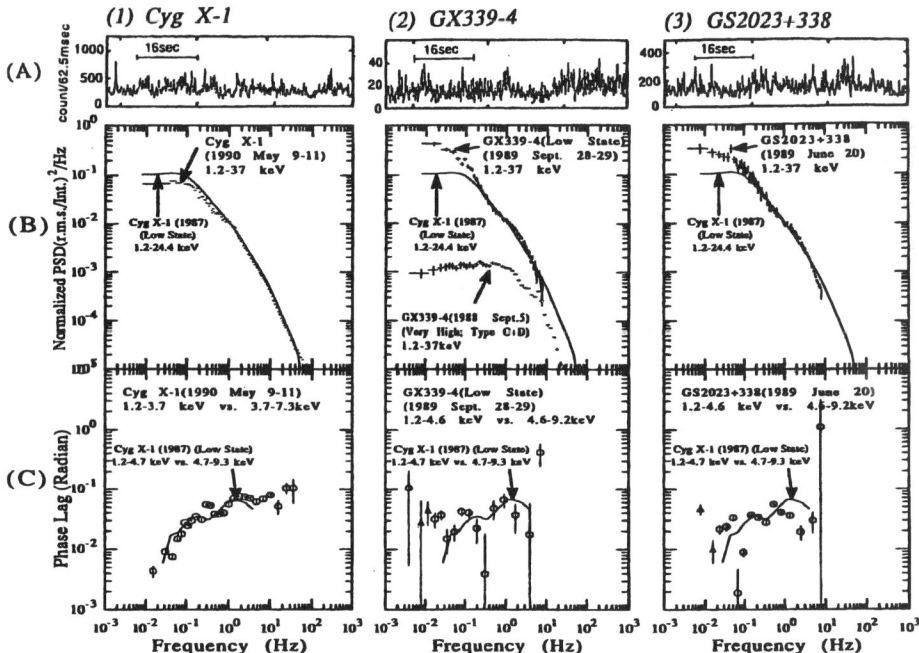

Figure 6. Time variation in the low state of three black hole candidates; Cyg X-1, GX339-4, and GS2023+338, after Miyamoto et al. (1992). Upper panels show an example of light curves. Middle panels show the power spectral densities (PSDs). The lower panels show the phase lag of the high energy component respect to the low energy component.

These properties indicate that the time variation can be described by one canonical phenomenon. This similarity is also confirmed by the similar shot structures of some black hole candidates found by Negoro (1998).

3.2. ANOTHER CANONICAL TIME VARIATION

Another canonical time variation in the high state was reported by Miyamoto et al.(1993). In the high state, the PSDs do not show same shape every time. Figure 7 shows typical PSDs of two black hole candidates and one neutron star binary. In the all sources, we can see two types of the shape; one is the power law shape and another is a flat top shape. Even in the neutron star binary; GX 5-1, it also shows similar power law shapes. There might be two components in the PSDs.

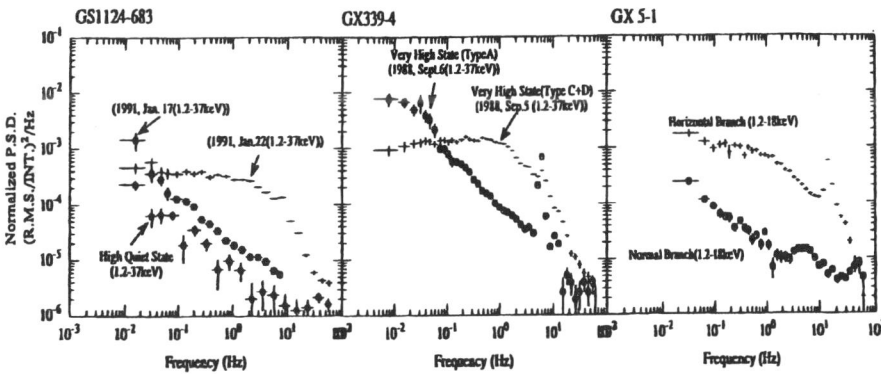

Figure 7. Power spectral densities (PSDs) in the high state of two black hole candidates; GS1124-683 and GX339-4 and one neutron star binary; GX 5-1, after Miyamoto et al. (1993).

4. Two Components

4.1. TWO COMPONENT IN ENERGY SPECTRA

We studied much more on the two component hypothesis. In the energy spectrum of the neutron star binaries, we can expect two emissions; from the accretion disk and from the surface of the neutron star (Mitsuda et al., 1984). In the case of the black hole binaries, we can expect the emission from only the accretion disk. Actually, the energy spectrum in the soft state is described by a disk blackbody model. The energy spectra of typical neutron star binaries can be roughly fitted by a 2 keV blackbody from the neutron star and disk blackbody with an inner edge temperature of about 1 keV. On the other hand the typical energy spectra of black hole candidate in their high state can be described by a disk emission only. This fact is an important support of the non-hard core at the center of the accretion flow in the black hole candidate.

However, the above discussion is not true. If we look at the high energy side above 10 keV of the spectra of the black hole candidates, there is an another component. The spectra of the neutron star binaries and of black hole binaries are shown in figure 8 (Tanaka and Sibazaki, 1996). The upper three panels show spectra of the neutron star binaries and the lower three panels show spectra of the black hole candidates. Spectra of neutron star binaries are well described by the sum of the neutron star component and the accretion disk component. While the spectra of the black hole candidates are roughly described by the accretion disk component but there

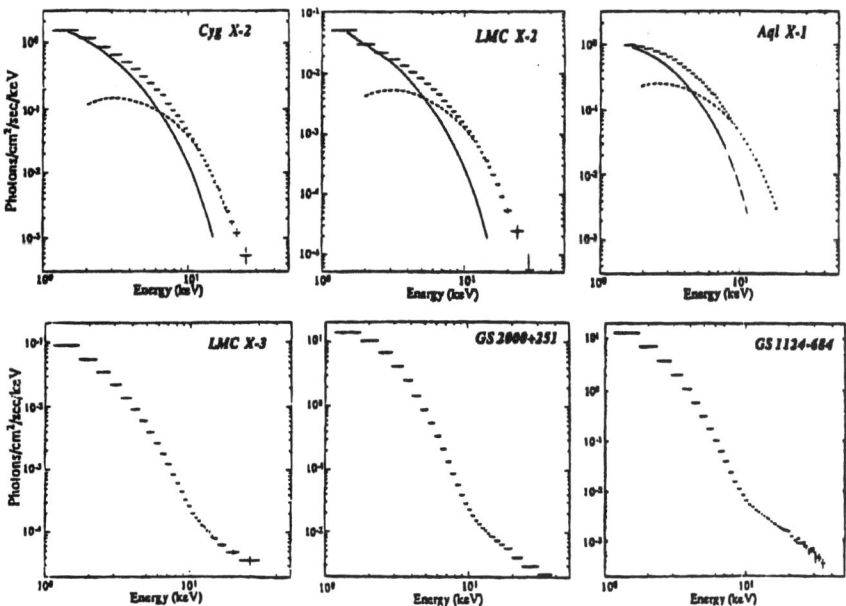

Figure 8. Energy spectra of three neutron star binaries and three black hole binaries, after Tanaka and Shibazaki (1996).

are another high energy component. We call them as a hard component. The energy spectra of black hole binaries can be decomposed into a disk component and a hard component.

4.2. TWO COMPONENT IN PSD

Figure 9 shows the collection of the PSDs of GS1124-64 (Miyamoto et al., 1994). Some have the flat top shape and some have power law shape. And some shows a combination of them. We studied the relation between this various power law shapes and the energy spectral shape. We introduce the PLF(fraction of the hard component); which is a ratio of the count rate of the hard component to the total count rate. The PLFs are indicated at the corner of the figures. If the PLF is large, the PSD shows the flat top shape. On the other hand, if the PLF(fraction of the hard component) is small, the PSD shows the power law shape. In summary; when the energy spectral shape has small hard component, the PSD shows power law shape and when the hard component is large in the energy spectrum, the PSD

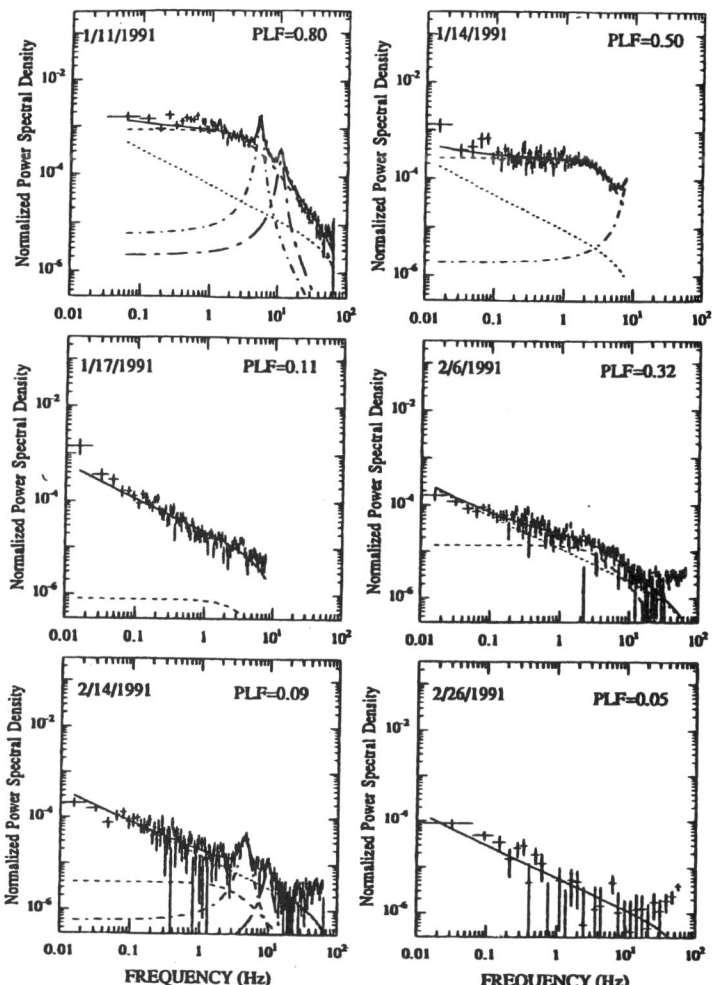

Figure 9. Collection of PSDs of GS1124-683, after Miyamoto et al. (1994).

shows flat top shape. This suggests that the PSD of the hard component has a flat top shape and the PSD of the disk component has a power law shape.

4.3. INHERENT TIME VARIATION OF THE TWO SPECTRAL COMPONENTS

Then let's assume that the flat top shape PSD is a variation of the hard component and the power law shape PSD is a variation of the disk component. Furthermore, we assume that both component varies independently and each have their own inherent value of the time variation.

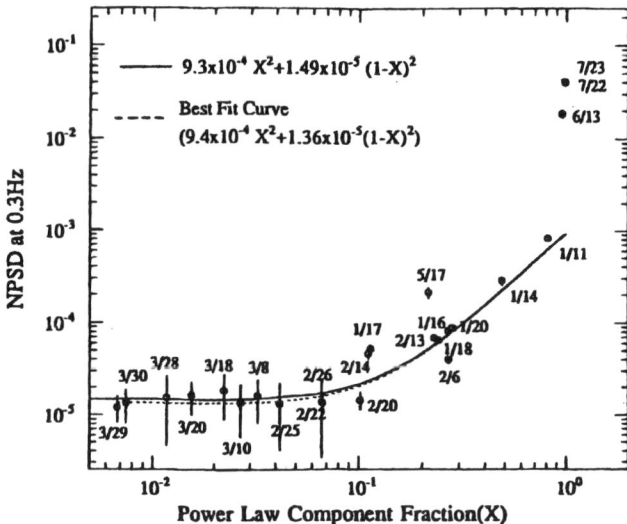

Figure 10. Normalized PSD at 0.3Hz as a function of PLF of GS1124-683, after Miyamoto et al. (1994).

Then the normalized PSD should be expressed by the following equation.

$$A \times X^2 + B \times (1-X)^2 \qquad (1)$$

where X is the PLF (fraction of the hard component), A is the inherent value of the PSD of the hard component and B is that of the disk component.

Figure 10 shows the normalized PSD values at 0.3 Hz as a function of PLF(X) derived from the GS1124-68 data. Except for the low state data, the data can be well explained by the assumed equation. The best fit values are $[9.4 \times 10^{-4}]$ (rms/int)^2/Hz and $[1.36 \times 10^{-5}]$(rms/int)^2/Hz for A and B respectively. Furthermore, if we decompose the energy spectra into the disk component and the hard component, and decomposed the PSDs into the power law shape component and the flat shape component, and normalize the PSDs by the each energy component, we got roughly constant values for the both components. These facts means that the assumptions are correct. Therefore, we conclude that the two spectral components varies independently and each has its own inherent value and shape of the PSD.

4.4. THE CASE OF NEUTRON STAR BINARY; GX 5-1

It is well known that the energy spectrum of GX 5-1 can be decomposed into two components; the disk component and the neutron star component

(Mitsuda et al., 1984). The PSD also shows sometimes power law shape and sometimes flat top shape (see figure 7). Therefore, according to the GS1124-68 case, we again fitted the normalized PSD as a function of the fraction of neutron star component(Kamado et al., 1997). First the normalized PSDs are plotted as a function of the fraction of the neutron star component. However, the square equation (equation (1)) does not fit well. Furthermore we normalized the each PSD components by its corresponding spectral component. The disk component seems constant during the small fraction of the neutron star component. It is interesting that the value of the disk component is very similar to that of the disk component of GS1124-68. However when the fraction of the neutron star component becomes large, the PSD value of the neutron star component increases steeply. Even the PSD value of the disk component becomes large. Thus we conclude that the decomposition of the two component may be correct but there is a correlation between the variations of the disk component and the neutron star component.

This is again important conclusion. In the case of neutron star binaries, the neutron star component is strongly coupling with the disk component. This is a natural result. However, in the case of black hole binaries, the hard component is independent from the disk component. We do not know the origin of the hard component, but it is not strongly coupling with the accretion disk.

5. Discussions; What is the black hole nature?

The difference between the neutron star binaries and black hole binaries is the center of the accretion flow;

> The black hole binaries have an event horizon.
> The neutron star binaries have a hard core, i.e. a neutron star.

From the observation of X-ray novae, we found a lot of interesting but difficult properties, among them, there are some observational support on the non-hard-core at the center of the accretion flow;

1. The spectral shape in the soft state shows no neutron star component. There is a hard component in the black hole binaries. But the hard component is not originate from the hard core, because the central hard core should directly couples with the accretion disk component.

2. The simple exponential decaying light curve can be explained by a simple cooling of the disk component. This also supports the non-central-hard core.

6. Acknowledgment

This review is based on the works with Porf. Miyamoto and other colleagues.

References

Barret, D., McClintock, J.E., and Grindlay, J.E. (1996) Luminosity Differences Between Black Holes and Neutron Stars, *ApJ*,**Vol. no. 473**, pp. 963–973

Ebisawa, K. et al. (1994) Spectral Evolution of the Bright X-Ray Nova GS1124-683 (Nova Muscae 1991) Observed with Ginga, *PASJ*,**Vol. no. 46**, pp. 375–394

Gilfanov, M., et al. (1995) Hard X-Ray Observations of Black-Hole Candidates, *in "The Lives of the Neutron Stars"* eds. M.A. Alpar et al., Kluwer Academic Publishers, Netherlands, pp. 331–354

Kaluzienski, L.J., Holt, S.S., Boldt, E.A., Serlemitsos, P.J., Eadie, G., Pounds, K.A., Ricketts, M.J., and Watson, M. (1975) The Light Curve of a Transient X-Ray Source, *ApJL*,**Vol. no. 201**, pp. 121–124

Kaluzienski, L.J., Holt, S.S., Boldt, E.A., Serlemitsos, P.J. (1977) Al-Sky Monitor Observation of the Decay of A0620-00, *ApJ*,**Vol. no. 212**, pp. 203–210

Kamado, Y., Kitamoto, S., Miyamoto, S. (1997) Time Variation of X-Rays from GX5-1, *PASJ*,**Vol. no. 49**, pp. 589–605

Kitamoto, S. (1992) Properties of X-Ray Emission from X-Ray Novae, *in proceeding of Fourth International Toki Conference on "Plasma Physics and Controlled Nuclear Fusion"* ESA SP-351, pp. 297–299

Kitamoto, S., Miyamoto, S., Tsunemi, H., Hayashida, K. (1991) Observation of a Very High State of GX339-4 with the ASM, *in "Frontiers of X-Ray Astronomy"*, edited by Y. Tanaka, and K. Koyama Universal Academy Press, Tokyo, pp. 321–322

Kitamoto, S. Tsunemi, H., Miyamoto, S., Yamashita, K., Mizobuchi, S., Nakagawa, M., Dotani, T., and Makino, F.(1989) GS2023+338; a new class of X-ray transient source? *Nature*,**Vol. no. 342**, pp. 518–520

Li, F.K., Sprott, G.F. and Clark, G.W. (1976) OSO-7 Observation of the X-Ray Nova 3U 1543-47 *ApJ*.**Vol. no. 203**, pp. 187–192

Matilsky, T., Bradt, H.V., Buff, J., Clark, G.W., Jernigan, J.G., Joss, P.C., Laufer, B., and McClintock, J.(1976) The Transient X-Ray Source A0620-00; Intensity Variations and Evidence for an 8 Day Periodicity,*ApJL*,**Vol. no. 210**, pp. 127–131

McClintock, J.E., and Reminard, R.A. (1986) The Black Hole Binary A0620-00, *ApJ*,**Vol. no. 308**, pp. 110–122

Mitsuda, K. et al. (1984) Energy Spectra of Low-Mass Binary X-Ray Sources Observed from Tenma, *PASJ*,**Vol. no. 36**, pp. 741–759

Miyamoto, S.(1995) Time Variation of X-Rays from X-Ray Stars. *in Proceedings of the IIAS Workshop on Mathematical Approach to Fluctuations (Vol.II)*, ed. T. Hida, World Scientific Publish Co.; Singapore, pp. 254–298

Miyamoto, S., Iga, S., Kitamoto, S., and Kamado, Y(1993) Another Canonical Time Variation of X-Rays from Black Hole Candidates in its Very High Flare State? *ApJL*,**Vol. no. 403**, pp. L39–L42

Miyamoto, S., Kimura, K., Kitamoto, S. (1991) X-Ray Variability of GX339-4 in its Very High State, *ApJ*,**Vol. no. 383**, pp. 784–807

Miyamoto, S., Kitamoto, S., Hayashida, K., Egoshi, W. (1995) Large Hysteretic Behavior of Stellar Black Hole Candidate X-Ray Binaries, *ApJL*,**Vol. no. 442**, pp. L13–L16

Miyamoto, S., Kitamoto, S., Iga, S., Hayashida, K., Terada, K. (1994) Normalized Power Spectral Densities of Two X-Ray Components, *ApJ*,**Vol. no. 435**, pp. 398–406

Miyamoto, S., Kitamoto, S., Iga, S., Negoro, H., Terada, K.(1992) Canonical Time Variations of X-Rays from Black Hole Candidates in the Low-Intensity State, *ApJ*,**Vol. no. 392**, pp. L21–L24

Narayan, R., Garcia, M.R. and McClintock, J.E. (1997) Advection-Dominated Accretion and Black Hole Event Horizons, *ApJL*,**Vol. no. 478**, pp. 79–82

Negoro, H. (1998) Are X-Rays Coming from Inside the Innermost Radius of Accretion Disks?, *in this volume*

Tanaka, Y., and Lewin, W.H.G. (1995) X-Ray Novae, *in "X-Ray Binaries"*, eds. Lewin, W.H.G., van Paradijs, J., and van den Heuvel, E.P.J., Cambridge Univ. Press, Cambridge, pp. 126

Tanaka, Y., and Shibazaki, N. (1996) X-Ray Novae, *Annu. Rev. Astron. Astrophys.*,**Vol. no. 34**, pp. 607–644

van der Hooft (1998) Rapid Hard X-Ray Variability in GRO J0422+32, *in proceeding of IAU S188 "The Hot Universe"* eds. K. Koyama, S. Kitamoto and M. Itoh, Kluwer Academic Publishers, Dordrecht, pp. 398–399

van der Klis, M. (1995) Rapid Variability in X-Ray Binaries; Unified Description, *in "The Lives of the Neutron Stars"* eds. M.A. Alpar et al., Kluwer Academic Publishers, Netherlands, pp. 301–330

Special Moments: (left to right) C. Haswell, S. Kitamoto & H. Negoro

Special Moments: T. Zannias & L. Ozernoy in the boat

A TRANSITION DISK MODEL FIT TO CYGNUS X-1

R. MISRA
Inter-University Center For Astronomy and Astrophysics
Ghaneshkhind Road, Pune, India

1. Transition Disk model and Cygnus X-1

During the last two decades, the X-ray/γ-ray spectrum of the black hole candidate Cygnus X-1 has been extensively studied by, e.g., HEAO1, EXOSAT, Ginga and the Compton GRO. This system settles into two distinct long term configurations, namely the so-called soft and hard states. These spectral states have also been observed in other black hole candidates. In the hard state, the X-ray spectrum is dominated by 2 – 200 keV emission, which has been interpreted by some to be the result of the inverse Compton scattering of cold external photons by a hot ($kT \approx 50$ keV) low density plasma. This plasma could be the innermost region of the accretion disk (Shapiro, Lightman & Eardley 1976), or it could be a corona overlying a relatively cold disk (Liang & Price 1977; Haardt & Maraschi 1993). An alternative model is one in which the emission is a sum of multi-temperature Wien peaks from the transition region of the disk (Maraschi & Molendi 1990; Misra & Melia 1996). All these models predict a similar spectrum which is more or less a power-law with an exponential cutoff.

The standard optically thick accretion disk is not self-consistent in its inner regions, since the effective optical depth is not as large as was assumed. Solving the disk structure without assuming a large effective optical depth, reveals that in the inner region the electron temperature is a rapidly varying function of radius. In a more recently developed model for the hard X-rays (Maraschi & Molendi 1990; Misra & Melia 1996), this region of the disk is viewed as the transition region since it is here that disk makes a gradual transition from a cold optically thick state to a hot plasma. The scattering optical depth in the transition is still large and the local spectrum emitted is saturated Comptonization (or a Wien peak). The sum of these multi-temperature Wien peaks reproduces the observed spectrum. The advantages of this model over the unsaturated Comptonization model are a) this model does not require a sudden transition from a cold to a hot phase which may not conserve energy (Misra & Melia 1996) and b) it explains the near invariance of the spectral shape to variations in luminosity. The main problem with the transition model is that it is locally unstable. The unsaturated Comptonization model is also thermally unstable but the high temperature there may make the disk globally stable by radial advection.

Apart from these theoretical arguments, it has not been possible to distinguish between the two models by the observed spectrum. The spectrum was consistent with a simple power-law with exponential cutoff and with the predicted spectra from both classes of models.

However, it may now be possible to differentiate between these models by combining the data from various high-energy instruments, such as OSSE, Ginga and EXOSAT, and thereby studying the broad band nature of this source. The initial steps in this direction were taken by Chitnis, Rao & Agrawal (1997) and Gierlinski et al. (1997). They concluded that a simple soft-photon Comptonization model does not adequately fit the data. Instead, significant systematic residuals were obtained, and it appeared that an additional *ad hoc* spectral component is needed to account for the observations.

Subsequently, Misra et al. (1997) showed that the transition disk model (Misra & Melia 1996) appears to fit the broad band data without invoking an additional component. However, this analysis used non-simultaneous observations by EXOSAT, XMPC balloon flights and OSSE. Since the source is known to be variable, the results of this analysis were not conclusive. Moreover, since only one set of observations was used, it was not clear whether the agreement with the transition model was a chance occurrence. An additional complication was the fact that a direct χ^2 comparison between the transition disk and soft photon Comptonization models could not be made since the latter was not fit to the same data. It was shown only that the transition model did not produce any systematic residuals whereas the Comptonization model (fit to a different data set) did give rise to these.

Misra et al. 1998 used four sets of simultaneous observations of Cygnus X-1 made with Ginga and OSSE. These data had already been analyzed with an isotropic Comptonization model by Gierlinski et al. (1997). They reanalyzed the data using the transition disk model and compared their results with those of the previous study. They established rather strongly that the transition disk model fits the data better than a pure soft-photon Comptonization model. In particular the χ^2 value for the transition disk model was in each case smaller than or comparable to that of the isotropic Comptonization model fit to the same data (Gierlinski et al. 1997). In addition, unlike the soft-photon Comptonization model, there are no system-

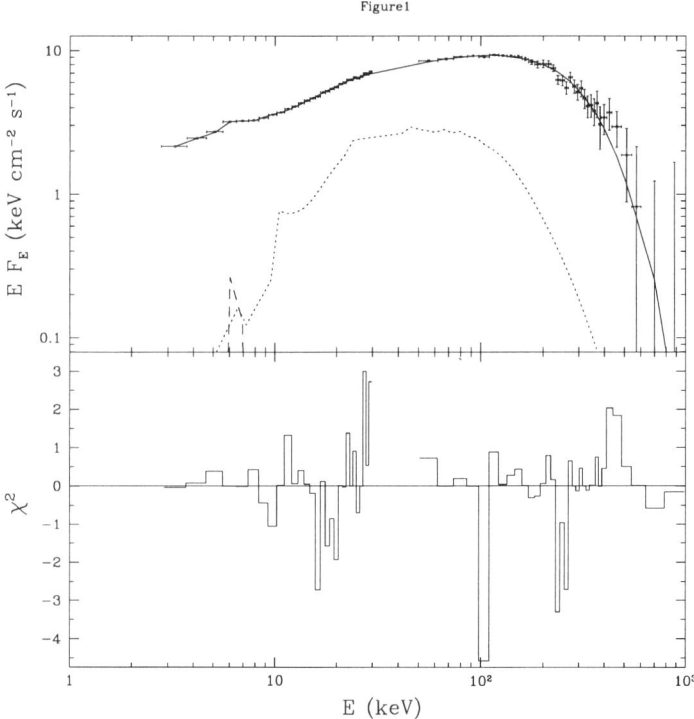

Figure 1. The deconvolved spectrum of Cygnus X-1 corresponding to the first data set, obtained from simultaneous observations with Ginga and OSSE. The contributions from individual model components, viz., the iron line and reflection, are shown separately. The residuals to the model fit are shown in the lower panel as a contribution to the χ^2.

atic residuals for the transition disk fit (Fig. 1). They have also determined the disk structure from the best fit temperature profiles, and through self-consistency checks, showed that their initial assumptions were valid.

It is still not clear whether more complicated soft-photon Comptonization model will be able to fit the data. In particular a multi-temperature Comptonization model may also fit the data as well as the transition model. However, the number of parameters of such a model will be larger than the transition disk model used by Misra et al. 1998. A better diagnostic test could be the timing analysis of Black hole candidates. It may be possible to confirm or rule out models with timing information like coherence, phase lag and power spectrum. If the radiative process involved in the production of X-rays in Black hole candidate systems is better understood, we most probably will be able to confirm the existence and probe deeper into the nature of black hole and their environment.

References

Chitnis, V., Rao, A.R. & Agrawal, P.C., 1997, *Astr. Ap.*, **331** 251.
Gierlinski et al., 1997, *mnras*, **288** 958.
Haardt, F. & Maraschi, L., 1993, **ApJ**, **413**, 507
Liang, E.P. & Price, R.H., 1977, **ApJ**, **218**, 247.
Maraschi, L. & Molendi, S., 1990, **ApJ**, **353**, 452.
Misra, R. & Melia, F., 1996, **ApJ**, **467**, 405.
Misra, R., Chitnis, V., Melia, F. & Rao, A.R., 1997, **ApJ**, **487**, 388.
Misra, R., Chitnis, V. & Melia, F., 1998, **ApJ**, **495**, 407.
Shakura, N.I. & Sunyaev, R.A. 1973, *Astr. Ap.*, **24**, 337.
Shapiro, S.L., Lightman A.P. & Eardley, D.M. 1976, **ApJ**, **204**, 187.

ARE THE X-RAYS COMING FROM THE INSIDE OF THE ACCRETION DISK OF THE BLACK HOLE ?

H. NEGORO

Institute of Physical and Chemical Research (RIKEN)
Hirosawa 2-1, Wako, Saitama 351-0198, Japan

1. Introduction

1.1. WHAT IS THE EVIDENCE FOR BLACK HOLES ?

Previous X-ray observations have revealed that about 20 X-ray stars show common X-ray properties distinct from neutron star systems and white dwarf systems. The central masses of about 10 of them were estimated from optical observations, and turned out to be in excess of the theoretical upper limit of the mass of a neutron star (e.g., Tanaka *et al.*, 1996).

Such massive X-ray stars commonly exhibit aperiodic rapid time variations in X-ray intensity on time scales of milliseconds, implying that the X-ray emission region is very compact, say 10^7 cm. The estimated masses and the compactness of the emission region strongly suggest that the X-ray stars are black holes. Those stars are, however, still called black hole candidates (BHCs) because we have not yet observed any property closely tied to black holes themselves.

Estimating the mass of invisible objects is the great first step to finding black holes as discussed by other authors in this volume. However, it does not tell us that the black holes we imagine really exist in the central regions of those objects. Since the black holes we imagine do not emit light at all, we cannot confirm their existence directly. Fortunately, relativistic theory also predicts that, on the inside of the innermost stable orbit, matter almost free-falls onto black holes and little energy is released. Thus, its confirmation will be the next step. In this sense, X-ray/γ-ray observations are of great importance because such high energy photons are thought to be emitted in the vicinity of the central object.

1.2. EVIDENCE FOR ACCRETION DISKS

Extensive X-ray observations have demonstrated the existence of accretion disks around massive objects. Ultrasoft spectral components have been observed in many BHCs, usually when the X-ray intensity is relatively high. The spectra are well represented by a multicolor disk blackbody model, which is a simple optically thick accretion disk model with the innermost radius r_{in} and the radial temperature dependence of $T \propto r^{-3/4}$, where r is the distance from the center and T is the temperature at r (Makishima et al., 1984). Most of the observations showed that r_{in} was almost *constant* even though X-ray flux of this component changed by a factor of more than 10, implying that r_{in} is the innermost stable orbit (e.g., Kitamoto, 1998).

The observed temperature and luminosity basically enable us to estimate the size of the innermost radius, which can be connected with the mass of the central object provided that $r_{in} = 3r_s$, where r_s is the Schwarzschild radius. The masses of BHCs, estimated in such a way, are systematically larger than those of low-mass X-ray binaries (e.g., Inoue, 1991).

ASCA detected a broad iron $K\alpha$ line from some AGNs. The line profiles imply that the emission region extends over $3r_s$ (Tanaka et al., 1995; Nandra et al., 1997). This can also be evidence for the accretion disk, though the disk structure has not been understood. However, neither of these observations give information on the inside of the innermost stable orbit.

1.3. X-RAY TIME VARIATIONS

BHCs show aperiodic time variations, which is one of their properties which is distinct from other systems. The nature of the time variations was, however, not well understood because the Fourier analyses of the time variations, which had long been utilized to evaluate the variations, did not give any useful information on the origin of the time variations. The time variations must have important information not only on the radiative process but also on the dynamical process. Therefore, they may contain a clue connecting the observed properties to the black holes.

Useful physical information obtained from results of previous timing analyses was the shortest time scale, ~ 3 ms, providing the maximum size of the emission region (Meekins et al., 1984), and the longest time scale, ~ 10 sec, suggesting that so-called X-ray shots arise from the mass accretion (Rothschild et al., 1974). The superposition technique is, however, a powerful method to obtain average profiles of the shots and their spectra without any ad hoc assumptions in contrast with the shot-noise models (Lochner et al., 1991 and references theirin). Obtained results can be closely related with the physical condition near the innermost radius and inside it. The results will be described next, focusing on the evidence for black holes.

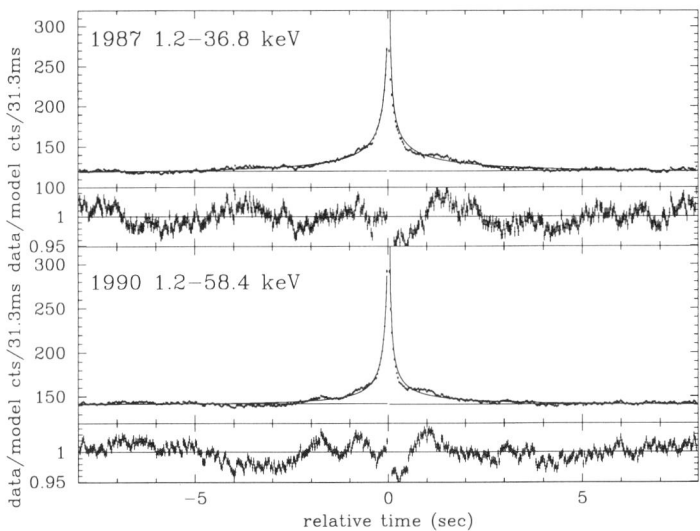

Figure 1. Average shot profiles of Cyg X-1 in 1987 and 1990 observations and the best fitting fractal function models (solid lines). Their ratios are also shown.

2. Data and Analyses

The superposition technique was first introduced in order to investigate average properties of the shots of Cyg X-1 using *Ginga* data (Negoro *et al.*, 1994). An average profile is obtained by superposing a number of shots with peak counts larger than a certain factor (typically 1.5-2.5) of the local mean number of counts. In superposing the shots, the events are aligned with their peak-bin positions in all the energy bands, not in each energy band (see Negoro, 1995a in detail).

All the results described here were obtained using *Ginga* data of Cyg X-1, GX 339-4, GS 2023+338 and GS 1124-68 in the low state (Negoro, 1997). In those data, selected shot events are statistically significant, but the peak-bin positions suffer from the statistical count fluctuations. Numerical simulations show that the accuracy of the alignment depends on the count rate, being 20-60 ms in the data used here.

3. Results

3.1. SHOT PROPERTIES

Figure 1 shows average shot profiles of Cyg X-1 in 1987 and 1990 observations (Negoro, 1995a). The other BHCs also show similar shot profiles (Negoro, 1997). Both the rise and decay profiles can be roughly represented by the following empirical functions. One is a sum of two exponential functions with time constants of about 0.1 sec and about 1 sec. A single exponential function gives poor fits to both the profiles due to long wings extending for more than 2 sec before and after the sharp peaks. The other is a *fractal* function proportional to $1/(\tau + |t|)^\alpha$ where t is the relative time from the maximum intensity time and τ and α are free parameters.

Due to the slightly asymmetric profiles, in most cases, the fractal function gives better fits to the rise profiles than to the decay profiles (see ratios in Fig. 1). τ and α obtained from fits to the rise profiles are 0.07–0.8 and 0.8–1.5, respectively. The obtained fitting parameters are different not only in different sources but also in the same source on different occasions (see Fig. 1), indicating that the shot profiles are not peculiar to each source. However, more detailed analyses show that the rise profiles do not change on time scales of hours or less, at least in Cyg X-1. The shot profiles also do not depend on the peak intensities, especially in the rise phases. Note that this independence, however, does not imply that all the shots have the completely same profile in the short time scales. Some selected shots have a similar profile to the average one, but others do not. The obtained profiles only present *average* profiles in some period.

Energy spectra of the shots give further information. The *Ginga* data used here have 12 or 48 spectral bins in the 1.2–37(58) keV band. All the sources show similar spectral evolution during the shots. The shots have softer spectra than the time averaged spectra, especially in the rise phases. The spectra suddenly harden at the peak intensities. The durations of the hardenings are 20–60 ms, which suffers from the mis-alignment in superposing shots as described before. Simulations show that a possibility of change in less than the minimum time resolution (7.8 ms) can not be ruled out. After the peak intensities, the spectra continue to be harder for 0.2-0.3 sec. After that, transient, slight softening is recognized in some sources.

3.2. COMPARISON WITH RESULTS OF FOURIER ANALYSES

Observed structures in power spectral densities (PSDs) and hard X-ray time lags below a few Hz are consistent with the above characteristics of the shots (Negoro, 1995a; Negoro, 1997). The PSDs of BHCs in the low state have two knees around 0.01–0.1 Hz and 1 Hz (Miyamoto *et al.*, 1992). The two knees reflect the shot profiles represented by the two time constants. The 'double-peaked' phase (time) lags are due to the gradual spectral hardening after the peak intensity and the following changes.

Structures above a few Hz in the PSDs and the time lags, however, can not be explained by the shots. This is not due to the fact that superposing a number of shots smears rapid (< 20–60 ms) time variations. Normalized PSDs showing the relative amplitudes of the variations indicate that, at those high frequencies, in contrast with the low frequencies, normalized PSDs in the low energy bands have lower values than those in the high energy bands, suggesting another high frequency component has a hard energy spectrum.

4. Discussion

4.1. ORIGIN OF THE TIME VARIATIONS AND RELATIONSHIP WITH BLACK HOLES

The gradual increase in intensity and the independence of the shot profiles from the peak intensities rule out shot occurrence models based on *local* disk instabilities and/or magnetic field reconnections. Because such models showed much shorter time scales (≤ 0.1 sec) for fluctuations. The duration and the magnitude of a fluctuation also depend on the radius where it occurs (e.g., Piran, 1978; Galeeve et al., 1979).

The general properties of the shots can be explained by a model in which the shots arise from density fluctuations of accreting matter drifting onto a black hole. If the radial velocity of a clump of accreting matter v_r ($= dr/dt$) and X-ray emission from it $l(r)$ are expressed by powers of the radial distance r such as $v_r = -C_1 r^{-\lambda}$ and $l = C_2 r^{-\mu}$, the expected time variation in intensity from the matter can be described as $a/(\tau_0-t)^{\alpha_0}$ where t is the time (set to be 0 when matter crosses the innermost radius r_{in} of the accretion disk), and $\alpha_0 = (\lambda+\mu)/(\lambda+1)$ and $\tau_0 = r_{in}^{\lambda+1}/(\lambda+1)C_1$ (Negoro, 1995b). This formula is exactly the fractal function used in describing the rise profiles.

This model is rather simple, but gives a relationship between the *observed* (*not assumed*) shot profiles and some physical parameters for the first time. Note that this function must (approximately) hold in any accretion model. Setting $\lambda = 1/2$ and $\mu = 2$, which are approximately correct in some accretion disk theories, gives $\alpha_0 = 5/3$, which is larger than the observed α. The observed shot profiles imply $\mu \sim 1$.

Thus, in this model, we just observe matter disappearing into a black hole as a shot. This was confirmed in a simulation based on the advection-dominated accretion flow (ADAF) model, in which some perturbation in the accretion disk gives rise to an intensity increase as it approaches the center (Mannmoto et al., 1996). If so, where does the emission after the peak intensity come from ? The rapid spectral hardening at the peak intensity suggests that the radiative process suddenly changes. The above simulation also shows that, before disappearing into a black hole, the perturbation is reflected around the sonic point as an outgoing sound wave.

This successful model and simulation, however, do not explain all the properties of the shots. For instance, there are significant sub structures in the observed shots (see the ratios of the data to the model in Fig. 1.). Thus, the obtained shot profiles have more information on the vicinity of the central object. Studying the obtained shot properties will allow us to understand whether the shock occurs or not (Lu, 1998), whether emission comes from the inside of the innermost region of the accretion disk or not, and to know whether the black holes we imagine really exist or not.

The author thanks Lev Titarchuk for his useful comments on this work in this meeting, and Brad Rubin for a critical reading of the manuscript.

References

Galeev, A., Rosner, R. and Vaiana, G. (1979) Structured Coronae of Accretion Disks *Astrophys. J.*, **229**, 318–326

Inoue, H. (1991) X-ray Study of Black Hole Candidates, Proc. *Frontiers of X-ray Astronomy*, 291–300

Kitamoto, S. (1998) *in this volume*

Lochner, J., Swank, J. and Szymkowiak, A. (1991) Shot Model Parameters for Cygnus X-1 through Phase Portrait Fitting *Astrophys. J.*, **376**, 295–311

Lu, J. (1998) *in this volume*

Makishima, K. et al. (1986) Simultaneous X-ray and Optical Observations of GX 339-4 in an X-ray High State, *Astrophys. J.*, **308**, 635–643

Mannmoto, T. et al. (1996) X-ray Fluctuations from Locally Unstable Advection-Dominated Disks *Astrophys. J.*, **464**, L135–L138

Meekins, J. et al. (1984) Millisecond Variability of Cygnus X-1, *Astrophys. J.*, **278**, 288–297

Miyamoto, S. et al. (1992) Canonical Time Variations of X-rays from Black Hole Candidates in the Low-Intensity State *Astrophys. J.*, **391**, L21–L24

Nandra, K. et al. (1997) *ASCA* Observations of Seyfert 1 Galaxies. II. Relativistic Iron $K\alpha$ Emission, *Astrophys. J.*, **477**, 602–622

Negoro, H., Miyamoto, S. and Kitamoto, S. (1994) Structure of X-Ray Shots in Cygnus X-1 in its Low State, *Astrophys. J.*, **423**, L127–L130

Negoro, H. (1995a) Time Variations of X-Rays from Cygnus X-1 and Implications for the Accretion Process, *Doctoral thesis, Osaka Univ., (ISAS Research Note. 616)*.

Negoro, H. (1995b) Shot Profiles of Cyg X-1, Proc. *Physics of Accretion Disks*, 153–158

Negoro, H. (1997) Long and Short-term X-ray Variabilities of Galactic Black Hole Candidates, *All-Sky X-Ray Observations in the Next Decade*, 97–102

Rothschild, R. et al. (1974) Millisecond Temporal Structure in Cygnus X-1 *Astrophys. J.*, **189**, L13–L16

Piran, T. (1978) The Role of Viscosity and Cooling Mechanisms in the Stability of Accretion Disks, *Astrophys. J.*, **221**, 652–660

Tanaka, Y. et al. (1995) Gravitationally Redshifted Emission Implying an Accretion Disk and Massive Black Hole in the Active Galaxy MCG-6-30-15, *Nature*, **375**, 659–660

Tanaka, Y. and Shibazaki, N. (1996) X-ray Novae, *Annu. Rev. Astron. Astrophys.*, **34**, pp. 607–644

List of Participants

P. Agrawal: Tata Institute of Fundamental Research, Homi Bhabha Road, Mumbai-400 005, India; pagrawal@tifrvax.tifr.res.in

P. Banerjee: IUCAA, Ghaneshkhind Road, Pune, India

S. G. Biswas: Viswabharati University, Shantiniketan, 731 235, India

S. Bhattacharya: Indian Inst. of Science, Bangalore, India; sudip@physics.iisc.ernet.in

G. Bisnovatyi-Kogan: IKI, Profsoyuznaya 84/32, Moscow, 117810, Russia; gkogan@mx.iki.rssi.ru

K. Borozdin: Space Research Institute, Moscow, Russia; kborozdin@kisa.iki.rssi.ru

A. Burinski: Nuclear Safely Institute, Moscow, Russia; grg@ibrae.ac.ru

S. K. Chakrabarti: SNBNCBS, JD Block, Salt Lake, Calcutta - 700091, India; chakraba@boson.bose.res.in

I. Chattopadhyay: SNBNCBS, JD Block, Salt Lake, Calcutta - 700091, India; indra@boson.bose.res.in

J. O'Connor: School of Mathematics and Statistics, La Trobe University, Bundoora, Melbourne, Australia; matjeo@luxor.latrobe.edu.au

T. K. Das: SNBNCBS, JD Block, Salt Lake, Calcutta - 700091, India; tdas@boson.bose.res.in

T. K. Das: ECRA, Inst. of Radio Physics & Electronics, Calcutta - 700009, India; tkdas@ecracu.ernet.in

M. K. Dasgupta: P 282, CIT Scheme VI M, Kankurgachi, Calcutta-700054, India

S. Dasgupta: Dept. of Physics, Pune University, Pune, 411 007, India; mohit@msclab.physics.unipune.ernet.in

J. Dey: Dept. of Physics, Presidency College, Calcutta, India; deyjm@giascl01.vsnl.net.in

M. Dey: Dept. of Physics, Presidency College, Calcutta, India; deyjm@giascl01.vsnl.net.in

M. Gierliński: Nicolus Copernicus Center, Poland; Marek_Gierlinski@camk.edu.pl

M. Gilfanov: Max-Planck-Institut für Astrophysik, Karl-Schwarzschild-Str. 1, 85740 Garching bei Munchen, Germany; gilfanov@MPA-Garching.mpg.de

C. Haswell: Astronomy Centre, University of Sussex, Brighton, BN1 9QJ, U.K.; chaswell@star.cpes.susx.ac.uk

L. Ho: Harvard-Smithsonian Center for Astrophysics, 60 Garden St., Cambridge, MA, 02138, USA; lho@coyote.harvard.edu

S. Kitamoto: Department of Earth and Space Science, Osaka University, 1-1, Machikaneyama-cho, Toyonaka, Osaka, 560, Japan; kitamoto@ess.sci.osaka-u.ac.jp

M. R. Kundu: Department of Physics, University of Maryland, USA; kundu@astro.umd.edu

S. Kuznetsov: Space Research Institute, RAS, Moscow, India; sik@hea.iki.rssi.ru

H. Mok Lee: Department of Earth Sciences, Pusan National University, Pusan 609-735, Korea; hmlee@uju.es.pusan.ac.kr

J. Lu: Center for Astrophysics, University of Science and Technology of China, Hefei, Anhui 230026, China; lujf@cfa.ustc.edu.cn

A. Maciołek-Niedźwiecki: Łódź University, Department of Physics, Pomorska 149/153, 90-236 Łódź, Poland; A_M_Niedzwiecki@camk.edu.pl

P. Magdziarz: University of Durham, Department of Physics, South Road, Durham DH1 3LE, U.K.; pawel_magdziarz@camk.edu.pl

C. K. Majumdar: SNBNCBS, JD Block, Salt Lake, Calcutta - 700091, India; ckm@boson.bose.res.in

S. Manickam: NCRA, Pune University Campus, Ganehkhind, Pune, India; sivman@abhijit.ncra.tifr.res.in

J. Mikolajewska: Nicolas Copernicus Center, Poland; Andrzej_Zdziarski@camk.edu.pl

R. Misra: IUCAA, Ghaneshkhind Road, Pune, India; rmisra@iucaa.ernet.in

A.P. Mitra: National Physical Laboratory, New Delhi, India; apmitra@doe.ernet.in

M. Miyoshi: National Astronomical Observatory Japan, 2-12, Hoshigaoka, Mizusawa, Iwate 023-0861, Japan; miyoshi@miz.nao.ac.jp

D. Molteni: Department of Physical and Astronomical Sciences, Via Archirafi 36, 90123, Palermo, Italy; molteni@gifco.fisica.unipa.it

R. Mukherjee: Barnard College & Columbia University, Dept. of Physics & Astronomy, 3009 Broadway, 506 Altschul, New York, NY 10027, USA; muk@astro.columbia.edu

B. Mukhopadhyay: SNBNCBS, JD Block, Salt Lake, Calcutta - 700091 USA; bm@boson.bose.res.in

H. Negoro: Inst. of Physical & Chemical Research (RIKEN), Hirosawa 2-1, Wako, Saitama 351-0198, Japan; negoro@craft.riken.go.jp

L. M. Ozernoy: Physics & Astronomy Dept. and Computational Sciences Inst., George Mason University, Fairfax, VA 22030-4444, USA; ozernoy@hubble.gmu.edu

B. Paul: Institute of Space and Astronautical Science, 3-1-1 Yoshinodai, Sagamihara, Kanagawa 229 Japan; bpaul@cygnus.tifr.res.in

A. Ray : SNBNCBS, JD Block, Salt Lake, Calcutta - 700091 India; aray@boson.bose.res.in

A. Roy: Viswabharati University, Shantiniketan, 731 235 India

A. Rozanska: Nicolus Copernicus Center, Poland; Agata_Rozanska@camk.edu.pl

D. Ryu: Department of Astronomy & Space Science, Chungnam Nat. Univ., Korea; ryu@canopus.chungnam.ac.kr

E. Salinas: Chalmers University of Technology, 412 96 Gothenburg, Sweden; ener.salinas@elkraft.chalmers.se

D. P. Sarkar: Jhargram Raj College, Medinipur, West Bengal, India

A. K. Sen: ECRA, Inst. of Radio Physics & Electronics, Calcutta - 700009 India; aksen@ecracu.ernet.in

A. Sillanpää: Tuorla Observatory, FIN-21500 Piikkio, Finland; aimosill@deneb.astro.utu.fi

L. O. Takalo: Tuorla Observatory, FIN-21500 Piikkio, Finland; takalo@deneb.astro.utu.fi

L. Titarchuk: NASA Goddard Space Flight Center, NASA/GSFC, code 660, Greenbelt, MD 20771, USA; titarchuk@lheavx.gsfc.nasa.gov

P. J. Wiita: Dept. of Physics and Astronomy, Georgia State University, Atlanta, GA, USA; wiita@chara.gsu.edu

T. Zannias: Institute de Fisica y Matematicas, Universidad, Michocana, S.N.H. Edificio C-3 Morelia Mich, Mexico; zannias@ginette.ifm.umich.mx

A. A. Zdziarski: Nicolas Copernicus Center, Poland; Andrzej_Zdziarski@camk.edu.pl

Author Index

E. E. Becklin	265
G. S. Bisnovatyi-Kogan	1
K. Borozdin	309
S. K. Chakrabarti	19
P. Charles	279
E. Churazov	319
T. K. Das	113
M. K. Das Gupta	241
T. Dotani	341
G. Gerardi	83
A. M. Ghez	265
M. Gilfanov	319
C. A. Haswell	293
L. C. Ho	157
S. Kitamoto	369
B. L. Klein	265
G. Lanzafame	83
H. M. Lee	187
J. F. Lu	61
A. Maciołek-Niedźwiecki	231
P. Magdziarz	231
R. Misra	385
M. Miyoshi	141
D. Molteni	83
M. Morris	265
R. Mukherjee	215
B. Mukhopadhyay	105
H. Negoro	389
B. Paul	351
M. Revnivtsev	309
D. Ryu	73
D. J. Saikia	247
C. R. Shrader	309
A. K. Sillanpää	209
R. Sunyaev	319
L. O. Takalo	203
L. G. Titarchuk	309
S. Trudolyubov	309
M. A. Valenza	83
P. J. Wiita	49
T. Zannias	123